The Study of PLANT COMMUNITIES

AN INTRODUCTION TO PLANT ECOLOGY

Second Edition

by Henry J. Oosting

PROFESSOR OF BOTANY, DUKE UNIVERSITY

W. H. FREEMAN AND COMPANY
San Francisco and London

Library of Congress Catalogue Card Number: 56-11029
Printed in the United States of America (*C10*)

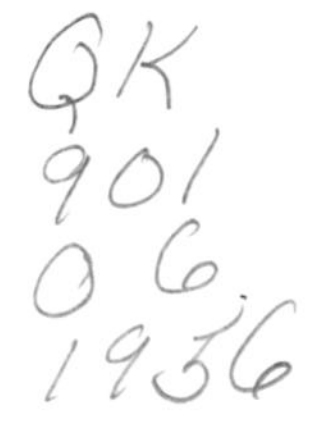

Preface to the Second Edition

AS WITH the first edition, this textbook is intended as an introduction to ecology using plant communities as a basis and the vegetation of North America as a primary source of illustrative material. The plan of presentation remains the same although the order of some subjects is rearranged with presumed improvement in logic. The attempt has been made to bring material up to date, in so far as it is practical to include the latest techniques and theories in an introductory textbook. Thus, there are revisions throughout with many sections completely rewritten. The chapter on succession and climax has been entirely redone with the intention of giving the student an impartial summary of both monoclimax and polyclimax theory and application. Because instructors have requested it and because of a rapidly expanding literature, the number of references cited has been very much increased. Nevertheless, where practical, books and review papers, usually American in origin and readily available to the student, have been cited in preference to foreign and original research publications.

To name all of those who have in some way contributed to this revision is impractical. Each will recognize his contribution and know that it is appreciated. A special acknowledgment is due to Dr. W. D. Billings and Dr. P. F. Bourdeau for reading the manuscript and contributing ideas and to Robert M. Linn who did the new line-drawings and adapted the vegetation map for the reduced reproduction necessary here.

H. J. O.

Durham, North Carolina
February, 1956.

Preface to the First Edition

THIS BOOK grew out of several successive reorganizations of an introductory course in plant ecology. Since it is intended as an introduction to plant ecology, effort has been made to make it as stimulating as possible while presenting basic information. From experience we know that this ideal is best achieved through study of plant communities with emphasis on field work. The plant community, therefore, is made the basis of this book.

The plan, in brief, proceeds from a consideration of the nature and variation of communities to methods of distinguishing and describing them. This is followed by a discussion of the factors which limit, maintain, and modify communities both locally and regionally. Thus the interrelationships between organisms and environment are emphasized and a foundation is laid for presentation of the concepts of succession and climax. Then the climax regions of North America become a logical consideration since they are illustrative of all that comes before. To answer the questions which must arise regarding the permanence of climax, a section is devoted to past climaxes and their study and reconstruction. Finally, the potentialities of the ecological point of view in practical considerations are emphasized by a survey of its possible and desirable applications in range management, agriculture, conservation, landscaping, forestry, and even human relations.

The intent has been to write a textbook with a wide usefulness. It was assumed that, in some instances, the text material might serve as the complete subject-matter of a course. To this end, the presentation aims at a fairly broad but solid foundation for ecological thinking and appreciation. At the same time there is no attempt at completeness, either in subject matter or bibliography, such as might be expected in a reference volume. Although controversial issues are not deliberately obscured, they are not emphasized. The assumption has been that a beginning student should acquire a working knowledge and appreciation of the field before he is introduced to matters that might confuse him.

A reasonable background of botanical and scientific experience is assumed so that, in general, college juniors and seniors might be expected to have the greatest appreciation of a course of this kind. A reasonable knowledge of plant physiology is expected, at least enough for comprehension of ordinary physiological processes. Although a student without some taxonomic training could hardly fully appreciate or enjoy an ecology course dealing with communities, he could use this book if he had some knowledge of plants. Both common and scientific names have been given regularly or at least the first time a species is mentioned. The plants which are named are almost without exception rather generally known species of long standing. It is not considered necessary, therefore, to include authorities with scientific names since they may invariably be found in standard manuals.

Suggestions for collateral reading may be found in the selected general references at the ends of chapters. Cited references are indicated in the text by number only and are listed in the bibliography at the end of the book. Citations are made where it seemed desirable to indicate the authority for or give credit for statements used in the text. Again, for those who may wish to go to original sources, references to survey and review papers are included. The bibliographies of these references are usually so extensive that the advanced student who uses them may quickly accumulate all the source material he needs.

Those who contributed directly or indirectly to the development of this book are too numerous to mention specifically, but I am deeply aware of my debt to former instructors, my colleagues, and my students. Many have given invaluable aid in the actual preparation of the book. A very special acknowledgment of assistance is due Miss Ruby Williams who, through a careful reading of the manuscript, did much to improve the mechanics of organization and to clarify and simplify the presentation.

The use of the book in mimeographed form provided a test of its value under a variety of conditions in different sections of the country. It was used in classes by Dr. W. D. Billings at the University of Nevada, Dr. M. F. Buell at Rutgers University, Dr. R. B. Livingston at the University of Missouri, and by Dr. J. F. Reed at the University of Wyoming, as well as at Duke University. The comments and suggestions derived from both students and instructors led to revisions and additions which are invaluable, particularly in their contribution to wider utility. It is truly with deep appreciation that the cooperation and assistance from all these sources is acknowledged.

Finally, although credit lines indicate the sources of illustrations, it is a real pleasure to acknowledge the courtesies and helpfulness of the numerous individuals and organizations involved. The excellent material they made available, sometimes with considerable trouble to themselves, often made it necessary to choose from several possibilities for a single illustration. It is regretted that not all the pictures could be used. The line-drawings were done by George A. Thompson and Robert Zahner whose assistance is gratefully acknowledged.

HENRY J. OOSTING

Durham, North Carolina
February, 1948.

Table of Contents

Part One: INTRODUCTION

Chapter 1 Ecology and Its Subject Matter

The scope and diversity of subject matter that may be included in the concept of ecology is so great that it cannot be expressed adequately in a single word. Nevertheless, the term *ecology* is surprisingly appropriate. Derived from the Greek words, *oikos*, meaning home, and *logia*, which may be translated "the study of," its implications seem quite right, when they are understood. A home implies that organisms are present and also that there are conditions linking them to their dwellings. These are the broad considerations of ecology—organisms, their homes, and the relationships existing between them. Included is knowledge of the organism, complete for all its processes, knowledge of the environment, especially of those conditions that might affect organisms, and above all, how the requirements of the organisms are satisfied or limited by the numerous variables of environment. All of these things are implied by the term ecology and by the commonly applied, simple definition, *the study of organisms in relation to their environment*.

Much of the basic subject matter of ecology has long been a part of scientific knowledge, usually acquired by specialists in other fields as unlike as physiology, geology, and climatology. The distinction and special function of ecology is to use any such information about organisms or environment and to integrate it for greater understanding and interpretation of the relationship between organisms and environment.

As a field of science, ecology is relatively new. Although the name first appeared in 1869, as "oecology,"[172] it did not immediately gain wide recognition, and acceptance of the philosophy progressed slowly. The greatest advancement has come during the present century, following the impetus supplied by the thinking and writing of a few men in the late 1890's. During the past fifty years ecology has gained a respected position as a biological discipline. Although it overlaps and draws upon the knowledge of other sciences, its effectiveness in integrating such knowledge for explanations of biological phenomena has amply demonstrated the need for, and advantages of, such a discipline. Even outside the sciences there is a rapidly growing appreciation of the meaning and objectives of ecology. The term now appears frequently, without explanation, in popular magazines and even in newspapers.

Life and Environment

An organism without environment is inconceivable,[185] for living things have certain requirements that must be satisfied by their surroundings if life is to continue. Their physiological processes, which, to sustain life, must all continue at rates above definite minima, are largely controlled by environmental conditions or substances. Most of the processes use water or require its presence; food manufacture is dependent upon carbon dioxide and like conditions; the universal process of respiration requires oxygen; and all processes are limited by, or vary with, temperature.

Since organisms must grow and reproduce to survive, they require energy, which they derive from food by respiration. Food, therefore, becomes a major consideration in explaining the activities of organisms. Green plants must be able to manufacture enough food for their needs with the supply of energy available to them. Also dependent upon this production are the numerous organisms that consume the products of green plants or feed upon each other. Food-chains are typical of the organisms in every environment. All these food relations may be considered in terms of energy as a common denominator for evaluating organism-environment relationships.

Every environment has a potential capacity to produce or support a population of organisms. Those organisms that are available and best capable of utilizing this potential will occupy the environment. Some will be there because of their ability to utilize the available light and nutrient energy to manufacture food under the existing conditions. Others, the consumers, will be present because, directly or indirectly, they can utilize the food and its energy which is provided by the producers, whether it is an excess they have stored or the producer itself, usually after death. The sequence forms a cycle, for the decomposing action of the consumer releases elements and simple compounds to the environment, making them available again for the producers. Energy may be lost anywhere in the cycle, and the productivity of particular environments varies according to the efficiency with which its organisms utilize energy under the existing conditions.

Such relationships, or energy cycles, exist in every environment that supports organisms. Thus we see that the basic relationship binding all organisms to each other and to the environment is one traceable to energy needs and uses; and, because the ultimate source of energy for both plants and animals is the sun, all organisms are mutually related to each other and to their environment.

If groups of organisms live together successfully, their demands and effects upon the energy cycle will not disrupt it. All the processes and activities taking place within the group will be in balance with the available supply of energy. A major concern of ecology, therefore, is to learn what that balance is and what controls it.

Scope of the Field of Ecology

The Environmental Complex. Environment includes everything that may affect an organism in any way. It is, therefore, a complex of *factors,* which may be: substances, such as soil and water; forces, such as wind and gravity; conditions, such as temperature and light; or other organisms. These factors may be studied or measured individually, but they must always be considered in terms of their interacting effects upon organisms and each other. The resulting complexity of environment and the array of subject matter encompassed suggest the necessity for drawing upon the knowledge of all fields of science for its understanding. Therein lie a complete justification of and explanation for ecology. Its special function is to consider such subject matter in terms of organisms. Any one field of science is relatively restricted to its own subject matter, whereas ecology brings together the knowledge of various sciences with the object of interpreting the responses of organisms to their environment.

Since all plants and animals, including man, are organisms, and since environment can at times include almost anything in the universe, the subject matter of ecology is almost unlimited. As a result, it is dependent upon the specialized fields of science for much of the knowledge it uses. It requires an understanding of the fundamentals of other sciences, an alertness to changes and new discoveries in various fields, and a constant consideration of the possibilities of using such information for interpreting or explaining the peculiarities, responses, and nature of organisms under the conditions in which they live.

The Organisms. The reference to organisms in the definition of ecology is intended to include all living things. It refers to all plants and all animals, including man. This means not only the larger, conspicuous organisms such as trees and deer of the forest, or grasses and bison of the plains, but any of the other lesser species in such environments. Some may be dependent on the larger organisms, some may be parasites, but all have relationships to each other and are using the resources available in their environments. The least of these, such as bacteria and protozoa, contribute to the breakdown of dead organic matter and the release of its components to be used again, fix nitrogen, or they may cause diseases. Knowledge of their contributions to and demands upon the environment may be extremely important for interpreting the success or failure of the larger organisms with which they are associated. All the organisms in an environment are subjects for ecological consideration; none can be ignored, for they may all affect each other in some way and all have relationships to the environment.

Areas of Specialization

The diversity of subject matter that may have ecological applications is discouraging in some ways but it is also exciting. The very fact that any new use of information from any source may help to solve ecological problems is a challenge to ingenuity and an invitation to attempt their solution by the application of special knowledge obtained from any source. It also emphasizes that an ecologist can profit by the widest possible scientific training and by having sufficient competence in related fields to recognize developments that can be turned to his use. Today, mastery of all sciences is impractical but an appreciation of the usefulness of other disciplines to ecology is not.

Every ecologist requires certain fundamental knowledge as a background. A basic biological foundation is, of course, a necessity, with a solid knowledge of taxonomy, genetics, and physiology as a prime requisite. Beyond this, the broader the biological experience the better, for if anatomy is not needed to solve a problem, pathology may provide the answer, or knowledge of microbiology may focus attention on a crucial matter that might otherwise be ignored.

The desirability of a basic understanding of physics and chemistry need hardly be emphasized, since both have their obvious uses in the interpretation of environmental conditions as well as in applications to physical and physiological problems. Some knowledge of geology is very useful; and, for certain types of work, a broad training in this field is a necessity. Soils are a constant concern of the ecologist both as to their origin and development and as to the paralleling vegetational characteristics as affected by water, aeration, and nutrition. The frequent recurrence of problems related to climatology suggests its desirability, and the increasing use of quantitative methods requires an appreciation of, if not actual facility in, the use of statistical methods and experimental design.

Because mastery of all these fields is obviously impossible, specialization is the natural outcome. Thus, the same or similar problems may, to advantage, be approached quite differently by ecologists with different backgrounds. If the objectives are the same and the approaches are ecologically sound, each may contribute to the solution of the problem because of his special knowledge. Again, the breadth of the subject allows for areas of specialization which scarcely overlap. Studies of root distribution in relation to soil moisture have little superficial similarity to those dealing with the fluctuating population of algae in a lake. Nevertheless, both are ecology, and the investigators are presumably both grounded in the same ecological principles. The wide diversity of ecological subjects is an invitation to specialization, a desirable result in so far as it contributes to the understanding of the interrelationships of organisms and environment and to clarification of the natural laws under which the complex operates.

Desirable as it may be that all ecologists should have a broad knowledge of basic biology, it is unfortunately all too common that investigators, teachers, and students are inadequately prepared to deal with the entire field. There is, as a result, a concentration on plant ecology or animal ecology as though the two were separate or distinct when, actually, they are often indistinguishable. This textbook, itself, represents this conventional but unnatural kind of specialization: it gives primary attention to the ecology of plants, although it does not ignore their relationships to animals. To concentrate attention still more, its emphasis will be upon natural groupings or communities of plants, particularly as they occur in terrestrial environments, and the reasons for their nature and occurrence. This treatment excludes much animal ecology as well as that of aquatic environments, which again involves specialization in limnology (fresh water), marine ecology, and oceanography. Ecological principles are equally applicable to all these environments, and all their organisms are subject to ecological controls. Regardless of the environment or the organisms, the natural laws are fundamentally the same. Restriction of the subject matter should not obscure the possibilities of wider application of the principles that are stressed.

The Origins of Plant Ecology

Early systematic botanists knew little of the geographic limits and ranges of species, but by the time of Linnaeus it became their practice to designate them as precisely as possible. This was, of course, a beginning in floristic plant geography which considers the origin and spread of species. As knowledge of floras grew, there developed a natural interest in explaining the distribution of species. An important contribution in this direction was made at the beginning of the nineteenth century by Humboldt, a German taxonomist who traveled widely and became impressed with the climatic correlations he observed. He presented his ideas so effectively[191] that the influence of his thinking is still apparent in the interpretations of climatic plant geography. Schouw,[346] one of Humboldt's students, was the first to attempt the formulation of "laws" regarding the effectiveness of light, temperature, and moisture in species distribution. A little later (1855), still another taxonomist, A. de Candolle, also considered the controlling influence of various environmental factors but put major emphasis on temperature. Numerous attempts to correlate the distribution of floras and types of vegetation with single factors continued for several years. Historically significant was the attempt by Merriam,[264] an American biologist, to show that all of North American flora and fauna are distributed in zones whose limits are fixed by temperature.

The geographer's preoccupation with climatic causes for the distribution of species was paralleled by another significant trend of interest initiated by the writings of Grisebach in the nineteenth century. He recognized groups of plants, or communities, as units of study and described the vegetation of the earth on this basis.[170] This was the first step in the direction of modern studies of plant communities. Although further expanded by the publications of Drude,[137] the trend received its greatest impetus from the writings of Warming, particularly his *Oecology of Plants*,[419] originally published in Danish in 1895. This publication marks the beginning of modern ecology as it is concerned with communities and the interrelationships of organisms and environment. Although Warming must be credited with recognizing the complexity of these relationships, he tended to place too much stress on water as a controlling factor. In 1898, Schimper published his monumental *Plant Geography upon a Physiological Basis*, which was later (1903) translated from the German into English. Its author followed the general plan of presentation begun by Warming but contributed substantially from his broad experience and travels. He came nearer to the modern interpretation of causes of distribution of vegetation by emphasizing the complexity of environment and the interaction of factors.

These, briefly, are the foundations of modern community studies and the philosophy of modern ecology. From them stem studies of the structure and classification of communities as emphasized by continental European ecologists particularly, intensive studies of habitat in the search for causes, and analysis and interpretation of vegetational change as developed by American and English workers. The history of modern ecology is so brief that the last of these developments can hardly be treated historically. They are the fundamentals of ecology today and, therefore, will be considered as part of the text material of this book.

Approaches to the Subject

It has been emphasized that ecology includes a variety of subjects and a multiplicity of areas of possible interest. It should be apparent that there are few set patterns for ecological studies, and that methods and techniques are adapted to circumstances and objectives as knowledge increases. However, there are general approaches to certain types of problems that can, to advantage, be broadly classified.

Autecology and Synecology. Certain problems can best be solved by working with individual organisms or species in the laboratory or in the field. Others can be solved only when the groupings of organisms are investigated as they occur naturally. Similarly, the environment may be analyzed one factor at a time or considered in its entirety as a complex of several factors. Each approach has its merits under condi-

Fig. 1. *Communities of contrasting life form as illustrated by vegetation on Roan Mountain in the southern Appalachians.* (1) *Deciduous forest of beech and maple.* (2) *Portion of a grassy bald in which grasses and*

sedges predominate. (3) *Portion of a shrub community made up largely of rhododendron (open coniferous forest in background).* (4) *Moss community in which young conifers are becoming established.* —Photos by D. M. Brown.

tions that should become apparent later. The two are distinguished as *autecology*—from the Greek root *autos* meaning self—which deals with individual organisms or factors, and *synecology*—from the Greek prefix *syn* meaning together—applied to studies of groups of organisms or to complexes of factors.

Autecology is not always distinguishable from other fields, such as physiology, dealing with processes, structure, and the functioning of the individual organism. There is no point in trying to distinguish autecology sharply from other related sciences; for autecology, in seeking to explain the requirements, tolerances, and responses of an organism, must draw upon and overlap the subject matter of several disciplines. Synecology is, however, clearly a field in itself whose objectives make it distinct from other fields of science. This is a partial reason for giving major consideration to synecology in this text and for bringing in autecology only when it contributes to the understanding of discussions of community problems.

Static and Dynamic Viewpoints. Plant communities may be studied as they are, without regard to what may have preceded them or to what their natural future may be. This leads to consideration of the abundance and significance of the species making up the community and permits detailed descriptions and precise classification of communities according to one system or another. It is typical of the work of several early continental Europeans, who, as a result, developed systems of classifying and describing communities and their structure. In America and England, the view was early adopted that a community is a changing thing whose origin, development, and probable future can be reconstructed or predicted. These two approaches have come to represent what are known as the *static* and *dynamic* points of view in community studies. The static approach is undoubtedly a product of the restricted areas of study in Europe, where civilization has long since destroyed or modified most natural communities. In the same way, the vast areas of virgin forest and grassland in America, with opportunities to observe natural variation on a large scale and under a variety of circumstances, must have contributed to development of the dynamic point of view. Undoubtedly each method has its place and usefulness. In fact, each has profited from the other, but, since the dynamic point of view has the broadest usefulness in both pure and applied ecology, it will be emphasized here.

Background for Community Study

Systems of description of vegetation that are based upon appearance or general nature of the plants have been used with some success, particularly by plant geographers. Such systems indicate size and form of plants; whether they are evergreen or deciduous, herbaceous or woody;[340] position of buds in the dormant season,[315] and various other characters classified under the general headings of growth forms or life forms. This makes possible the visualization and superficial comparison of otherwise unfamiliar vegetation and likewise may serve to bring out certain characteristics of communities that otherwise might not be apparent. Such systems are either based upon previous detailed studies of the species, or they may be a means of superficially characterizing vegetation of which the taxonomy is still inadequately known. They can only supplement studies based upon taxonomy, since description of a community, to be adequate for all purposes, must be based upon species. The field ecologist must, therefore, have a thorough working knowledge of taxonomy and, preferably, some experience with the flora of the region of his studies.

Just as the study of vegetation must remain more or less superficial without a solid knowledge of the flora, so will interpretations and explanations be limited by the amount of autecological information available about the species and their environments. Physiological-ecological investigations, in the field and under natural conditions, constantly modify synecological conclusions that have been made deductively, or they suggest new interpretations and investigations. The quality of community studies, therefore, depends upon certain fundamentals, which include a knowledge of the individual species and their requirements and responses.

*General References**

L. J. HENDERSON. *The Fitness of the Environment.*

E. RÜBEL. Ecology, plant geography and geo-botany; their history and aim.

* See *References Cited* on page 402 and following for complete listings.

A. F. W. SCHIMPER. *Plant Geography upon a Physiological Basis.*

P. B. SEARS. *Life and Environment.*

A. G. TANSLEY. The early history of modern plant ecology in Britain.

E. WARMING. *Oecology of Plants.*

Part Two: THE PLANT COMMUNITY

Chapter 2 Nature of the Community

Recognition of a plant community or distinguishing one community from another is probably simpler than recording the characteristics by which the community is recognizable. To refer to a stand of pine, a grassy field, or a lowland forest is, in a sense, recognizing communities, and most of us have done this from childhood. Such communities are the basic vegetational units of the ecologist, and, therefore, their specific and general characters should be stated to insure agreement as to concepts.

Definition

A good working definition is as follows: A community is an aggregation of living organisms having mutual relationships among themselves and to their environment. This applies to the specific example which one has in mind or which one is observing—that is, the concrete community or *stand.* At the same time, it does not exclude the possibility of visualizing an abstract community synthesized from several or many concrete examples or stands. Thus a particular stand of pine would be a concrete community, and the community in the abstract would include all the stands of that species.

A stand need not be limited to trees. Any group of plants satisfying the definition of a community may be so termed—a mat of lichens on a rock, covering only a few square inches, an algal mat on a pond, or a forest of fairly homogeneous composition extending over a thousand acres or more.

Mutual Relationships Among Organisms

Competition. Mutual relationships among the individuals of a community include all the direct or indirect effects that organisms have upon each other. Foremost among these is *competition,* which results whenever several organisms require the same things in the same environment. The intensity of competition is determined by the amount in which the demands exceed the supply. Competition may

Fig. 2. *A stand of mixed conifers in Idaho.*—U. S. Forest Service.

occur between individuals of the same species. Because they are alike, their demands are identical, and, if the supply of water or nutrients or light is insufficient to satisfy the needs of all, then some will be handicapped or eliminated. This is particularly noticeable in young, crowded forest stands but is equally true among roadside weeds or in a vegetable garden. All plants may survive for a time in a stunted condition; then some individuals are gradually eliminated. In the forest, thinning to reduce competition usually pays with more or better lumber. The justification for eliminating garden weeds is the improved crop.

In an unmanaged community the numbers of individuals and the species are naturally adjusted by competitive demands to use the resources of the available environment to near capacity and increasingly so as the stand grows older. This results in mature communities becoming so stabilized in their make-up that a new species can only invade after disturbance unless it has exceptional competitive efficiency

Fig. 3. *A stand of moss* (Hypnum crista-castrensis) (*note coin*) *in northern Wisconsin. Although this species is a dependent within the forest community, it forms a stand nevertheless.*—Photo by L. E. Anderson

or its needs can be satisfied from some incompletely used part of the environment.

Stratification. Usually there are several species involved in competition within a stand. If plants of several species that start simultaneously make the same demands upon the habitat, they may survive in about equal numbers and occupy the same position in the community. Those whose requirements differ will affect each other less but will most certainly not be of equal importance in the community.

Fig. 4. *Very much overstocked stand of naturally seeded, eight-year-old, loblolly pine. Although many individuals will die in the next few years and thus produce natural thinning, the remaining trees will remain spindly and growth will not be satisfactory. Artificial thinning to reduce competition is apt to pay dividends in such stands.*—Photo by C. F. Korstian.

Fig. 5. *Young loblolly pine stand, which was overstocked (left) for best growth. The stand was thinned experimentally soon after it was photographed. Same stand (right) only two years after thinning, shows marked increase in size in the reduced number of trunks. The increase in rate of growth will be apparent for a number of years.*—Photo by C. F. Korstian.

A tall-growing species outgrows a potentially short one under the same conditions. If the latter then survives, shaded by the first, it does so because its light requirements are not great. Thus the one tends to occupy a higher level than the other and to form an overstory. In this way *stratification* may develop in a stand in which the upper stratum of plants usually includes the species that control and characterize the community. These are termed the *dominant* individuals. If they are removed for any reason, as by selective cutting or disease, dominance is usually assumed by other species, and the character of the community is changed completely. This change does not occur when

Fig. 6. *Stratification in an oak-hickory forest community as seen in spring when the subordinate tree stratum is especially marked by flowering of dogwood and redbud.*—Photo by H. L. Blomquist.

lesser species in *subordinate* strata are removed, for, with the dominants intact, the same type of community can regenerate itself.

Stratification may likewise be seen among the shrubs and herbs beneath the trees, since some may be tall and some low. The lowest exposed stratum is made up of mosses, lichens, and sometimes algae, which may form a mat or ground cover on the forest floor, and a final stratum of fungi, bacteria, and algae in the upper layers of the soil can also be recognized. The species making up these lesser strata probably offer little direct competition to the trees above them. Most of these plants have appeared here, and are able to survive, because of conditions provided by the tree strata. Indirectly, however, they may offer serious competition to the continued dominance of the trees; because, if the trees are to maintain themselves in the community, they must be able to reproduce themselves. If the seedlings of tree species cannot meet the competition of lesser species, whether it be in the herb or shrub stratum, such trees must eventually disappear from the community. Thus permanent or true dominance involves the ability to compete successfully in all strata of the community. The effects of competition are most apparent in the lesser strata, and undoubtedly competition is greatest between the seedlings of species of all strata, since all must start small and in the same restricted environment of the forest floor.

Some ecologists maintain that each of these strata is itself a community (synusia), which should be considered as a distinct unit of vegetation. Whether or not the strata are so recognized, they cannot be neglected in any study of communities. Often an understanding of the community as a whole is possible only after information is complete on the individual strata.

Dependence. Within any community some species, although a part of the community, are at the same time dependent upon the whole for their survival. To a great extent, these are inconspicuous organisms, which, at first glance, might well be overlooked or ignored. Most of the bryophytes and thallophytes, as well as some vascular plants, require the special conditions provided by larger seed plants; shade and moisture are usually of greatest importance to their survival. Such dependent organisms would soon disappear if the dominant vegetation were removed.

Epiphytes grow on the trunks, the branches, and even on the leaves of the larger plants. In subtropical and tropical forests they may be conspicuous because of both size and abundance. In forests of temperate zones they may be easily overlooked, for they are usually mosses, liverworts, or lichens. These may be restricted to certain communities, and sometimes individual species will grow only on specific trees. Fungi, including bacteria, make up an important part of many communities, especially forests. Here they may be parasitic and cause diseases that may at times become so serious as to destroy a stand or even to eliminate a community. Other saprophytic fungi, living in the soil or litter of the forest floor, although dependent upon the community, likewise contribute to its perpetuation through their activities in decomposition of organic matter. Still others, again often

Fig. 7. *Indian pipe* (Monotropa uniflora), *a saprophyte* (*on left*), *and squaw root* (Conopholis americana), *a parasite dependent upon the shade, moisture, and nutrients available only on the floor of a mature forest.*—Indian pipe photo by H. L. Blomquist.

host-specific, live in an association with the roots of vascular plants in a relationship termed mycorhiza (see Fig. 97).

Finally, animals, largely as dependents but often at the same time as influents (having effects on vegetation), are likewise a part of the biotic community. Large species such as deer, which move about freely, are not necessarily associated with a single community. However, many

Fig. 8. *Spanish "moss"* (Tillandsia usneoides), *an epiphytic flowering plant, growing on live oak, North Carolina coast.*—Photo by H. L. Blomquist.

smaller, less widely ranging species are definitely restricted to single communities, and even some birds and flying insects may be constantly associated with certain types of vegetation. Many beetles, borers, moths, etc. are extremely destructive parasites, while other similar small animals live on the remains of dead plants. The animals are apt to be related to the community through food requirements and, if present in large numbers, may have extremely destructive effects.

Mutual Relationships to Environment

Plants must be adapted to the environment in which they live if they are to survive for long. Some can withstand heat, some cold; some require a large continuous supply of moisture, others require only a small amount, which need be available only periodically. Thus the climate of a region controls the kinds of plants that may grow there. The general vegetation type or growth form, such as grassland, desert, or forest, is a product of the complex of climatic factors effective in a region and can be used as a generalized basis for evaluating the climate. For example, knowing something of the growth forms able to survive under the extreme conditions of moisture and temperature associated with a desert, one may automatically accept a repetition of these growth forms anywhere else in the world as indicative of desert conditions. The scrubby broad-leaved evergreens (chaparral) that cover much of southern California are a product of the climatic conditions peculiar to the area. The same growth form is repeated in a few widely separated regions of the world, where, although made up of quite different species, it exists in a similar complex of climatic conditions. In the same way the vast expanses of deciduous or coniferous forests in the temperate regions of the world each occur where climatic characteristics fall within definite limits, similar throughout.

General Climate and Vegetation Type. Within the general vegetation type, certain variations may be expected. Species may differ although the growth form is uniform for all. Such differences are most pronounced when a type of growth form extends over a wide latitudinal range. In the arctic flora, which has an otherwise uniform physiognomy, the number of species declines steadily northward. Within the grassland areas of the Middle West, there is obvious uniformity of growth form from Canada to Texas, yet some species found in the south are not found in the north, and other species may be found only in the north. Even those species that seem to range from one limit of a growth form to another may likewise include *ecotypes,* that have certain characteristics, probably physiological but genetically controlled, which limit the extent of their area of favorable growth. Recently it has been shown that certain grasses that seem to range throughout the latitudinal extent of the prairie cannot be satisfactorily used to reseed northern areas when the seed has been obtained in the south. Foresters, too, recognize that it is advantageous to replant with seedlings grown from locally produced seed.

The more extreme the climatic conditions, affecting plants adversely, the less diversity can there be among the species and the fewer the species there will be, because not many will have the adaptations

Fig. 9. *Transition zones between stands of two life forms. The forest at right (mostly buckeye) shows the usual gradual transition from a closed stand to scattered, widely spaced individuals over a wide band —such as is typical of most transitions from one community to another. The abrupt transition from beech forest to grassland (at left) is unusual. Roan Mt. in the southern Appalachians.*—Photo by D. M. Brown.

necessary for their survival. The numbers of species in a general vegetation type are by no means constant throughout, especially nearing the limits of the type. Here it might be expected that conditions would be something less than optimum and that some species would be less well adapted to the extremes than others. The same can be said for numbers of individuals of a species. As conditions favoring a species vary from their maximum, the number of individuals may be expected likewise to fluctuate, and, near the limits of the range of a growth form, the numbers of individuals of that growth form would also decline. In the same sense, but in the opposite direction, this marginal area would support a few species and individuals of the contiguous growth form; thus *transition zones* (also called ecotones) between communities are characteristic. Sometimes these transitions are wide, sometimes relatively narrow, but rarely does one community, large or small, have a sharp line of demarcation between itself and its neighbor.

Local Habitats and Species Differences. Climatic areas are of considerable extent and usually include local diverse conditions of soil or topography. Often these variations are so great that some habitats will

Fig. 10. *Aerial view of the forest that extends along the meandering Souris River far into the grassland of Nebraska.*—U. S. Forest Service.

have localized environments quite unfavorable to the species and even to the growth form of the region as a whole. Often the conditions may be so much more favorable than those of the general climate that a growth form from a neighboring region can compete successfully. This is well illustrated by the trees and shrubs extending far into the prairie along the streams, where the favorable soil moisture is sufficient for them to live successfully in a grassland climate.

In the northern hemisphere, a south-facing bluff forms a habitat that is almost always warmer and drier than the average for the region, while a north-facing bluff is cooler and moister. Barren exposures of rock or high, rocky ridges represent one extreme in local habitats, while flood plains of streams, lakes, and lake margins represent the other. Because the vegetation of such habitats is controlled much more by the local conditions than the general climate, it will include species that can live only under these special circumstances, and even their growth forms may differ from those of the region as a whole. These local variations may be extremely restricted in area, scarcely affecting the general physiognomic picture, as would be true of the vegetation around a spring, or on a boulder in the woods; but they may also be so extensive as to be misleading. Cypress or cypress-gum swamps in some sections of the southern states are so extensive that they might appear to be as much related to general climate as to a special habitat. Many of the pine forests of New England and the other northern states lie in a climatic region where spruce and fir should eventually predominate. They are so extensively distributed that some ecologists have recognized them as the ultimate growth form for the region and as controlled by climate, whereas, pine dominance within the climatic area is initiated after disturbance, is maintained for long periods on light, sandy soils, but is subject to natural replacement, eventually.

The kinds of plants, as to form and appearance, that can grow in a climatic region are, therefore, determined by the overall climate. Because the species within the general growth form have genetically controlled variations in their requirements and amplitude of environmental tolerances, the species and their proportions may vary from one limit of a climatic area to another. Local habitats may have such marked differences in growing conditions that not only will species differ but even the growth form may not be that of the climate of the region.

The Community and the Ecosystem

A machine, no matter how elaborate, is a combination of wheels, gears, levers, or other devices, designed to convert energy from one form to another so that ultimately a product results, either tangible or otherwise. All the parts are related to each other and the operation of the whole is dependent upon the least as well as the greatest. Its capacity to produce is a function of energy put in and the efficiency with which it is used.

A community, in a broad sense and including all its organisms, is like a machine in that it is a system of interacting parts, using and transforming energy. As such, its efficiency and productivity may be rated, but this is not a simple matter. The community is only a part of a larger energy system that includes environment, the medium in which it operates and the basic source of the energy and materials it uses. The amount and availability of these requirements control the functioning of the community, which is, in effect, both the producer in, and the product of, the system. Unlike that of a commercial machine, the product, in a natural community, is regularly returned to the system as organisms die, to be broken down and used again, often with little loss of energy. Such a complex system of organisms, forces, substances and conditions as are encompassed by the community, its environment, and their interactions, has been appropriately termed an *ecosystem.*[394] This concept of an ecosystem is a useful one, for the problems of community ecology are almost invariably problems of the whole system, not of its parts. Understanding of the parts involves a sound conception of their relationships to the whole.

General References

J. Braun-Blanquet. *Plant Sociology: The Study of Plant Communities.*

A. F. W. Schimper. *Plant Geography upon a Physiological Basis.*

E. Warming. *Oecology of Plants.*

J. E. Weaver and F. E. Clements. *Plant Ecology.*

Chapter 3 Vegetational Analysis Quantitative Methods

Bases for Characterizing a Community

The fallacy of doing detailed physiological studies with an unnamed plant is obvious. If the physiologist does not know the species, and sometimes the variety or strain, with which he is working, his conclusions will be limited to the particular group of plants he is using in his experiments. The studies of taxonomists, floristic geographers, and geneticists represent an accumulation of information and data upon which the physiologist can draw and which he can use to make generalizations and comparisons. All this information is connoted by the scientific name of the plant being studied.

The ecologist, although working with communities, deals with problems similar to those of the physiologist when he sets up theories, attempts to find causes, to draw conclusions, or to formulate laws. However, before the ecologist can proceed to an investigation of causes or to experimental considerations in a community he must know its make-up in the same way that a physiologist knows about a species from the work of taxonomists and others. At present, most of the larger, regional, climatic vegetation types are so well known that their concepts are probably as distinct to the ecologist as are those of most common species to the taxonomist. For lesser communities, however, this is not true. Furthermore, identification of such a community in terms of a specific concept requires more than a superficial examination. Perhaps an ecological classification of plant communities will never be achieved with the same degree of perfection found in taxonomic classification; such perfection is not necessary. It is necessary, however, that there be means of characterizing a community with sufficient accuracy to permit identification at any time, to compare it with other similar communities, and to have an adequate permanent record of its nature and occurrence. This is an important first step in the study of plant communities. If such work is well done, it is justified on its own merits as a phase of ecological investigation.

If the major interest in a community is an experimental one and the preliminary analysis and description of the vegetation have not previously been made, the experimenter must first learn and record the characteristics of the community with which he intends to work. Again, after experimentation or treatment, whether it be of the community as a whole or of individual species, it often becomes necessary to evaluate the results in terms of the community as a whole. There must also be a means of comparing the original and the resulting communities at the beginning and at the end of each experiment or treatment. The relationship of the individual species to the community and the responses of the individual species can best be interpreted when the constitution of the entire community is positively established.

It is illogical to proceed with explanations when the subject itself is indefinite or unknown. Therefore, the first objective in ecological work is to learn the composition and structure of the community under consideration. Then, and only then, logically follows a search for causes, experimentation, and interpretations based upon a firm foundation.

Quantitative Data a Necessity

In the early days of ecology, observation and description were considered adequate for recording the characteristics of a community, but few observers see the same thing in the same way, and few writers have the ability to translate exactly into words the things they have seen. Thus, as in other sciences, ecology has become more precise as it has developed and, with its concern for greater detail, has demanded accurate measurement and precise records of vegetation. This demand has led naturally to quantitative methods and terminology, which are becoming more uniform and, therefore, more useful. Their use permits positive statements concerning the numbers and sizes of individuals as well as their spacing and distribution within a stand. With such data, it is possible to make comparisons of species or groups of species within a stand or between stands. Likewise, the data constitute a permanent record, which can be referred to again if the same stand or similar stands are studied later. Also, as a permanent record, they are subject to reconsideration by other investigators, who may reinterpret them in the light of additional experience or information.

Sampling

The need for quantitative records has made it necessary to give serious consideration to methods of sampling. Usually the members of an entire community cannot be counted or measured, and even if this were done, the information would be no more useful or significant than an adequate set of data acquired by proper sampling. Since this is true, it becomes of prime importance to determine what constitutes an adequate sample in terms of the community as a whole and how to obtain such a sample with a minimum of effort. At best, sampling for vegetational data is tedious and time-consuming; often it may be extremely hard work. Nevertheless, sampling conserves both time and labor as compared with an attempt to analyze a whole community, and its results are much more significant than those obtained by mere observation.

In this connection it should be emphasized that the early procedures of observation and reconnaissance are still of extreme importance in determining where, how, and what to sample. These activities are still a necessary part of community study, although they cannot be substituted for detailed analysis. They serve to form a basis for theories or ideas that may in turn be substantiated by quantitative evidence obtained by sampling. Preliminary reconnaissance may likewise help to reduce the effort expended in sampling. No sampling should be done without considerable knowledge of the history, physiography, and vegetation of the region as a whole.

The potential usefulness of aerial photographs in all these connections deserves special mention. They often provide information available from no other source or which would require much time to obtain on the ground. Much of the United States and Canada has been photographed, and prints are available through several government agencies or may be examined at various state and county offices, especially those operated under the U. S. Department of Agriculture. Many of the uses of aerial photographs for ecological field work are so obvious that they need not be detailed. Of the various publications discussing uses and limitations, one on forestry,[386] a field in which aerial photography is now a recognized tool, would perhaps be most useful.

Prior to sampling, examples of the community should have been observed repeatedly in different parts of its range and more particularly under any varying conditions in which it may exist. Finally, the specific stand should be observed thoroughly to determine its obvious variations, its extent, limits, and transitions to contiguous communities. Then, knowing all this, together with the size of individual plants, the

Fig. 11. *A small quadrat laid out with meter sticks, pinned at corners and ready for list-count. This permanent quadrat, relocated by paint markings on boulders, was used at Glacier Bay, Alaska, for the study of early development of vegetation on raw morainic soil.*[100] *Thirty-seven years before picture was taken this area was covered with ice.*—Photo by W. S. Cooper.

strata present, and the purposes for which the sampling is to be done, one may plan his procedure in terms of the desired results, the desired degree of accuracy, and the time available for doing the work.

Ecologists call a sample unit or plot a *quadrat,* and the method of sampling by the use of plots is commonly called the *quadrat method.* The use of the sample plot is by no means restricted to ecology, but its application in the sampling of natural vegetation has led to methods peculiarly adapted to the ecologist's needs. The quadrat has almost unlimited applications and has been used in a great variety of ways.

Fig. 12. *One of a series of permanent experimental plots treated as chart quadrats. Note iron pipes used as markers and the grid pattern of string to facilitate accurate charting. One of controls in background.*—Photo by C. F. Korstian.

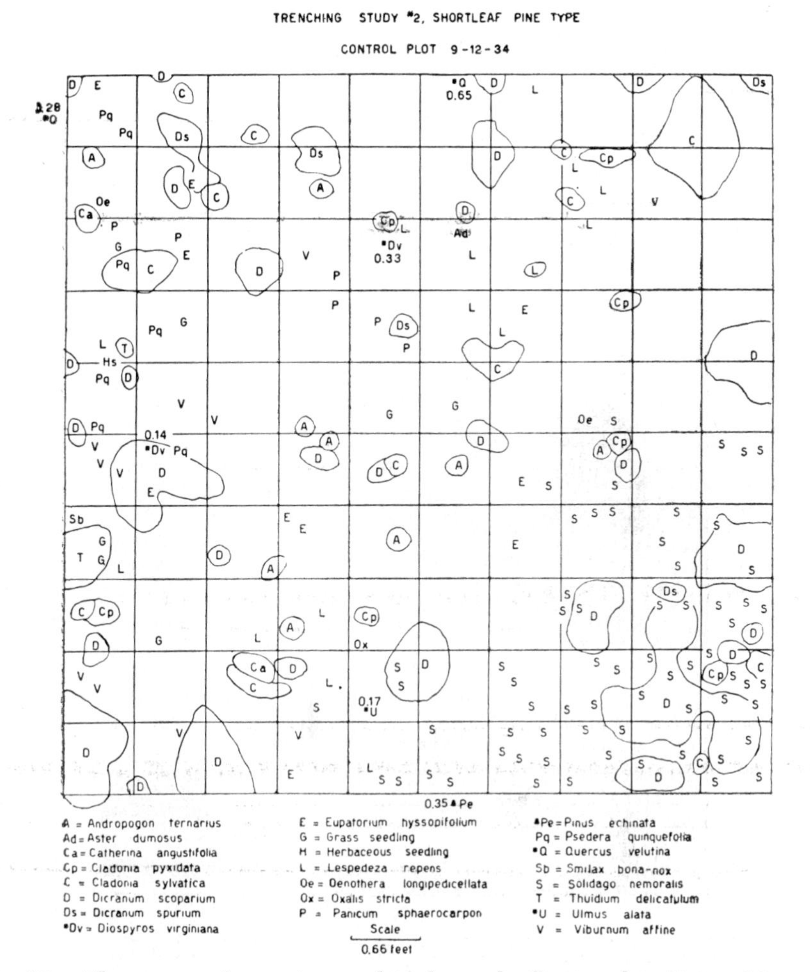

Fig. 13. *The system of mapping used in the study illustrated in Figure 12. Such a procedure is adaptable to many situations.*[219]

Kinds of Quadrats. The *list* quadrat, a simple tabulation of species present, is adequate for certain purposes but the *list-count* quadrat is most commonly used. With this the species are recorded and their numbers determined by count. This method is subject to many modifications depending upon circumstances. For trees, the individual trunk diameters or areal spread of crowns might be recorded and later used for segregating size classes, or perhaps for computing basal area (indicative of dominance) for species. Bunch grasses, too, are often measured across the base to obtain a basal area figure, which, combined with the count, will give a better expression of the relative importance of species. With herbs it is sometimes desirable to have additional information on the weight of tops, which must, therefore, be removed for each species. In any event, the species are listed and tabulated by number, weight, or size.

A *chart* quadrat is a more detailed record of the individuals present, giving their size and distribution within the area. This is time-consuming, even on small quadrats with a relatively simple arrangement

Fig. 14. *Mapping a quadrat by the use of a pantograph, which reduces all details to scale.*—U. S. Forest Service.

and few species. It does, however, permit study at a later date—an advantage not to be ignored under many circumstances. Small quadrats may be photographed with considerable success if proper equipment can be brought to them conveniently. Camera stands of various sorts have been designed that permit vertical views, and the photographs can be studied at leisure. Fairly accurate coverage for individual species can be determined from the prints with a planimeter (a mechanical device for determining the area of a surface with irregular boundaries). Such records are particularly useful when the areas are to be studied over a period of time and when they are subject to treatment. When a high degree of accuracy is desired for small plots, a pantograph[297] can be used with a drawing board, or sketching on coordinate paper may be quite satisfactory, especially if the quadrat itself is marked off into a grid pattern, as with strings. For small quadrats of low or matted vegetation, a rigid frame permanently rigged with fine cross wires to form a grid (see Fig. 12) can be used to advantage since it can be moved from place to place, thus saving the time of marking off each new quadrat.[35] Small quadrats in relatively tall herbaceous vegetation or among shrubs and saplings can be laid out more easily with rods or wooden strips cut to proper length than with tapes (see Fig. 11). At times, in rapid surveys, estimation of cover by inspection of each plot may have to suffice. After a little experience, especially when using a grid, considerable accuracy can be attained; but, because of personal bias, all observations should be made by one individual.

The use of *permanent quadrats* has been advocated by many ecologists, but few have followed their own excellent advice. Whenever there is a possibility that a sampling area may again be visited for further study, the quadrats should be marked permanently, for surprisingly worth-while results may be obtained by restudying identical areas after a period of years. Such results are often valuable out of all proportion to the effort required, especially when compared to the initial study. Most quadrat studies are planned for immediate results and to help solve problems of the moment, but with little extra effort they could be used to yield returns over a period of years. Permanent quadrats should always be considered when an area of study has a long-time potential freedom from major disturbance as well as when effects of changes of usage might be of interest. When Dr. W. S. Cooper made his now widely known study of vegetation on Isle Royale in Lake Superior, he photographed his sampling areas and carefully marked the spots even though he had no definite plan for restudying the area. Seventeen years later he was able to relocate these points exactly, and he obtained a striking series of matched pictures illustrating the development of each of the vegetation types on the island.[99] A number of similar illustrations could be mentioned, but they are far too few. Because vegetation does change and memory sometimes fails, a warning is in order that permanent plots should be permanently marked (as in Fig. 11 or otherwise) and a careful record made for relocating them.

Quadrats originally set up for permanent study are usually of an experimental nature. Perhaps they are to be subject to a treatment of some sort, as, for example, different degrees of grazing, watering, or thinning. For acceptable results these must always be laid out in pairs so that an untreated plot can be used as a check or control on the treated area. Usually it is desirable to replicate the pairs one or more times, and this must be given serious thought in terms of the extent of the stand and uniformity of conditions. Such experimental areas are often established near at hand and in easily accessible places, for

Fig. 15. *Paired pictures illustrating slow development of vegetation on rocks on Isle Royale. Lower picture taken seventeen years after upper.* —Photos by W. S. Cooper.[99]

they are to be visited regularly. With plans made in advance, materials for permanent marking are among the first equipment to be assembled. Substantial lengths of old pipe or scrap metal, when driven into the ground leaving a few inches protruding, are permanent and very satisfactory markers. If they are painted conspicuously and marked with numbers, there can be no confusion.

Experimental quadrats are of many types. Studies of competition and survival may involve thinning of stands, eliminating undesirable species, or introducing other species, either by seeding or planting seedlings, the object being to observe effects on the community or the introduced species. Newly exposed bare areas may be studied to follow the natural development of vegetation, or areas may be denuded and attempts made to produce artificial communities. Perhaps the quadrats are used to evaluate the effects of some controlled factor such as artificial watering or shading or the application of a fertilizer. Again, animals may be the factor under consideration, and then exclosures of the vegetation or the animals will be necessary, depending upon objectives.[119] Exclosures should not be considered lightly, for their installation may require considerable time and labor. Also certain types of materials may be surprisingly expensive, especially if plots are replicated. If the effects of grazing are to be studied, a barbed-wire fence will keep out cattle, but rodents must also be considered. They may be attracted by the very things that flourish within the exclosure after the cattle are kept out. Again, small plots may be fenced for rabbits and yet permit squirrels or birds to come in over the top. Then the entire plot must be covered. Lesser rodents may go through or tunnel under the wire, and suitable precautions must be taken to check them.

The effects of the exclosure itself upon the vegetation should not be ignored since it may serve as a windbreak, which may reduce transpiration and intercept snow, soil, and seeds. Small plots completely screened over will have quite a different microclimate from unscreened areas. To hold constant a single variable within an exclosure is difficult, but it can be approached by having exclosures as large as possible, by insuring a liberal transition or isolation strip around the margin, which will not be used in sampling, and by having the barriers as low and as open as possible within the limitations of the experiment.

Sampling Procedure. Because it is often desirable to subject sampling data to statistical analysis, it is well to know something of the requirements and limits of sampling procedure for this purpose. Any standard textbook on statistics will include information on sampling methods to use for making "estimates" of various quantitative characteristics of a population, as well as estimates of error, which serve as a measure of accuracy. Invariably, the statistical treatments assume that the data are the result of random sampling and that they are free of bias, either personal or introduced by the method. An acceptable sampling pro-

Fig. 16. *A large grassland exclosure (left), which shows range recovery in absence of grazing such as has continued on deteriorated grassland (right) where burroweed and snakeweed predominate.*—U. S. Forest Service.

cedure must, therefore, satisfy these requirements. Desirable as it might be, a detailed discussion of these matters is impractical here. Reference can advantageously be made to any generalized presentation of sampling theory, or one of the few introductions written specifically for the ecologist may be consulted.[199a]

Lest the novice be carried away by the possibilities of applying statistical methods to community study, it is advisable to establish a point or two. Statisticians who have interested themselves in phytosociological problems have emphasized that statistical analysis can never be a substitute for good judgment either in selection of communities for quantitative study or in the interpretation of data obtained. The subjective element will continue to be important in community study. Statistical analysis has its greatest usefulness in interpreting community structure and very little for classifying communities.

Seldom can all the information an ecologist seeks by sampling communities be evaluated statistically, and often such statistical evaluation is unnecessary. Regardless, it will be apparent in following sections that many of the statisticians' requirements and recommendations have been incorporated in the methods used by ecologists, especially when they contribute to greater efficiency and probable accuracy of information.

Because vegetation is so variable, generalizations cannot be made to fit all situations; and because objectives are rarely the same, methods quite satisfactory in one instance may not be so in another. Actually, the unit sample may be any size or shape, and any number may be used in a variety of ways, depending upon circumstances and objectives. As one soon learns, the major concern is to get adequate information with a minimum of effort. Set rules for sampling are not advisable, but certain generalizations may well be considered in the light of experience.

Shape of Quadrat. The term, *quadrat,* implies a square, and this shape has been more commonly used by ecologists than any other. This is probably a matter of habit, for other shapes are just as usable and sometimes more efficient. When Raunkiaer[315] was making his pioneer studies of frequency, he at first used a square frame for marking his sample areas but later used a circle exclusively. Seeking data from many small quadrats that were randomly distributed, he contended that "The most convenient forms and sizes of the unit areas are the best." With low vegetation, he found that circular plots had a distinct advantage, laying them out by means of a set of hoops or rings of proper size tossed in all directions from a central point. Larger circular plots can be quickly and accurately marked with a string attached to a free-turning ring on a central axis. Obviously, this method is not satisfactory where vegetation is more than waist high.

Although it was shown some time ago that stability of results could be obtained with less sampling by use of a strip or rectangular plot than a square,[195] there was not an immediate concern over the relative merits of shapes of sampling units. Now, however, it is rather generally accepted that rectangular units are significantly more efficient than squares of equal area, since they will tend to include a better representation of the variation in a stand. For low, herbaceous vegetation, it has been shown by statistical studies that plots $\frac{1}{4} \times 4$ m. were the most efficient in size and that to secure the same amount of information with squares as with strips, nearly twice as large an area would have to be observed. Short strips (1:4) gave less variable data than squares but more variable than long strips (1:16).[77] The same general conclusions were reached after studies of certain types of sagebrush-grass range sampling.[298] In a forest study, the best estimate of tree population was obtained with strips 4×140 m. and 10×140 m. with a fourteen percent sample, when the long axis of the plots crossed the contours and vegetational banding.[37] In terms of effort and time, it is of interest that when alternate segments, no longer than the width of the strip, were systematically omitted, the precision of the sample was hardly impaired, although the total sample was reduced by half.

A variation of the usual sample plot is the point quadrat, which has been used effectively for sampling low, dense, or matted vegetation. A frame supporting a vertical row of equally spaced, slender pins is used as a sample unit. When lowered over the vegetation, only the species are recorded whose leaves are first touched by each pin point. The point method is useful for determining cover values and relative abundance of species. Adaptable to most sampling manipulations possible with quadrats, it is rapid and usable for survey or intensive work.[52] In extensive grassland surveys, a pair of rimless wheels with pins for spokes and drawn like a cart has been used effectively.

Currently, plotless sampling is receiving attention because of its relative rapidity. Most widely tested is the random pairs method,[103] which makes use of two-tree samples at each of a predetermined number of points spaced on compass lines. Granted the assumption that spacing of individuals in a stand deviates at random from a theoretical condition in which they would all be equidistant, it is only necessary to determine the average distance between randomly selected pairs of individuals to permit estimates of the population.

By the present recommended procedure,[104] the individual nearest each sample point is taken as the first of a random pair; the second is the individual nearest the first but outside the 180° sector bisected by a line from the point to the first. The average of the determined distances between pairs (d) permits calculation, by an empirical formula, of mean area per individual (ma) and individuals per acre: $ma = (0.8d)^2$ and trees per acre $= 43560 \div ma$.

The method, now frequently used, has been tested against plot procedures and in known populations with good results. Certain initial discrepancies are eliminated by the present corrected formula. Others may have resulted from use of an inadequate number of pairs. At least forty are recommended, and unless thirty individuals of a species are sampled, its relative density values will not be dependable.

It will be noted that the method is designed primarily for forest stands, although successfully used elsewhere, and deals with dominants (over 4 in. d.b.h. for trees) only. Lesser vegetation must still be sampled by some supplementary plot system which can be carried along at the same time.

The Bitterlich variable-radius plotless method[171] assumes that the probability of a tree being included in a point count is proportional to its basal area. An angle-gauge is used to determine which trees are tallied. Although developed for forestry, the method holds promise for ecological sampling,[363] particularly basal area,[322] and merits mention because it almost certainly will be used and tested.

Size and Number of Quadrats. A community is rarely homogeneous throughout as to species and their distribution. Newly formed habitats, such as sandbars or tidal flats, where often only a single species is a pioneer, may support a nearly homogeneous stand, but the usual community will have some variation. If there were no variation, a single relatively small sample would always be sufficient. Since variation is the rule, it becomes necessary to have samples large enough or numerous enough to include the variation and to have it fairly represented in the data. There is thus always a question of how large

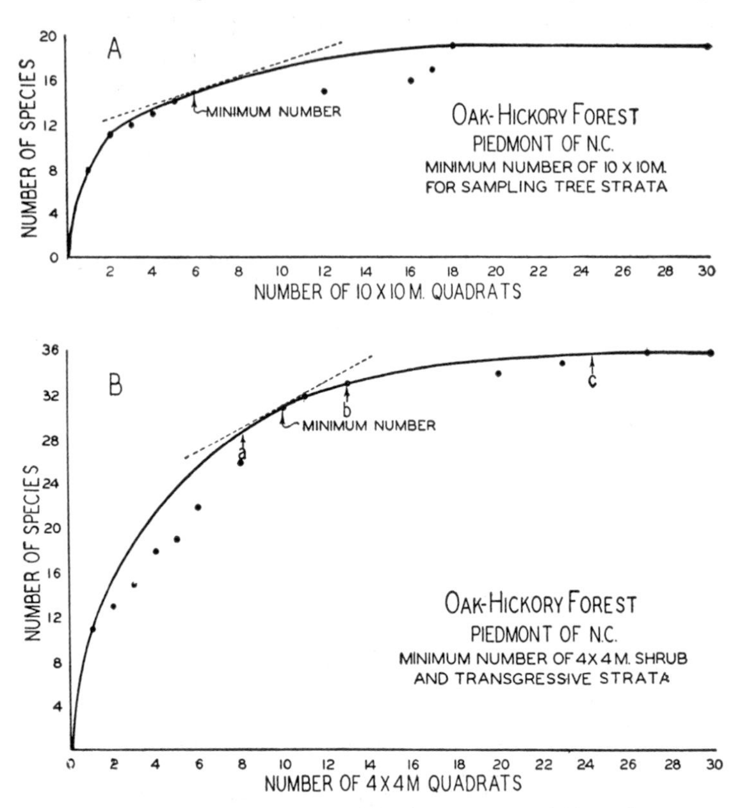

Fig. 17. *Species:area curves for an oak-hickory forest, (A) indicating a minimum of six 10 by 10 m. quadrats for sampling the arborescent strata, and (B) a minimum of ten 4 by 4 m. quadrats for sampling the transgressive and shrub strata. (C) A dune grassland community re-*

and how numerous the quadrats should be for adequate sampling.

The literature dealing with this problem is too extensive to review here, and the statistical technicalities would be inappropriate in such an introductory discussion. Ecological sampling is receiving so much attention and recommendations of procedures for specific purposes are sufficiently divergent to assure that methods will not be standardized for some time to come. There are available papers that consider the problems of sample size and number[60, 66] and summarize conclusions for a variety of circumstances.[13, 164] Others report tests of the usefulness and application of currently favored procedures.[113, 300] Reference

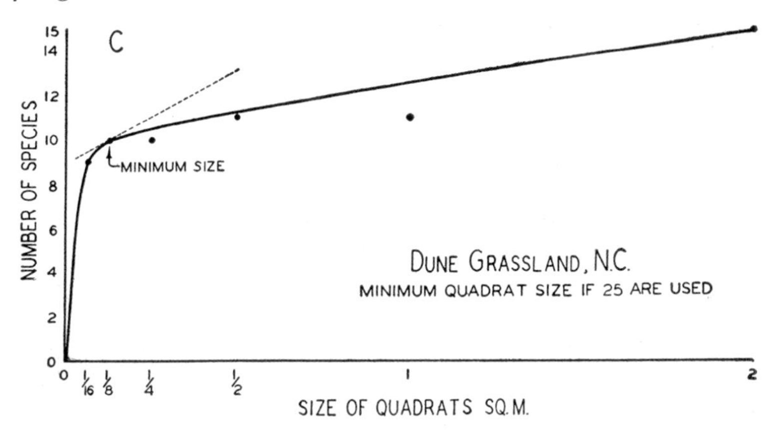

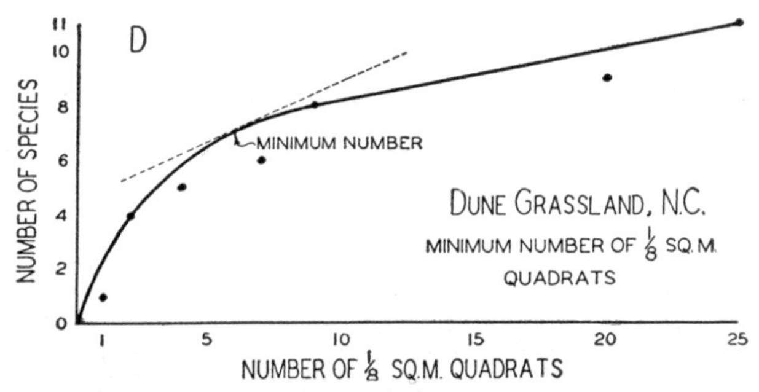

quired a quadrat of not less than ⅛ sq. m. and (D) a minimum of six such samples. The lines tangent to the curves were put in using the triangle method described on page 46. In (B), point a *is equivalent to the average increment per sample, at point* b *the yield is only one-half this increment, and at point* c *only one-quarter the increment.*

to such papers and their extensive bibliographies will make it clear why only generalities are presented here.

Species:area curves have been applied to ecological sampling so frequently that their use and their limitations should be summarized. Originally used by European ecologists to determine the "minimal area" constituting a stand, they have since been used primarily to help decide on the number and size of samples for adequate sampling of a population or an individual community.

The number of species (absolute or percent of total) accumulated in the sampling is plotted on the y axis against the numbers or sizes of samples on the x axis. When the points are joined the characteristic curve first rises abruptly, because many species occur in the first samples, and then tends to level off as fewer species are added with increased sampling (Fig. 17, A, B). Since the break in the curve represents the point beyond which added sampling effort yields diminishing returns, the need for information about the added species must be weighed against the extra effort required to get it. It is generally agreed that sampling is not adequate unless it somewhat exceeds that producing the break and flattening of the curve. The deceptive simplicity of this idea needs certain qualifications.

The form of the curve depends upon the ratio of the x and y axes, and consequently the position of the break cannot be determined by inspection. Therefore, a method of evaluation applicable to any curve is desirable. It has frequently been assumed that sampling was adequate when a 10 percent increase yielded additional species equal only to 10 percent of the total present.[63] This point can be located mechanically on the curve. A line is drawn through zero on the graph and the plotted point representing 10 percent of the species and 10 percent of the sample area. Then, a line parallel to the first and tangent to the curve will have the 10 percent relationship centered at the point of the tangent, regardless of the form of the curve. If a right triangle is placed with its hypotenuse on the first line and then moved upward along a ruler held against its upright side, the hypotenuse, when tangent to the curve, will mark the point. For more intensive sampling the point may be set at the 5 percent rise for a 10 percent increase of sample. Another method of evaluation permits sampling to stop at any point.[66] The average increment of species per plot is calculated from the total, and this point is located on the curve (Fig. 17, point a). Then, since further sampling yields progressively less than the average, in the region of point b, a sample yields only one-half, and at point c, only one-fourth, the number of species obtained by a sample at a.

If, in practice, a series of quadrats of an arbitrarily set size is run in a stand, a species:area curve constructed from the data will be indicative of the minimum number of such quadrats it would have been necessary to use to include in the sample a desired proportion of the species present. The minimum plot size necessary for this same purpose can also be determined from such a preliminary series of data, provided each quadrat is subdivided into successively smaller plots (Fig. 18, B & C) for which records are kept separately.[66] To construct a curve, the data for the small areas are combined to give a figure for the next largest and so on to the largest unit that includes all the smaller ones. The number of species obtained is then plotted against size of quadrat and the usual curve is formed (Fig. 17, A, D), which may be evaluated as before.

It is pertinent to know that by the nature of the species:area curve, addition of sampling units causes the 10 percent point to shift to the right.[321] Also, the number of plots used affects the average increment and an evaluation on this basis. Furthermore, the curve indicates only the approximate proportion of species represented in the sampling. This is certainly valuable knowledge for the ecologist and is probably the chief justification for use of the curve. It does not in any way tell him how adequately the phytosociological characteristics of these

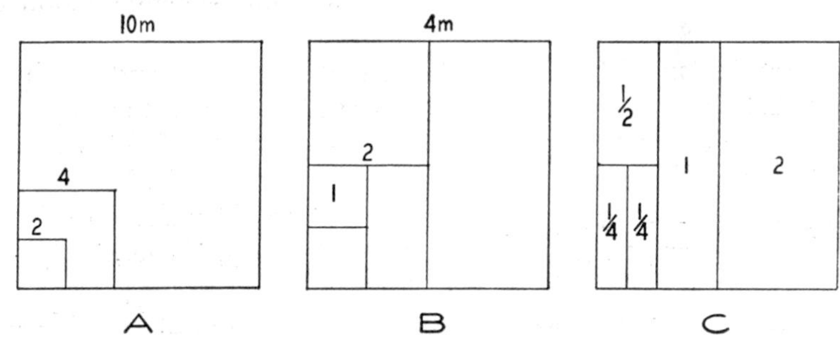

Fig. 18. *Nested quadrats. (A) shows a plan used successfully for sampling the several strata in forest stands. (B) and (C) show systems of dividing plots of any size for accumulating data to be used in determining the desirable size of plot by means of species:area curves.*

species will be represented in the data, although this has often been assumed. Actually, it is desirable that the sample be substantially larger than indicated by a species:area curve. It is generally agreed that adequate sampling for all purposes requires more quadrats,[105, 413] but because of the variety of information sought in a single procedure, recommendations differ considerably.

When vegetation is stratified, plots large enough to sample the dominants will certainly be large enough for all plants and strata. The work involved in measuring or counting the lesser vegetation in such plots, however, would be unnecessarily great. It, therefore, becomes advisable to sample each stratum separately with an appropriate size of plot for each. These plots can be "nested" one within the other, and the work thus materially reduced. The system is adaptable to circumstances. Sampling forest vegetation has often been done satisfactorily by using 10 × 10 m. plots for trees, 4 × 4 m. plots for all other woody vegetation up to ten feet tall, and 1 × 1 m. plots for herbs.[280] By separating the data for trees into overstory and understory individuals and by recording separately those woody plants less than one foot tall and those from one to ten feet tall, five strata can be distinguished. More might be necessary or advisable under some conditions.

In general, it may be said that small plots require less work than large plots, both in the laying out and in the obtaining of data, even though more small plots than large ones are needed for complete sampling. At the same time, there is a further saving of effort in that the total area sampled by small plots may usually be less than that sampled by large plots and yet give comparably valuable information.

Distribution of Quadrats. When the size, shape, and numbers of quadrats have been determined, there still remains the question of how they are to be placed efficiently and in such a fashion that they will give representative data for the stand as a whole. If a stand had a perfectly homogeneous composition, it would make no difference where the sampling was done, but this is rarely, if ever, true. Differences in the soil, drainage, and topography are usually present and are reflected in the vegetation. These variations must be fairly represented in the sample. It becomes necessary, therefore, to distribute the quadrats throughout the stand, and a plan that will eliminate the human factor in placing the individual plots is desirable.

The statistician prefers a sampling system that gives him data obtained at random.[348] This demands a division of the entire stand into possible sampling areas; and then, to assure that every area has equal probability of being included, a selection of actual sampling areas determined strictly by chance. Under such conditions, the statistician is able to express mathematically how good his sampling may be. Such a method frequently brings several sampling areas into close proximity at the same time that wide areas are left unsampled. Within these wide areas, there are very likely to occur a number of infrequent or unusual species in small numbers, which would be of little concern in a statistical treatment but whose presence could be of great interest to the ecologist. He would usually prefer to insure having as many of the variations as possible represented in his data, because they are subject to interpretation in terms of experience and the nature of related communities. For such purposes, statistical methods are often of little help. It is, therefore, probable that quadrats distributed systematically throughout the stand as evenly and widely as possible are quite satisfactory for much ecological sampling. In fact, systematic sampling may be better than random sampling for certain ecological purposes.

Any method that will insure wide and even distribution of samples should be satisfactory. The limits and extent of the stand must first be ascertained, and sampling plans made accordingly. Once the plan is made, it should be followed rigidly unless some previously unknown irregularity, like a swamp or an outcrop of rock, should fall within a sample.

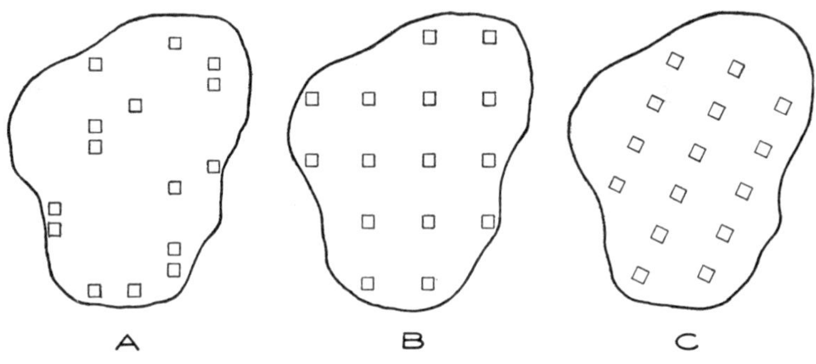

Fig. 19. *The distribution of quadrats in a stand according to three different systems. (A) Random distribution as determined by Tippett's numbers.*[401] *(B) Spaced as widely and evenly as possible by survey and measurement. (C) Distributed evenly along lines run by compass or sighting; spacing determined by pacing.*

In small stands it is possible to plan a grid pattern and to sample at regular intervals in this pattern. When stands are large but of reasonable uniformity, it is common practice to run one or more lines across the greatest extent and to space the quadrats evenly along these lines. It would appear that the more widely the plots are spaced in an area to be sampled the greater the efficiency of the sampling unit, provided the spacing is not so great as to make correlation negligible between adjacent plots.[298] Under some conditions, it may be desirable to run the lines with a surveyor's transit, although a compass line will usually suffice, and in open country it is possible to run them by sighting on some landmark. The spacing may sometimes require accurate measurement, but pacing may serve quite satisfactorily. The important thing is to avoid any method bordering on personal judgment in placing the plots, once the sampling is under way. This admonition should be remembered, particularly when the sampling is being done to prove or disprove a point. Under such conditions, there is often a strong temptation to shift a plot a few feet or more to include or exclude a desired or undesired species or condition.

A random sampling procedure incorporating desirable characteristics of systematic sampling and giving the same kind of information without adding to the work could often be used to advantage. The following method, designed as a result of studies of sampling efficiency,[37, 38] has been very satisfactory when used in several types of forest.

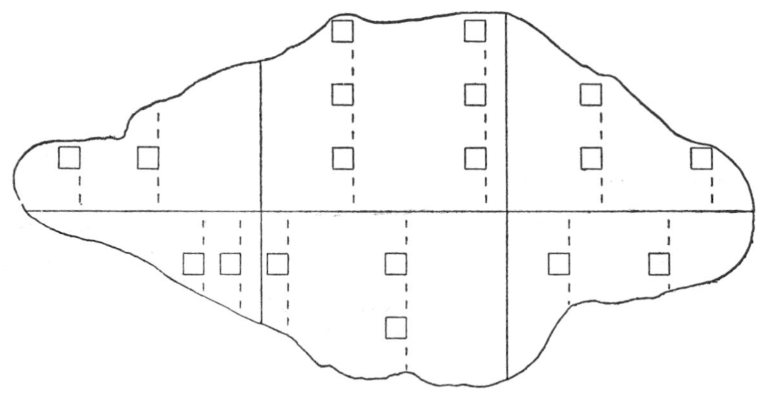

Fig. 20. *An illustration of stratified random sampling in which plots are equally spaced on each of two randomly selected lines in the "blocks" of an irregularly shaped stand.*

In each stand a square area, 100 m. on a side (2.5 acres), is roughly delimited as are the included ten possible strips, 10 × 100 m., oriented across the contour lines. Two of the strips are randomly selected and each is divided into ten equal units (10 m. square). By using five alternate units in each strip as quadrats, ten 10 × 10 m. samples are employed and a 10 percent sample (1000 m.^{2}) of the original area is obtained.

This method does not increase the work in the field. It gives the same information and retains some of the advantages of systematic sampling. It incorporates the increase in sampling efficiency attributed to strip samples but without the work of tallying an entire strip. It permits statistical evaluation if it is desirable.

In extensive stands with possibilities for considerable variation, this method can be expanded into the statistician's "stratified random sampling" procedure. This method, instead of sampling the entire population completely at random, divides it into blocks within which sampling is at random. Obviously, this assures relatively wide distribution of plots.

For ecological purposes an adaptation of such a block system designed for timber cruising[348] has real merit. A base line is established on the long axis of the stand and, any number of blocks of diverse area but equal width are delimited on each side by equidistant lines, perpendicular to the base. In each block two lines are run from randomly selected points on the base to the boundary, and quadrats are equally spaced along the lines at intervals determined by the desired sampling intensity. Any number of blocks may be used; the area of the stand can be calculated; the sampling intensity is known, and the data may be analyzed statistically. The field procedure is adjustable to circumstances and requires no more time than systematic sampling.

Transects. A *transect* is a sampling strip extending across a stand or several stands. It is most often used when differences in vegetation are apparent and are to be correlated with one or more factors that differ between two points. From a flood plain of a river to the adjacent upland there would be marked changes in moisture conditions, and in such a place a transect can be useful for determining the range of moisture requirements of individual species. Transects are also useful in altitudinal studies and in any situation where transitions between communities occur.

Sizes of transects, just as sizes of quadrats, will be determined by conditions. A transect reaching from one small community to another, across a transition zone, might need to be only a few meters long and perhaps a meter or less in width. Transects from lake margins across the several marginal girdles of vegetation that are usually present might be much longer. One reaching from high-tide mark across seaside dunes might be several hundred meters long. A study of the zonation of vegetation on the Sierra Nevada was made by mapping a transect seven miles wide and extending across the mountain range for a distance of eighty miles.[217]

When it seems desirable to map an entire transect in detail, it is advisable to do so by blocks. Values for each block may then be conveniently used as quadrat data, an additional means of analysis and expression of results. A variation of the transect is the method of

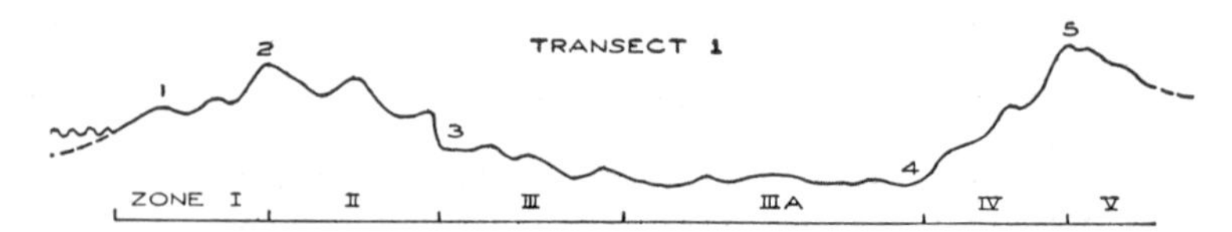

Fig. 21. *Diagrammatic profile along a transect on the dunes at Ft. Macon, N. C. Physiographic-vegetational zones are indicated. Transect was 110 meters long and horizontal scale is one-half the vertical.*[286]

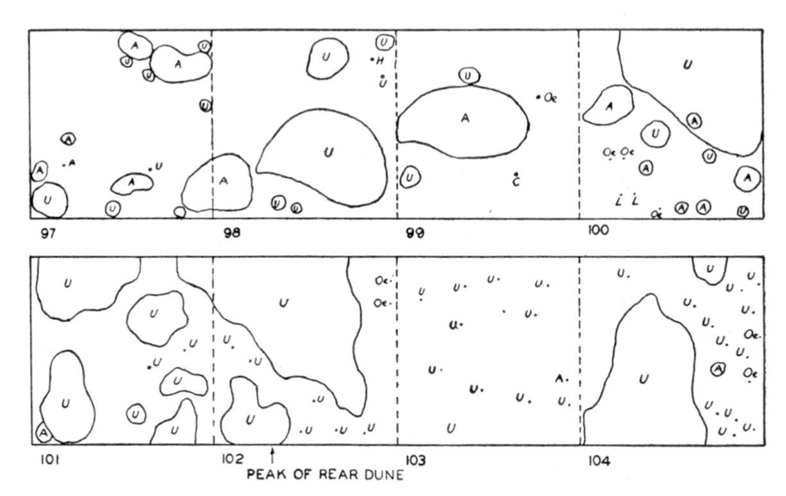

Fig. 22. *Portion of field-mapped transect along profile shown in Figure 21 from 97 m. through 104 m. across the transition from Zone 4 to Zone 5, where dominance changes from* Andropogon *to* Uniola. *The symbols indicate* A—Andropogon, U—Uniola, H—Heterotheca, C— Cenchrus, Oe—Oenothera, L—Erigeron. *Such a map gives accurate quantitative data for each species as well as a visual record of changes in vegetation associated with habitat. See Table 1.*

sampling a unit area at regular intervals along a line. These intervals may be determined by distance or altitude. Such records taken on several lines are particularly helpful in mapping several vegetation types that intergrade irregularly over an extensive area. In the early land surveys of the northern and midwestern states, it was required

Table 1. *Average density (D) and cover (C), by zones, of principal species mapped on a transect from high tide to the crest of the rear dune at Ft. Macon, N. C. (see Fig. 21). Both cover and density values show the predominance of* Uniola *in exposed zones;* Andropogon *in protected ones. The distribution is correlated with amount of salt spray reaching the plants from the open sea.*[286]

		Transect 1					
	Zones	*I*	*II*	*III*	*IIIa*	*IV*	*V*
Uniola paniculata L.	D	11.8	7.5	3.3	0.4	4.0	10.6
	C	2.7	6.9	0.3	0.03	9.1	18.9
Andropogon littoralis Nash	D	2.1	2.5	5.7	6.6	4.1	0.9
	C	2.2	7.1	7.3	15.4	5.5	2.3
Oenothera humifusa Nutt.	D	1.1	1.8	7.2	1.8	6.5	4.5
	C	0.06				0.9	
Heterotheca subaxillaris (Lam.)	D	4.3	0.5	0.5	0.2	1.1	0.8
Britt. and Rose	C	0.5	0.1				
Erigeron canadensis L.	D	0.4	1.2	15.1	0.06	6.2	5.1
Euphorbia polygonifolia L.	D	0.1	0.3	0.3		0.2	
Fimbristylis castanea (Michx.)	D			1.4	11.6	1.9	
Vahl	C			0.2	16.6	0.1	
Myrica cerifera L.	D				0.5		
	C				14.7		

that the characteristic trees be listed in the records for definite intervals along the lines run by the surveyors. Since the county and township lines they established still stand, it has been possible to reconstruct with considerable accuracy the composition of the forests as they then existed as well as the limits of forest and grassland.[210] These surveyors' "transects" were some of the first and longest ever run.

Sometimes there is an advantage in the use of "line transects" (also called the "line-intercept" method), in which the species are tabulated as they occur along a line. The method is adaptable to the determination of numerical abundance, frequency, coverage, and other characteristics. It is rapid and gives accurate information, if the vegetation has the same growth form and the same average crown diameter throughout. It is particularly useful in dense stands of scrubby vegetation,

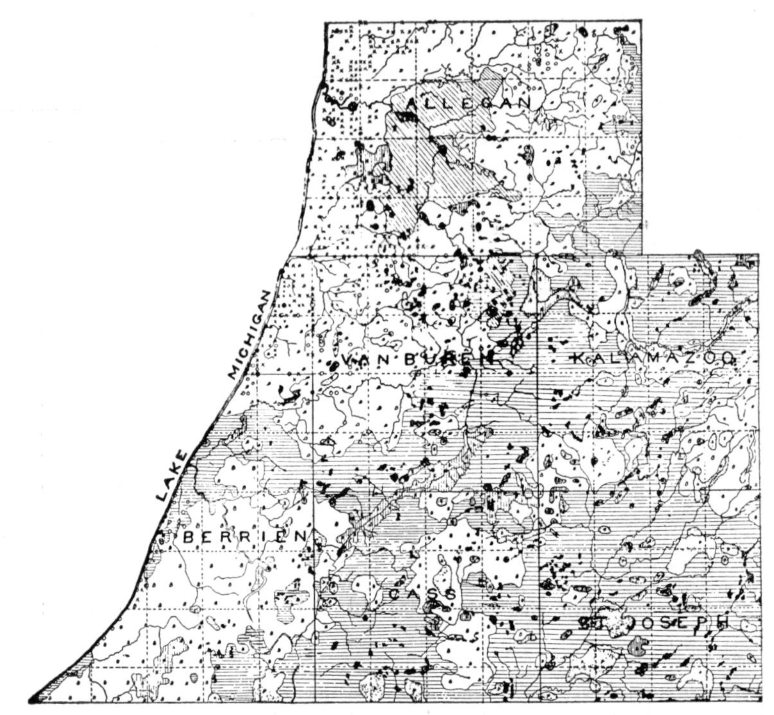

Fig. 23. *Forest associations of southwestern Michigan as reconstructed from the field notes of the old land survey. Unshaded areas, marked* B, *beech-maple forest;* X = *hemlocks, along lake shore, a codominant with beech and maple;* O = *white pines (a mark for each locality of occurrence noted in the survey); horizontally shaded areas, oak-hickory forest; obliquely shaded areas, oak-pine forest; stippled areas, dry prairies; and vertically shaded areas, swamp associations.*—From Kenoyer.[210]

which would be very difficult to sample with quadrats. Determinations of cover in dense chaparral using line transects gave results that compared very favorably with those obtained by complete charting, although the transects were made in a small fraction of the time required for the detailed procedure.[17]

Bisects. These are variations of transects in that they are sample strips aiming to show the vertical distribution of vegetation. Thus they may include stratification and layer communities from dominant trees to seedlings on the forest floor. Graphically reproduced, the resulting profiles may be highly instructive for making comparisons between stands or types of stands, and for interpreting community dynamics. The most valuable are those that also show root distribution. Knowledge of the stratification of parts below ground, particularly with respect to the soil profile, its physical characteristics and moisture

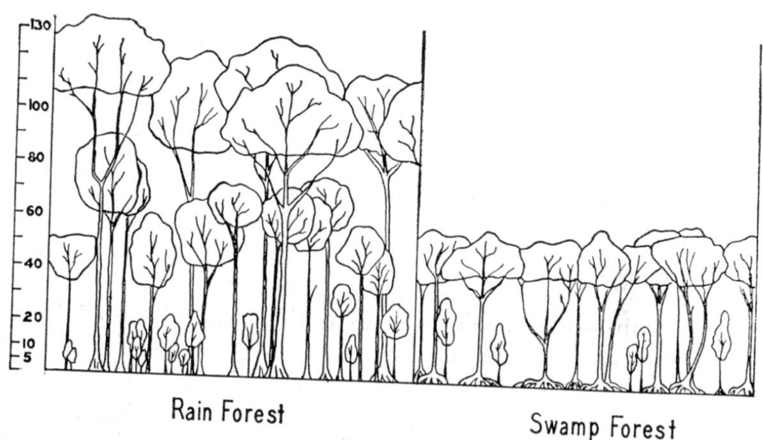

Fig. 24. *Profile diagrams (bisects) of two types of tropical forest. Note that difference in height of trees and in form of trunk is well shown and that rain forest has three distinct strata of trees but swamp forest has essentially one.*—After Beard.[19]

relations, can be very valuable for interpreting competitive abilities of species, their presence or absence in some habitats, or their abilities to survive together. Although the work involved has probably restricted studies of this kind, those that have been made were most rewarding.

General References

Dorothy Brown. *Methods of Surveying and Measuring Vegetation.*

S. A. Cain. Sample-Plot Technique Applied to Alpine Vegetation in Wyo.

F. X. Schumacher and R. A. Chapman. *Sampling Methods in Forestry and Range Management.*

Chapter 4 Vegetational Analysis Phytosociological Objectives

European workers, who first used the term *phytosociology,* have long been interested in detailed structure, precise description, and systems of classifying communities. As a result, the term phytosociology has frequently been reserved, in error, for such studies only. Actually, the foundations for detailed community analysis and classification are creditable to the phytosociologists, but their concept of plant sociology includes in addition, community development, interrelationships, and distribution, as well as all causal factors involved.

Let it be clearly understood, then, that this section deals with the phase of phytosociology related to vegetation analysis. There is no intent to imply that this is the limit of sociological studies. The development of this field was marked by (1), growth of concepts and systems of terminology with which the characteristics of a community could be adequately expressed, and (2), the testing and refinement of methods for obtaining quantitative data on the structure and composition of a community to support the systems of description.

Both methods and terminology have become progressively more standardized, but, as yet, there is not complete agreement among workers. The problems to be resolved are still of the same nature as those of earlier days as is illustrated by a recent characterization[300] that groups them into two categories: (*a*) the size and number of quadrats to be utilized and (*b*) the conditions to be investigated. The first we have discussed at some length as a part of quantitative methods in community analysis. It should be remembered that the development of these methods has been strongly influenced by phytosociological interests. Although the quadrat method in ecology probably has various origins, its adaptation and refinement for complete analysis and description of communities must be largely credited to European workers.

The phytosociological values necessary for an adequate characterization of a community would hardly be agreed upon by all workers even today. Through the years this has been the subject of much debate. Some early workers attempted to describe communities on the basis of a single value (e.g., frequency) for each species. Today such a simple system would not be recommended by anyone, and, regardless of objectives, several values are now used in all phytosociological analyses. Methods of sampling and objectives have always influenced each other; and, therefore, it is not surprising that early workers had widely different approaches, which led to somewhat different conclusions. Several centers of thought and research naturally grew up, which still influence our thinking and procedure. Presently, the ideas of the so-called Zurich-Montpellier school, under the leadership of Dr. J. Braun-Blanquet, have gained the greatest support. These will, therefore, be used as the framework of this chapter in which phytosociological concepts and their application will be summarized.

Phytosociological methodology is far from standardized but its terminology is stabilizing noticeably. As appreciation of the values of sociological data has grown there has also come about a need for uniform expressions to convey ideas and concepts precisely. Although the methods were initially developed to characterize communities for purposes of classification, they are being used to advantage in many other ways. Whenever communities must be described or the significance of individual species in a community must be evaluated, phytosociological concepts and methods are applicable and usually with distinct advantages. This means that the methods are useful in experimental studies of communities, for comparing one community with another, for showing changes in a community from year to year, and, in fact, whenever information is needed about community structure and the part contributed by various species. Applications are almost unlimited. To illustrate, various of its methods have been used to advantage in such diverse problems as correlating the progressive changes of vegetation and soil on abandoned fields,[26] showing the effects of different intensities of fire on the structure of pine stands,[281] and for demonstrating variability of virgin forest with changes of site.[285]

Structural Characteristics

The sociological characters of an individual stand or concrete community may be conveniently grouped in two categories: quantitative and qualitative. Quantitative characters, obtained by quadrat methods, indicate numbers of individuals, their sizes, and the space they occupy. Qualitative characters indicate how species are grouped or distributed, or describe stratification, periodicity, and similar conditions, often based upon knowledge derived from long familiarity and observation of the community.

Quantitative Characters.

Numbers of Individuals. Under some circumstances, it may not be practicable to make actual counts, but plentifulness may rapidly be estimated according to some scale of *abundance* similar to the following:

1. very rare
2. rare
3. infrequent
4. abundant
5. very abundant

Such estimates are particularly useful when several similar stands of uniform composition are to be surveyed within a limited time. When there is time for adequate sampling, the determination of actual numbers by counting is of greater value, because it permits the expression of *density*, which is abundance by number on a unit-area basis.

Density is the average number of individuals per area sampled. Since it is an absolute expression, the significance of density in interpretation may be overemphasized unless one remembers that it is an average value. Not all species with equal densities are of equal importance in a community, or need they be similarly distributed. If ten individuals of a species are counted on a series of ten plots, the density is "one" regardless of whether they are all found in one plot or one in each of the ten plots. It becomes necessary, therefore, to interpret density values or to specify other characters that, combined with density, serve to complete the picture. One such value is *frequency*.

Table 2. *Portion of a list of species occurring on the east coast of Greenland at fourteen localities ranging from (A) 70° N latitude, southward to (N) 65° N latitude. Both presence and degree of importance of the species in each locality is indicated by the field-assigned numbers according to the following scale:*
5—very common (important constituent of several closed communities); 4—common (more scattered occurrence); 3—here and there; 2—uncommon; 1—rare; + —present.
The listed species were selected to show how the system of values indicates range limits and progressive changes of importance with latitude. From Böcher.[33]

	Localities													
	A	B	C	D	E	F	G	H	I	J	K	L	M	N
Cystopteris fragilis	4	+	3		4	+	3	4	4	4	4	?	4	4
Cerastium alpinum	4	4	5	4	5	+	4	4	5	4	4	4	4	5
Minuartia biflora	4	4	4	4	4	+	4	4	4	4	4	4	4	4
Silene acaulis	5	4	4	4	4	+	4	5	4	4	5	4	5	5
Sedum roseum	4	+	4	2	4	+	1	5	5	4	3	4	5	5
Oxyria digyna	5	5	5	4	4	4	4	5	5	5	5	5	5	5
Polygonum viviparum	5	5	5	5	5	5	5	5	5	5	5	5	5	5
Salix herbacea	5	5	5	5	5	5	5	5	5	5	5	5	5	5
Potentilla tridentata										1		3	?	3
Polystichum lonchitis								1					2	3
Alchemilla filicaulis				1				3	5	4		?	5	5
Sagina intermedia		+				+	3	3	4		4	+	3	3
Draba rupestris	?		3				3	3	4	4	4	4	4	4
Empetrum nigrum	3	2	2	1	2	+	5	5	5	5	5	5	5	5
Salix arctophila	+	+	+	+	+	+	4	4	5	5	5	5	5	5
Epilobium arcticum	3													
Potentilla pulchella	3	+												
Rununculus sulphureus	3		2											
Draba lactea	4	+	+	+	+									
Dryas octopetala	5	+	2		2									
Draba alpina	4	4	4	4	3		3							
Cassiope tetragona	5	5	5	5	4				1	1	4			1

Frequency. This value is an expression of the percentage of sample plots in which a species occurs. In the example above, the plants that were all found on a single plot would have a frequency value of 10 percent, whereas, if they had occurred in every plot, the value would be 100 percent. Thus frequency becomes a very useful value, when used in combination with density, for then not only the number of

Table 3. *The effect on frequency of increasing size of quadrat as illustrated by data on Alpine fell-field vegetation in the Rockies. Quadrat sizes in sq. m.* From([66]).

	1/10	1/4	1/2	1/1
Arenaria sajanensis	100	100	100	100
Selaginella densa	100	100	100	100
Trifolium dasyphyllum	80	100	100	100
Eritrichium argenteum	80	80	90	90
Sieversia turbinata	50	50	60	80
Polemonium confertum	40	40	40	50
Phlox caespitosa	30	50	50	60
Sedum stenopetalum	30	50	50	60
Paronychia pulvinata	30	30	50	50
Silene acaulis	20	30	30	70
Potentilla nelsoniana	20	20	30	30
Potentilla quinquefolia	20	20	20	30
Potentilla sp.	10	20	30	30
Polygonum bistortoides	10	20	20	20
Artemisia scopulorum	10	10	20	20
Sieversia ciliata	....	10	20	30
Arenaria macrantha	....	....	....	10
Erigeron compositus	....	....	....	10
Total species	15	16	16	18
Average frequencies	42	45.6	48.7	52.2

individuals is known but also how widely they are distributed in the stand. Knowledge of these two quantitative characteristics, in combination, is fundamental to an understanding of community structure. The use of frequency as a single determination in analytic procedure has proven unsatisfactory, although numerous attempts have been made to show its adequacy.

It should be emphasized that frequency values cannot be compared unless determined with plots of equal size. The larger the plots, the higher the frequency.

Frequencies may conveniently be grouped into classes, for example, A 1-20%, B 21-40%, C 41-60%, D 61-80%, E 81-100%. Raunkiaer[315] used these five classes and, on the basis of more than eight thousand frequency percentages, found that Class A included 53 percent of the species; B, 14 percent; C, 9 percent; D, 8 percent; and E, 16 percent.

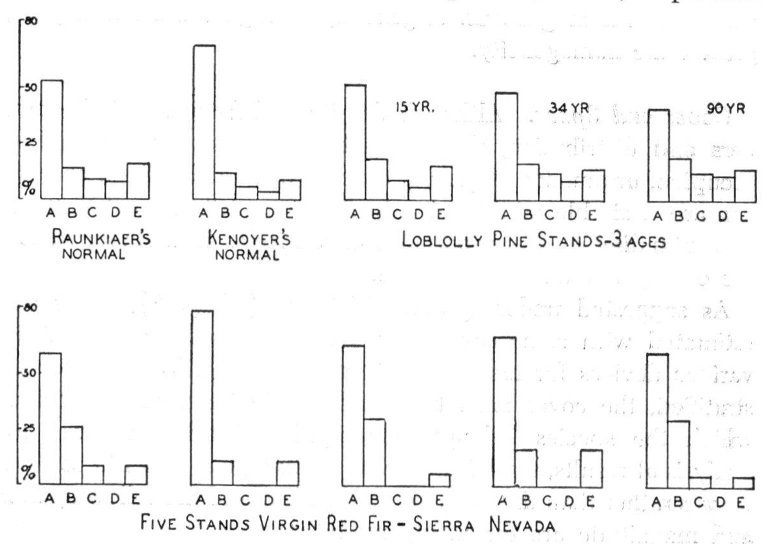

Fig. 25. *Frequency diagrams of pine stands of different ages and of virgin red fir stands compared with Raunkiaer's and Kenoyer's normals. The pine stands were all relatively homogeneous but became slightly less so with age as the total number of species increased by 25 percent and the accidentals declined. Class E, the dominants, remained essentially constant throughout the series. All the virgin red fir stands were extremely homogeneous in spite of a high proportion of incidentals occurring sporadically. The stands were also similar to each other although widely distributed along the Sierra Nevada.*

From these data he drew his "Law of Frequency," which states that class $A > B > C \gtreqless D < E$. This led to numerous investigations to check on the validity and universality of the principle of frequency distributions in plant communities.[208] The results have been in essential agreement regardless of the vegetation type. Class A will normally be very high because of the numerous sporadic species to be found with low frequency in most stands. Class E must always be relatively high because of the species that dominate the community. If quadrats are enlarged, within limits, classes A and E will enlarge and the classes between will decrease accordingly. Frequency classes, therefore, are comparable only when based upon samples of the same size.

A frequency diagram is suggestive of the homogeneity of a stand since floristic uniformity varies directly with the proportionate size of classes A and E. When classes B, C, and D are relatively high, the stand is not homogeneous. In general, the higher Class E may be, the greater the homogeneity.

Cover and Space. Although density and frequency indicate numbers and distribution, they do not indicate size, volume of space occupied, or amount of ground covered or shaded. These characteristics are desirable additional values that contribute materially to an understanding of the importance of a species in a stand, since they are closely related to dominance.

As suggested under Quadrat Methods (Chap. 3), cover can be estimated with some success or may be accurately determined by various devices for measurement and recording. When vegetation is stratified, the cover must be considered in terms of the stratum to which the species belongs. For rapid estimation, as well as for analysis of results, there is a distinct advantage at times in using cover classes rather than the specific values. Classes of the following number and magnitude are commonly used:

1. covering less than 5% of the ground surface
2. covering 5% to 25%
3. covering 25% to 50%
4. covering 50% to 75%
5. covering 75% to 100%

In studies of grassland, estimates and measurements of cover are extremely useful because the variations in size and form of grasses make counts difficult and of little value. For expressing cover, sometimes as area of coverage, sometimes as basal area of clumps, range ecologists frequently use the term, *density*. This usage is, of course, at variance with the usual phytosociological application and, consequently, leads to confusion of interpretation unless it is known, for example, that a "density list" [142] applied to grassland, refers to area or cover for each species, and that "square foot density" [390] also indicates coverage evaluated by a different method.

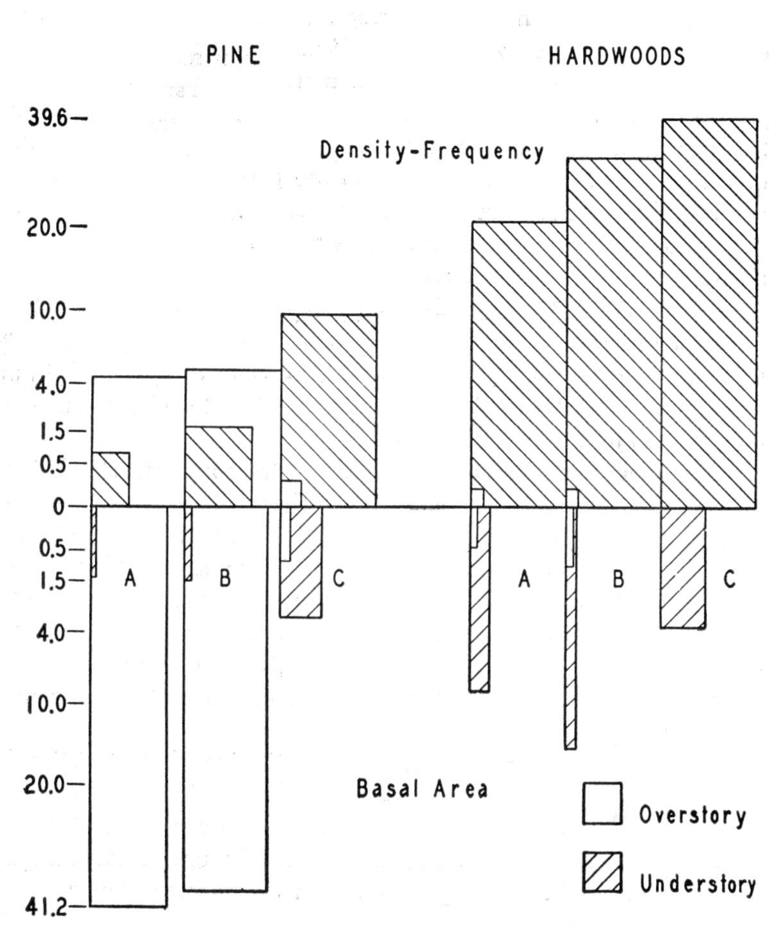

Fig. 26. *Bar diagrams of density, frequency, and basal area to compare pine and hardwood development in an unburned pine stand (A), with portions previously subjected to surface fire (B), and crown fire (C). Densities are indicated by the height of the columns above the zero line and frequencies by the width of the columns. Basal areas in square feet are indicated by the length of the columns below the zero line, and the width of these columns indicates percent of total basal area in the stand. Values for density and absolute basal area are reproduced on a sliding scale because of their wide range.*[281]

Determination of the volume of space occupied by species is difficult and has not been widely done. When all plants are small, cover alone serves very well, especially when strata are distinguished. With grasses, as in pasture studies, clipping and weighing the tops is sometimes necessary for accurate comparisons. In forest studies, the estimate of volume of standing timber as used by foresters can be used to advantage, but a more useful value is basal area. Diameters can be determined accurately and quickly with a diameter tape, and basal area, easily obtained from standard tables, can add much to an evaluation in terms of size and bulk that cannot be visualized through the other quantitative characters. This provides a particularly useful means of comparing the relative importance of species of trees and, in addition, permits analysis in terms of size or diameter classes among the sapling and understory individuals.

There is a growing feeling that cover of trees in forest stands is even more expressive of relative dominance than is basal area. Determination of cover involves measurements of crown projection at ground level, which can be slow and awkward unless a sighting level of some kind or a device with a combination of mirrors[55] is used to locate the limits of each crown. Very satisfactory results have been obtained in several types of forest by combining what is essentially a line intercept procedure with strips or plots of different widths for trees, shrubs, and herbs.[237]

When measurements of several quantitative characteristics have been made, their evaluation may often be facilitated by combining them graphically. Bar diagrams (Figs. 26, 110) are frequently helpful and phytographs[246] (Fig. 27) are especially useful for comparing characteristics of several species within or between stands.

Qualitative Characters. These characters, which include sociability, vitality, stratification, and periodicity, are mostly not derived from quadrat studies but from observation of, and wide experience with, the community. They describe the plan and organization of its components, which have been evaluated previously in terms of measurements and counts. When the quantitative analysis has been fairly complete, especially including density or cover in conjunction with frequency, and when strata have been analyzed separately, the qualitative characters are already largely included in the quantitative picture.

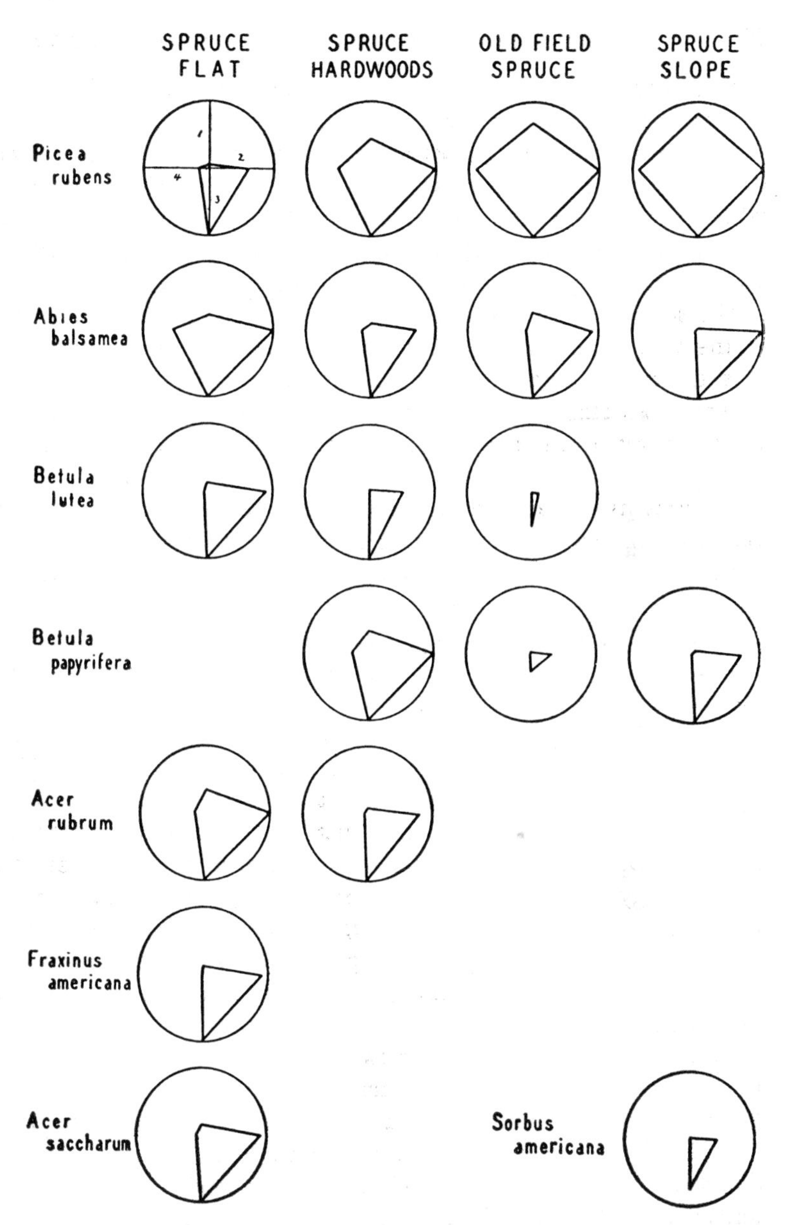

Fig. 27. *Phytographs showing the relative importance of the dominant species of trees in four types of pulpwood forest in northwestern Maine. Radius 1, percentage of total dominant density; Radius 2, percentage frequency; Radius 3, percentage of total size classes represented; Radius 4, percentage of total basal area. Each radius is scaled from 0 at the center to 100% at the circumference.*[291]

Sociability. This character evaluates the degree that individuals of a species are grouped or how they are distributed in a stand. It has also been expressed as *gregariousness* or *dispersion.* Each of the various scales used to indicate degree of sociability include expressions which range from plants occurring singly, as one extreme, through intermediate conditions (patches, colonies, or groups), to large colonies, mats, or pure stands at the opposite extreme.

The sociability of a species is not a constant, for it is determined by the habitat and the resulting competition of the species with which it is associated. Since habitat conditions are not constant and since communities change, especially in plant succession, the sociability of a species, even in the same locality, may change considerably.

Dispersion is a statistical expression that has been applied to sociability. If dispersion is normal, it implies a randomized distribution such as might be expected by chance. In hyperdispersion there is irregular distribution, which results in crowded individuals in some areas and their complete absence from others. Hypodispersion means that the arrangement is more regular than would be expected by chance, as, for instance, the plants in a cornfield. All of these conditions are recognizable in natural communities and, when density-frequency values have been determined, are noticeable in the data.

This phase of phytosociology that deals with the relationships of species in populations, as to their numbers, groupings, and arrangement, has become an area of intensive statistical study. Although the methodology of the statistician has served well for estimating error in sampling techniques and determining the degree of variance in data obtained by various methods, its greatest contributions have been, and probably will be, in clarifying the nature of community structure.

Approached on a statistical basis the nature of sociability or dispersion has been shown to be most complicated and far from understood. The many facets of population structure that are being explored are producing an abundant and highly technical literature, which cannot be appropriately reviewed in an introductory presentation. Randomness,[267] contagious distribution,[126] dispersion,[337] distribution patterns,[11] spatial patterns,[189] and abundance of species[312] are examples of expressions used in the titles of some recent publications whose bibliographies will lead quickly to many other studies of a similar nature. Actually, all have the same general objective, namely, a mathematical model that fits the distribution and association of species in natural populations.

Vitality. Not all species found in a given stand need belong to the community. Unless the plants are reproducing, they are not completely adapted to the conditions and may disappear entirely. Even species constantly present in a community may be derived from seeds produced elsewhere and transported by wind or some other agency. It becomes necessary, therefore, to know something of the vigor and prosperity of the species before classifying it as a true community member.

Vitality need not always be listed for all species, but it must be considered in evaluating their importance, whether it is done systematically or not. Vitality classes or degrees of vitality include (1) ephemeral adventives, which germinate occasionally but cannot increase, (2) plants maintaining themselves by vegetative means but not completing the life cycle, (3) well-developed plants, which regularly complete the life cycle.

Changes in the vitality of species are often indicators of community change or plant succession. Dominants decreasing in numbers and reproducing feebly indicate future radical changes. Rapidly increasing numbers of a species previously of little importance may suggest the new dominants to come.

Stratification. The necessity for recognizing the strata of a community becomes obvious when sampling is attempted. The several strata that may occur were described under sampling procedure. Diagrams of stratification combined with cover are often used effectively to show the relative significance of the several layers in a stand. The physical and physiological requirements of species in different strata can be appreciated fully only when the stratification both above and below ground is clearly worked out. Then the microenvironments of these strata may be considered in terms of cause and effect.

Periodicity. The conspicuous rhythmic phenomena in plant communities are those related to seasonal climatic change, and, of these phenological changes, the obvious ones have been given most attention. Fowering and fruiting periods have been noted for so long that they are fairly well known; in fact, phenology is often thought of as

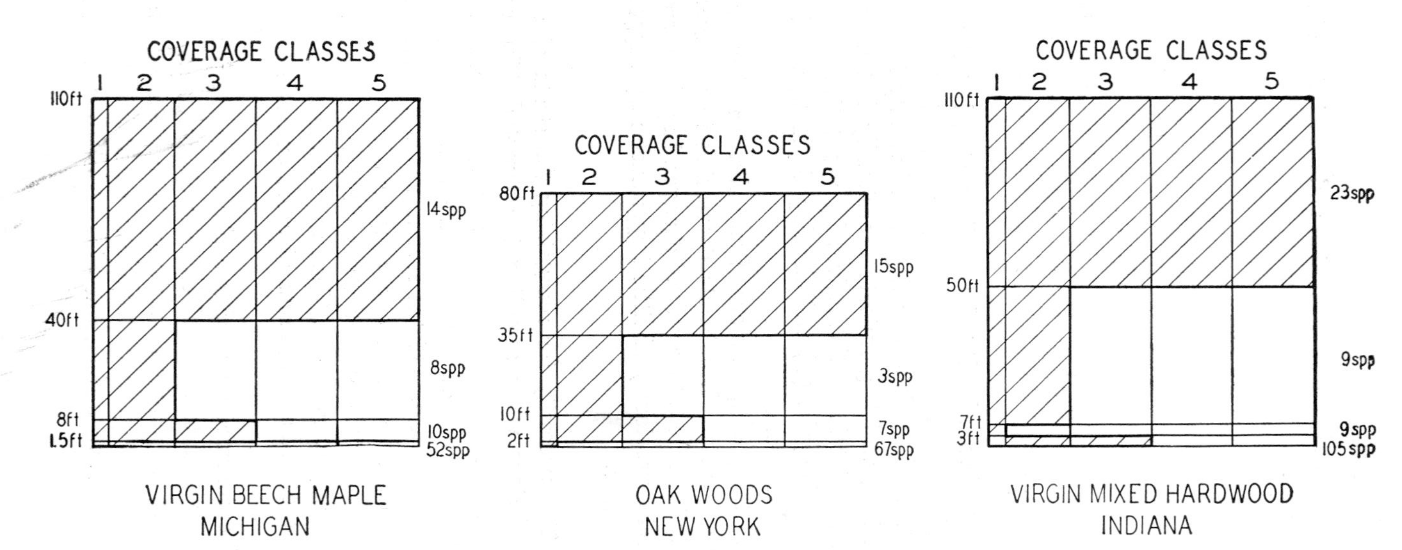

Fig. 28. *Coverage-stratification diagrams comparing types of hardwood forest. Number and height of strata are graphically shown as well as the coverage class and number of species involved in each stratum.*—Adapted from papers by S. A. Cain.

referring only to these phenomena. In community studies the terms *aspect dominance* and *seasonal dominance* have been used to describe situations in which a species or group of species appears to be dominant for a portion of the year, usually because of conspicuous floral characters.

Of equal importance to the community is the seasonal development of vegetative parts. The seasonal aspect of the individual may proceed through several phases, including a leafy period, a leafless period, a

Fig. 29. *Aspect dominance as illustrated by chandelier cactus* (Opuntia arborescens) *in a mixed prairie community* (Bouteloua-Hilaria). *El Paso County, Colo. The cactus makes up only 8.9 percent of the total cover.*—Photo by R. B. Livingston.

flowering period, a fruiting period, an embryo period, and perhaps others. Rarely will all the species of a community have these periods strictly coinciding. Consequently, in temperate climates the community as a whole usually has seasonal aspects, which are termed *vernal, estival, autumnal,* and *hibernal.* The structure and species of a community are strongly influenced by the extent to which periodic phenomena in the individuals are adjusted to each other.

Light and moisture conditions on the floor of a deciduous forest during the vernal period permit the growth and maturation of numerous herbs before the estival period. When the trees and shrubs are in full leaf, these herbs are already declining to a fruiting or resting condition and are unaffected by the reduced light and moisture available to them. These vernal herbs are a part of the community and must be so considered.

Another illustration of a periodic phenomenon that may be important in sociological relations is the time of growth. Height growth has been

Fig. 30. *Vernal aspect dominance of atamasco lily* (Zephyranthes atamasco) *in a low North Carolina meadow where only grasses, rushes, and sedges are visible a few weeks later.*—Photo by H. L. Blomquist.

systematically studied for numerous woody species, but the periods of root elongation are rarely known. Studies of loblolly pine[319] showed, surprisingly, that it makes some root growth in every month of the year. Even in the winter months, its roots are constantly coming in contact with new supplies of soil water, which fact may partially explain its ability to thrive in the southeastern states where transpiration may at times be fairly high during the winter.

Periodicity may be controlled by a variety of factors. Length of day affects the time of flowering, some species requiring long days, some short. The fall of leaves in autumn is a response not to temperature but to length of day. Desert vegetation may flower or not depending upon precipitation, and semidesert plants regularly flourish during the brief seasonal rains and exist in an almost dormant condition for the remainder of the year. Arctic and alpine areas usually receive little rain. The melting snow provides the moisture for vegetation. In situations where little snow accumulates or where it melts and disappears quickly, the vegetation is sparse and takes on a hibernal aspect very quickly. Where snow patches remain well into the summer and provide a water supply by melting gradually, the estival aspect remains for several weeks after plants in less favorable sites near by have gone to seed. Plants deeply buried under snow may not be exposed until so late in the season that conditions are unfavorable for flowering, and, as a result, they produce no fruit or seed.

Synthetic Characteristics

It has previously been pointed out that it is often desirable as well as practical to consider a community in the abstract as well as in the concrete sense. When a community is studied on this basis, it becomes necessary to observe numerous stands and to determine whether they actually do belong to the same community and to what extent they vary from each other. It is desirable also to know which species, singly or in combination, may be taken as indicators, which species are only incidental, which ones are always present, and which ones occur only when a stand develops under a given set of conditions.

Thus, for a complete synthetic analysis, it is desirable to have information on as many stands as possible, or at least enough stands to be representative of the whole. These should be distributed throughout the range of the community and under all the variety of conditions in which they develop. Again, to make a proper analysis, only those stands should be employed that are in a comparable stage of development or maturity and that are extensive enough to include all the important species and most of the anticipated variations.

Presence. A most useful synthetic character involves merely the degree of regularity with which a species occurs in the stands ob-

Table 4. *Portion of a presence table compiled from sixteen stands of virgin red fir* (Abies magnifica) *forest in the Sierra Nevada. Only* Abies magnifica *and* Pinus monticola, *of the trees, are constantly present (Class 5). Only one shrub,* Ribes viscosissimum, *is a constant, others falling in Class 3 or lower. Five herbs are constants, eight are mostly present (Class 4), and five are often present (Class 3). Eleven herbs of Class 2 (seldom present) and 46 of Class 1 (rare) are not listed.*[287]

Species	1	2	3	4	5	6	7	8	9	10	11	12	13	14	15	16
TREES																
Abies magnifica	x	x	x	x	x	x	x	x	x	x	x	x	x	x	x	x
Pinus monticola	x	x	x	x	..	x	x	x	x	x	x	x	x	x	x	..
Pinus contorta	..	..	x	..	..	x	x	x	x	x	x	x	x	x	x	x
Tsuga mertensiana	..	x	..	..	..	..	..	x	x	x	x	x	x	x	x	..
Abies concolor	..	..	x	x	x	x	x	x	..	..	..	..	..	x	..	x
Acer glabrum	..	..	x	x	..	..	..	..	..	..	..	..	..	x	..	..
SHRUBS																
Ribes viscosissimum	..	..	x	x	x	x	x	x	x	x	x	x	x	x	..	x
Symphoricarpos rotundifolius	..	..	x	..	x	..	..	x	x	..	x	..	..	x	..	x
Ribes montigenum	..	..	..	..	..	..	..	..	..	x	x	x	x	x	..	x
Sambucus racemosa	..	..	..	..	x	..	..	..	x	..	..	x	..	x	..	..
Ribes cereum	x	..	..	..	..	..	x	x	x	..	..	..	..	..	..	..
Spiraea densiflora	..	..	x	..	..	x	..	..	..	x	..	..	..	..	x	..
Arctostaphylos nevadensis	x	..	x	..	..	x	..	x	..	..	..	..	..	..	..	..
Symphoricarpos mollis	..	..	x	..	..	..	..	..	..	x	..	..	..	..	..	x
Lonicera conjugialis	..	..	..	..	..	..	x	..	..	x	x	..	..	..	..	..
Quercus vaccinifolia	..	..	..	x	..	..	..	..	..	x	..	..	..	..	..	..
Amelanchier alnifolia	..	..	x	x	..	..	..	..	..	..	..	..	..	..	..	..
Rubus parviflorus	..	..	..	..	x	..	..	..	..	..	..	..	..	..	..	..
HERBS																
Chrysopsis breweri	x	x	x	x	x	x	x	..	x	x	x	x	x	x	x	x
Monardella odoratissima	x	x	x	x	x	..	x	x	x	x	x	x	..	x	x	x
Gayophytum ramosissimum	..	x	x	..	x	x	x	x	x	x	x	x	x	x	x	x
Pedicularis semibarbata	..	..	x	x	x	x	x	x	x	x	x	x	x	x	x	x
Pirola picta	x	x	x	x	x	x	x	x	..	x	x	..	x	x	..	x
Phacelia hydrophylloides	..	..	x	x	x	..	x	x	x	..	x	x	x	x	x	x
Poa bolanderi	x	x	x	..	..	x	x	x	..	..	x	x	x	x	x	x
Arabis platysperma	x	..	x	x	x	..	x	x	x	..	x	x	..	x	x	..
Corallorrhiza maculata	..	x	x	x	x	..	x	x	..	x	x	..	x	..	x	x
Thalictrum fendleri	..	x	x	..	x	x	x	x	x	..	x	x	x	x	..	..
Kelloggia galioides	..	..	x	x	x	x	x	x	x	x	x	..	..	x	..	x
Erigeron salsuginosus var. *angustifolius*	x	x	..	x	..	x	x	..	..	x	x	..	x	..	x	x
Hieracium albiflorum	..	x	..	x	x	x	x	..	..	x	..	x	x	..	x	x
Lupinus andersoni var. *fulcratus*	x	x	x	x	x	..	..	..	..	x	x	..	..	..	x	x
Viola purpurea	..	..	x	x	..	..	x	..	x	..	x	..	x	x	x	x
Chimaphila umbellata	x	x	x	x	x	x	x	x	..	..	..	..	..	..	..	..
Penstemon gracilentus	..	..	x	x	x	x	x	x	x	..	..	..	..	..	..	..
Plagiobothrys hispidus	..	x	..	..	..	x	x	..	x	..	..	..	x	x	..	x

served. When the species present in each of the stands have been tabulated, the presence of each is expressed by the percentage of stands in which it occurred or by a five-degree scale of presence classes.

1. rare (1-20% of the stands)
2. seldom present (21-40%)
3. often present (41-60%)
4. mostly present (61-80%)
5. constantly present (81-100%)

The number of stands necessary for a study of presence as well as the necessary extent of stands cannot be arbitrarily stated. Mature, relatively homogeneous, undisturbed stands of virgin forest would require the observation of only small portions of individual stands and a relatively small number of stands to give dependable information. In younger, less stable vegetation, more stands and a wider observation would be necessary so that variation would be represented, and so that those species seldom present or rare would fall into their proper classes. What this minimal area should be and what the minimum

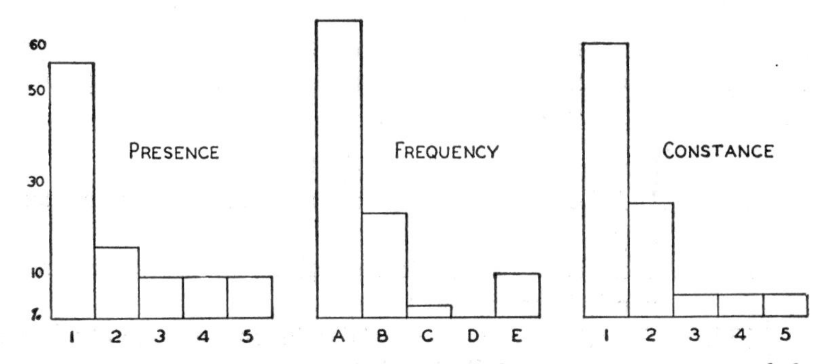

Fig. 31. *Presence, frequency, and constance diagrams for Sierran red fir forest, based on sixteen stands. The presence diagram is normal, especially in the absence of a second maximum. The constance diagram is constructed from regular quadrat data rather than a constance sample. Compared to a presence diagram it should show a material decrease in Class 1 because of the greater odds on discovery of a single plant of an accidental species in a restricted area. Surprisingly, with only forty species, it has the same form as the presence diagram (ninety-seven species) although the high constance classes are reduced. The frequency graph is normal, and typical of relatively homogeneous stands.*

number of species might be for the community must largely be determined by experience and familiarity with the community.

Constance. When a unit area in each stand instead of the entire stand is used for listing species, as for presence, the values are termed *constance.* There is thus no fundamental difference between presence and constance. The latter has the advantage of eliminating discrepancies resulting from sampling stands of unequal size. The lower classes of constance are more uniform than those of presence, for the larger the area examined the greater the number of incidental species encountered.

Constancy bears a relationship to the abstract community very similar to that of frequency in the concrete community. The problems of minimal area are similar; and species: area curves, as used in frequency determinations, may be helpful. Nevertheless, experience may have to be the basis of decision. Both concepts are concerned with homogeneity, the one with that of the stand, the other with that of the abstract community. If constance values are divided into five classes and these are diagrammed as for frequency, the results are quite different. Instead of two maxima as in frequency, only the classes representing irregular occurrences are high, and each succeeding class is apt to include fewer species.

Fidelity. This character is indicative of the degree with which a species is restricted to a particular kind of community. Species may be grouped into five fidelity classes.

Fid. 1. Strangers, appearing accidentally
Fid. 2. Indifferents, without pronounced affinity for any community
Fid. 3. Preferents, present in several communities but predominantly in one of them
Fid. 4. Selectives, found especially in one community but met with occasionally in others.
Fid. 5. Exclusives, found completely, or almost so, in only one community

Species with fidelities 3-5 are termed *characteristic species* in a community. Positive establishment of which species are characteristic is possible only after all communities of a region have been studied sociologically. Approximations can, of course, be made by those of wide experience, but even then the assigned values must be considered with skepticism. If fidelity values could be accurately determined, they would contribute strongly to the recognition and classification of a community. However, complete regional sociological studies are so rare that precise statements of fidelity for species are usually impractical for most communities. As a result, constance, an absolute value determined within the community in question, is commonly used as a criterion of the sociologically important species. It may be, as some believe, that constance has greater significance than fidelity for this purpose. It should be noted that characteristic species are more responsive to habitat variations and are consequently of greater indicator significance than are, in general, the species of high constance. The availability of both values is desirable. That fidelity values may yet be made precise is suggested by a new statistical approach that deserves testing in the recognition of characteristic species. An index of fidelity and an index of indicator significance, both based on frequencies (or constancies), are proposed which make possible the rating of species as to their significance and the distinguishing of communities as well.[165]

Physiognomic Characteristics

Life-form. Of the several systems for classifying life-forms of plants[69] the most widely used is based primarily on the position of and degree of protection afforded the perennating bud during unfavorable growing seasons.[315] On this basis, five major classes are recognized:

Phanerophytes (P)—trees and shrubs, buds least protected; predominating in tropical floras.
Chamaephytes (Ch)—buds at ground level during unfavorable seasons; compose increasing proportion of flora with increasing distance from tropics.
Hemicryptophytes (H)—dormant buds just beneath or in soil surface; important in temperate climates.
Cryptophytes (Cr)—buds deeply buried and food-storing (e.g. bulbs); adapted to extremes of environment.
Therophytes (Th)—depend only on seeds for survival; best protected of all groups against cold and heat during dormancy.

Phanerophytes are often subdivided on the basis of height, and ferns and fern allies are commonly put in a separate category. Other classes sometimes considered are epiphytes and microscopic organisms.

The proportions of species in these classes are determined for a particular flora, and comparison is made with the proportions in the *normal spectrum,* a standard based on a group of species randomly selected from the flora of the world. Every regional spectrum will have at least one class whose percentage is substantially higher than that of the normal, and it is taken as an indicator of the climate and environments of the region involved. In general, deserts have therophytic climates, the tropics are phenerophytic, alpine and arctic climates are chamaephytic, and most temperate regions are hemicryptophytic.

Table 5. *Life-form spectra for the four major climatic types and the normal spectrum.*[315]

Locality	No. spp.	Life-form Classes				
		Ph	Ch	H	Cr	Th
Phanerophytic:						
St. Thomas and St. Jan	904	**61**	12	9	4	14
Therophytic:						
Death Valley, Calif.	294	26	7	18	7	**42**
Hemicryptophytic:						
Altamaha, Ga.	717	23	4	**55**	10	8
Chamaephytic:						
St. Laurence Is., Alaska	126	0	**23**	61	15	1
Normal Spectrum	1000	46	9	26	6	13

The system has been used for comparing life-form characteristics of different regions and for demonstrating progressive changes with both altitude and latitude. It has also been used to compare specific communities in the same and different areas and the life-form changes associated with succession. When dealing with specific communities for which sociological data are available, more realistic comparisons are possible. Instead of using a floristic basis in which a species counts as a unit, regardless of its abundance, each species count is weighted according to its frequency[68] or its cover[389] to give it a value in proportion to its ecological significance. Evidence is sufficient to show that properly handled life-form statistics reflect differences in general climates and microclimates as well.

Leaf-size. There has been little systematic study of leaf-size although there is consistent variation within and between communities, and vegetation of clearly differing environments has obviously distinct leaf-size characteristics. A system of classification devised by Raunkiaer[315] could be used in combination with his life-form concepts. It has six size-classes with 25 sq. mm. for the lower limit and succeeding classes each nine times larger than the last:

Class 1—Leptophyll — 25 sq. mm.
2—Nanophyll — 225 sq. mm.
3—Microphyll — 2,025 sq. mm.
4—Mesophyll — 18,222 sq. mm.
5—Macrophyll — 164,025 sq. mm.
6—Megaphyll — larger than Class 5.

The leaves are first grouped as being evergreen or deciduous, simple or compound, and then assigned to classes by size and their proportions calculated. These proportions can certainly be related to environments, and more knowledge of leaf-sizes might well contribute to a better understanding of physiological processes in community relationships.

Coefficient of Community. When comparing two communities or the vegetation of two regions, a mathematical expression of the similarity of lists of species may be useful. If community X is compared to Y, the number of species common to both, expressed as a percentage of the total number for X plus Y has been termed the *coefficient of community.* The same principle can be used for evaluating variation or similarity among several stands of an abstract community. Then, however, each must be compared with a standard or list of the characteristic species of the community as a whole.[197]

A similar index,[226] which incorporates quantitative data for the species, goes beyond floristic comparison and has greater meaning in community analysis. Measurements (e.g. frequency) are expressed as percentages to have comparable data, and the proportion of similarity between two communities is expressed by the ratio of percentages shared to total percentages. The coefficient is calculated by the formula $2w/a + b$ times 100. For species occurring in both stands the lowest of each pair of percentages is summed (w) and doubled to represent the degree that the two communities share the measured characteristic. This value divided by the sum of all species percentages in both stands $(a + b)$ gives the coefficient of similarity. If several stands are to be compared, a coefficient is calculated for each one paired with each of the others to permit arranging them in the order of their similarity.[111]

Objectives Determine Procedure

If these several sociological concepts are grouped systematically in tabular form, their relationships become clearer (Table 6). Such a

Table 6. *A summary of sociological concepts that permits presentation of the important data for a community in a single tabulation. The quantitative data (1) are derived from quadrats; the analytic data (A) from the study of some one community; the synthetic data (B) from the study of several different examples (stands) of the same community.*—After Cain.[59]

	SOCIOLOGICAL SUMMARY												
	I Organization											II Physiognomy	
	A Analytic								B Synthetic				
	1 Quantitative				2 Qualitative								
	a	b	c	d	a	b	c	d	a	b	c	a	b
SPECIES LIST	Abundance	Density	Dominance	Frequence	Sociability	Vitality	Periodicity	Stratification	Presence	Constance	Fidelity	Life Form	Leaf Size
Classes	1-5	—	1-5	1-5	1-5	1-3	1-4	1-4	1-5	1-5	1-5	1-10	1-6

grouping has the further usefulness of presenting tabulation of values obtained in the field in compact and logical order for interpretation.

When the objective is merely to describe a community as completely as possible, it might well be desirable to have such a table completely filled out. In studies involving the application of phytosociological methods to special problems it is frequently only necessary to use a few of the values. This does not mean that not all are of significance, or that some can be ignored entirely. Rather, it suggests that each has its uses, and that some are applicable where others are not.

The limitations and possibilities of usefulness of the several concepts become increasingly understandable after one has had some experience with them. Nevertheless, selection of the most useful values for study and application to a particular problem always remains a matter for serious consideration. The concepts to be used must be selected in terms of their contribution to the object of the study, the time available, and the labor involved.

General References

J. Braun-Blanquet. *Plant Sociology: The Study of Plant Communities.*

S. A. Cain. Concerning Certain Phytosociological Concepts.

C. Raunkiaer. *The Life Forms of Plants and Statistical Plant Geography; Being the collected papers of C. Raunkiaer.*

Part Three: FACTORS CONTROLLING THE COMMUNITY: THE ENVIRONMENT

Vegetational analysis gives the information necessary to describe and name a community and provides data that can be used to compare it with other communities or with itself after a lapse of time or an experimental treatment. This in itself is worth while, but the ecologist has the added objective of correlating the vegetational record so obtained with the environment. To interpret the vegetational statistics, and to explain them in terms of cause and effect, leads to an analysis of the environment and its relationships to the community.

The environment is a complex of many factors interacting upon each other and the vegetation. It is possible and appropriate to think of it as a whole that, together with the vegetation, forms an energy system of forces and conditions, an ecosystem. However, our knowledge of the system is so incomplete and its complexities are so great that a discussion on this basis is, as yet, somewhat impractical. The effects of specific individual factors, under controlled conditions, are not all known, and the interactions of factors under natural conditions are even less well understood. There appears, therefore, no alternative but to segregate major or general factors of the environment and to consider their independent effects at the same time recognizing that plant responses are a part of the entire inter-related system. Knowledge of the operation of single factors yields many explanations and a greater appreciation of the interacting variables. The chapters of this section, therefore, deal successively with climatic, physiographic, and biological factors as each may operate in the complex of factors termed environment.

Chapter 5 Climatic Factors: Radiant Energy, Light, and Temperature

The sun is the source of the earth's radiant energy (insolation). This energy, radiating as waves, includes those wave lengths of the visible spectrum that we term "light" and those that lie just beyond the visible spectrum, called "heat" if slightly longer, or "ultraviolet light" if shorter. The amount of insolation reaching the earth is always reduced because of absorption by the atmosphere (6-8 percent), and as much as 40 percent may be reflected by clouds. The remainder reaching soil or water on the earth may be further varied by such factors as distance from the sun at different seasons, duration of radiation, and the angle of the rays with the earth's surface. The last determines the amount of air through which the rays pass, modifies the amount of reflection and absorption, and likewise controls the amount of energy falling on a unit area simply by spreading or concentrating a given amount of energy over more or less space. With these things in mind, one can more easily explain insolational variation with latitude and topography.

Insolation varies only slightly at the Equator, because the angle of the sun's rays never exceeds 23½° from zenith, and the days are uniformly twelve hours long. Twice a year, on March 21 and September 21, called the equinoxes, the sun is at zenith at the Equator at noon, and its circle of illumination exactly reaches the North and South poles simultaneously. After March 21, because of the tilt of the earth's axis, the North Pole comes progressively nearer to the sun until June 21, after which the shift is reversed to bring the pole back to the equinox position by September 21. The North Pole's movement away from the sun continues until December 21, after which it starts its shift back to the June position again. The shifting of the pole toward the sun causes the circle of illumination to extend far beyond the pole and results in continuous insolation at the pole during the June solstice; but, since the December solstice results in diametrically opposite con-

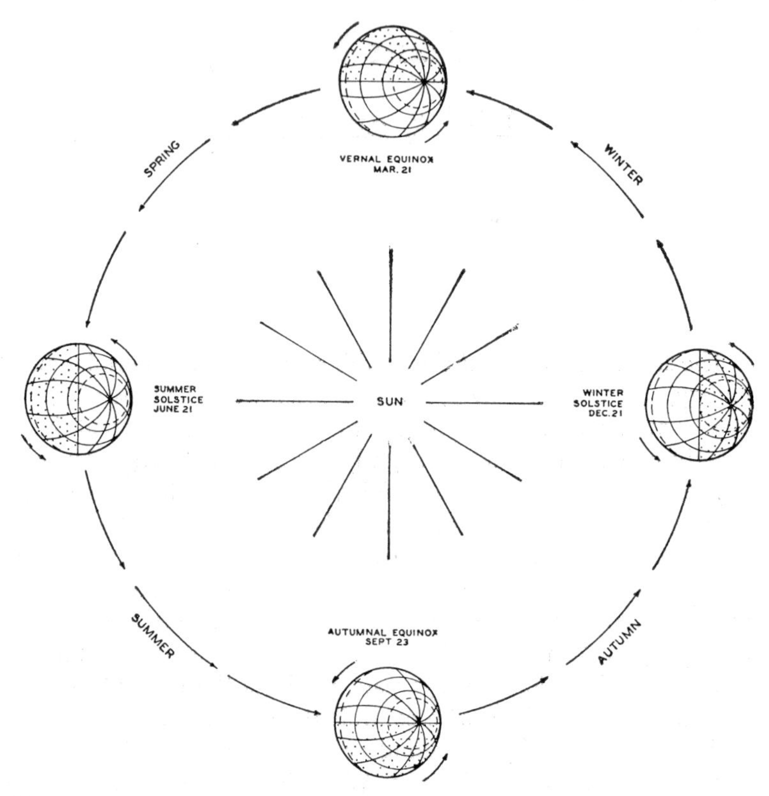

Fig. 32. *Diagrammatic representation of the changing position of the earth with respect to the sun and its relationship to insolation and change of seasons in the Northern Hemisphere.*—Adapted from Trewartha.[408]

ditions, it represents a period without insolation. Conditions in the Southern Hemisphere are, of course, always exactly reversed.

Thus, because of differences in insolation, we have seasons marked by variation in length of day and temperature. Since the periodic differences in insolation become more marked with distance from the Equator, the seasons likewise become more distinct with increasing latitude. The greatest total insolation, however, occurs at the

Equator and decreases with distance from the Equator in spite of the increasing length of day. Toward the poles, intensity of insolation is reduced because of the increasing angle of incidence.

These introductory statements refer to insolation as a whole. We may now more conveniently consider separately the visible portion of the spectrum known as light, and those longer, invisible wave lengths

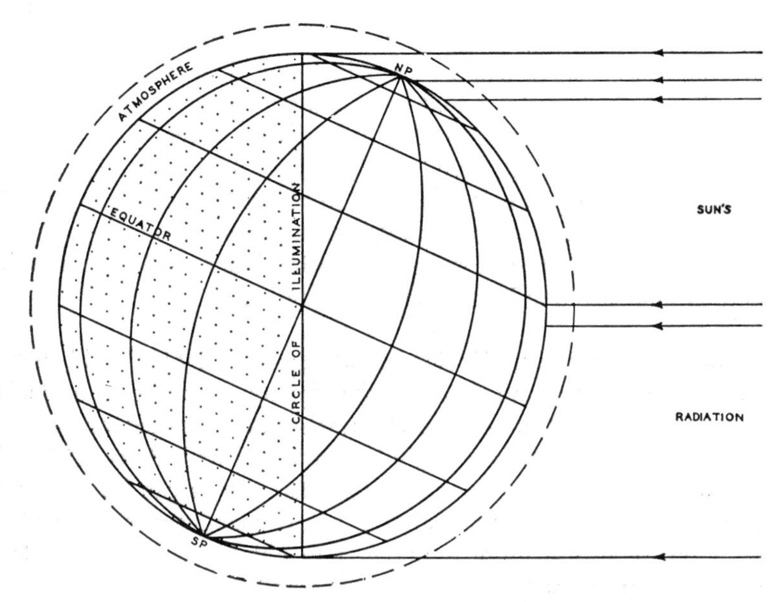

Fig. 33. *Circle of illumination, areas of daylight and darkness, angles of sun's rays at different latitudes, and differences in areas affected and thickness of atmosphere penetrated at time of summer solstice.*—Adapted from Ward and Powers.[418]

known as heat, whose presence or absence are expressed as temperature.

Light

That portion of the sun's radiant energy that forms the visible spectrum and that we commonly term "light" strikes the earth in quantities far in excess of the apparent needs of plants. Although green plants, with very few exceptions, are the only organisms that can directly convert this energy to their own use, they actually change to potential energy only about one percent of the light energy they receive. It has been estimated that, of the total solar energy falling upon a given field of corn during a growing season, only 0.13 percent can be "stored" as potential energy. However, this also suggests that, to function normally, plants require much more light energy than they actually use. Not all wave lengths are used equally. Green light is mostly reflected or transmitted. The longer wave lengths, in the red end of the spectrum, are much more effective in photosynthesis than are the shorter wave lengths of yellow and blue. Not all species are equally efficient under equal illumination. Some require a specific limited range of intensities, either high or low, while others have wide tolerance. Some need certain lengths of day and thus function normally only in particular seasons. To add to the difficulties of interpreting plant-light relationships, it is not always possible to distinguish between light effects and those of total insolation, which include heat and its influence on physiological processes.

Light Measurements. Ecological studies of light should not be casually undertaken in spite of the apparent simplicity of making measurements with modern instruments. As suggested above, plant responses and light values rarely bear a simple and direct relationship to each other. Whether or not these relationships can be interpreted may depend upon proper planning before making measurements. In addition, there are problems related to obtaining measurements for ecological purposes that must be considered.

Chemical, illuminating, electrical, and heating effects of light are measurable, and for each a different type of instrument is used.[371] Field ecologists have largely abandoned the first two approaches in favor of electrical measurements since the perfection of compact, sturdy photoelectric instruments with which accurate and rapid determinations can be made. These are sensitive to approximately the same portion of the spectrum as is the human eye. Since they are selective instruments, there may be some question of the advisability of generalizing as to plant responses in relation to the measurements they obtain. In most field studies this does not become a serious limitation, because the usual objective is to compare relative intensities of light in two or more situations or habitats. For this purpose, the photoelectric method is quite usable.

The method has, however, other limitations, and its use requires certain precautions. Preferably two or more instruments should be

Table 7. *Light measurements, in foot candles, made with a Weston photometer in a mixed pine-hardwood stand between 12:00 and 1:00* P.M. *when full sunlight was 9,500 foot candles. Readings taken along three lines, at three-foot intervals, at a height of three feet. After completion of a line, the measurements were repeated at the same points. Note the great variation in readings at the same points at different times (sun flecks) and that some points are apparently much less shaded than others.*

	Line 1		Line 2		Line 3	
	300	500	500	400	100	100
	300	500	500	300	200	400
	200	300	300	400	300	100
	200	300	400	200	200	600
	300	300	300	200	200	200
	300	200	300	100	100	200
	200	200	400	3000	200	300
	100	200	3600	2400	200	400
	2400	200	4400	3600	300	300
	300	200	400	200	100	200
	300	200	2000	400	200	200
	200	2400	400	400	100	200
	500	200	400	300	100	200
	500	200	600	400	100	100
	500	200	1000	800	100	200
	300	200	200	300	200	200
	300	200	200	400	1600	200
	4200	200	300	200	100	100
	200	200	1200	5000	100	100
	200	2100	1600	2400	200	100
	200	600	300	4400	100	100
	300	300	300	300	100	100
	400	500	100	200	300	100
	200	200	300	200	200	200
	400	200	800	2000	200	100
Aver.	532	432	832	1140	224	200

Average for the stand = 560 ft. candles. 5.9% of full sunlight.

available, and the readings should be made simultaneously. Even so, readings should be made only on a clear day and, when periodic observations are made, at the same time of day. Results should be expressed as percentages of full sunlight at the time when each observation is made. At sea level this would be approximately ten thousand foot candles on a clear day at noon, but values as high as twelve thousand foot candles have been obtained in the clear air of high mountains. If for any reason the readings in the open are low on a given day, no further observations should be made.

Because of its flat sensitive surface, the instrument can be operated in only one plane at a time. If readings are made simultaneously at noon, with the instrument in a horizontal plane, many complicating factors are automatically eliminated. The instruments are extremely sensitive to slight variations in light, and this necessitates numerous readings to arrive at average conditions. The slightest air movement shifts the position of leaves and permits bright sun flecks to come through a forest canopy. These flecks come and go, first at one point and then at another, and cannot be ignored in evaluating light in a stand. Their inclusion is best accomplished by making observations at a rather large number of uniformly or randomly distributed predetermined points and averaging the results. In all instances, the instrument should be in the same position relative to the observer and the ground.

A sensitive surface of spherical form is usually more desirable than a flat one. Where reflected light is appreciable, a sphere will record from all directions. If a continuous record is to be obtained, the sphere records accurately, because one-half its surface always faces the sun regardless of its position. Several radiometers, which measure heat effects and are nonselective of wave lengths, are spherical in form and are advantageous in other respects. If a photoelectric cell is given more than a short exposure to strong light, the current it generates falls off because of solarization, but the radiometer can be exposed indefinitely without such effects. It is, therefore, adaptable to continuous operation with a recording device. Various instruments for continuous recording of duration or intensity are available commercially[266] and less expensive adaptations have been devised.[384]

Such equipment is not always available to the field ecologist, but, even so, some form of measurement is far more dependable than an estimate. Good approximations of light intensity may be obtained with photographic light meters even though they are not calibrated in foot candles. Useful values are obtainable by exposing black and white bulb atmometers (page 123) in pairs. When one pair is exposed in the open, and differences from pairs in near-by habitats are expressed as percentages of the value in full sunlight, the results may be quite as satisfactory as with more elaborate equipment. Since the atmometers would be operating continuously, they might even be more meaningful in terms of the vegetation.

Light Variations. The biologically important variations of light are those in intensity and quality. These occur periodically, recurring seasonally and daily to a degree that is determined by latitude[211] as discussed under the general heading of insolation. Of course, altitude modifies the regional variations, and topography results in more localized variation through the effects of angle of slope and direction of exposure. The principles are discussed later (p. 109) in some detail. It should be sufficient here to present an illustration of how slope and exposure affect light in the southern Appalachian Mountains (Fig. 34).[58]

The principles involved here have been developed into a system for classifying habitats and vegetation in semi-arid, once-forested, Palestine.[42] On the assumption that limiting factors of specific environments will be related to solar energy in a region with uniform precipitation and geology, insolation totals were determined for various slopes and exposures to arrive at an I-E (inclination-exposure) factor for any site. Because the amplitude of I-E tolerance varies widely, this factor can be used to typify any component of the complex vegetational mosaic. Likewise, it gives an effective evaluation of sites, especially for afforestation or the introduction of exotics.

Variation in quality of light is not so obvious as variation of intensity. Quality, however, is variable, largely because of the same factors that modify intensity, for the amount of absorption and diffusion by the atmosphere determines what wave lengths reach the earth. Clouds, fog, smoke, dust, or even atmospheric moisture alone, increase diffusion and absorption; and, as a consequence, dry regions receive more light than humid ones, and open country receives more light than smoky or smog-afflicted cities. The greater the diffusion, the higher the percentage of red light and the lower the percentage of blue reaching the earth.

A local variation of far greater general ecological importance is that produced by vegetation of one stratum upon that of a lesser stratum beneath it. Because plants growing in the shade of others

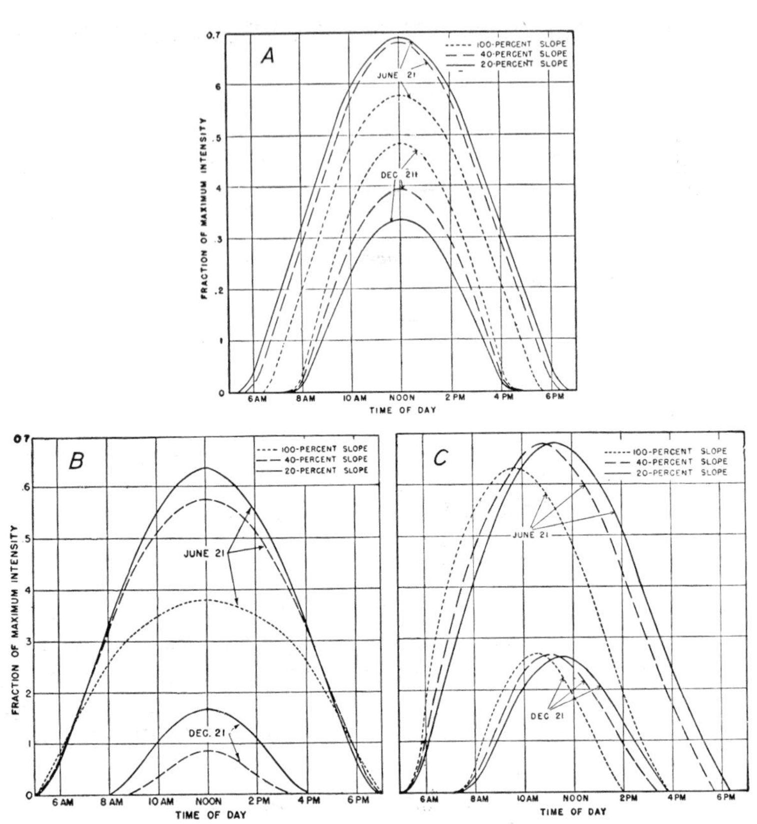

Fig. 34. *Intensity of radiation received at different times of day on* (A) *south,* (B) *north, and* (C) *east slopes in the southern Appalachians, on June 21 and on December 21. For S. exposure, in summer, the 20 percent slope receives greatest radiation because it forms an angle of almost 90° with the sun's rays at noon. In winter, when the sun is low, the 100 percent slope receives more radiation than the 20 or the 40 percent slope. For N. exposures, in summer, 20 percent slopes receive almost as much radiation as 20 percent south slopes. In December, 100 percent N. slopes are in complete topographic shade but 100 percent S. slopes receive 48 percent of maximum radiation at noon. Curves for west slopes would be mirror images of those for east slopes.*—From Byram and Jemison.[58]

receive only the light that is not absorbed or reflected, they must be adapted to functioning with reduced light intensity (often reduced to 15 percent or less) of somewhat different quality (reduced red and blue light) than those in full sunlight receive. Consequently, there are species representing a wide range of tolerance to shade, for no forest is so dense that nothing can grow beneath it, even when there is a reduction to 1 percent or less of full sunlight, as under some tropical forests. The reduction of light intensity under a forest canopy is probably of more ecological importance than the change in quality.

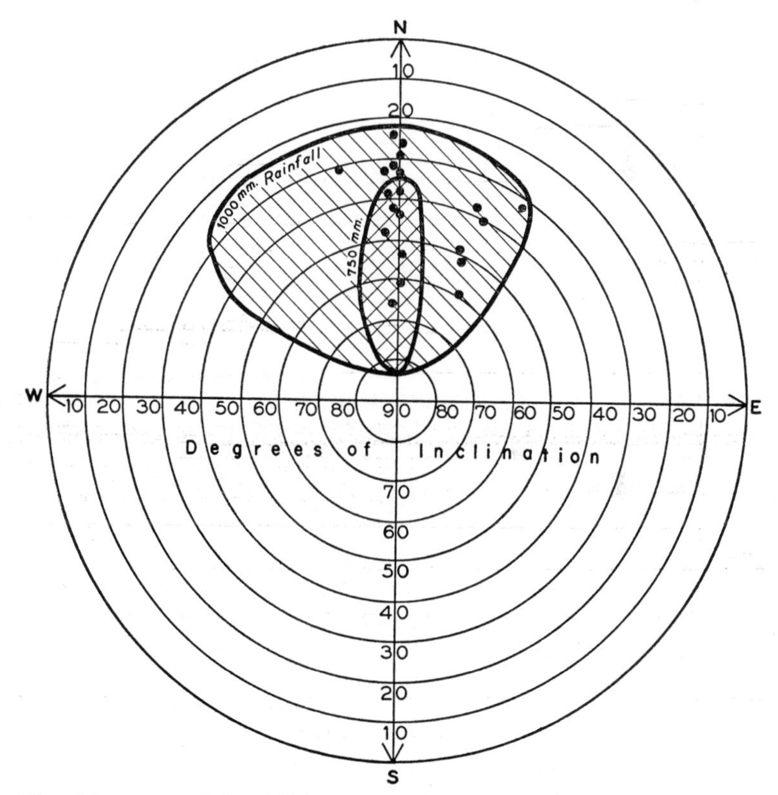

Fig. 35. *The range of the I-E factor to which the Kermes oak-laurel community is adapted. Each circle represents an angle of slope (0° level terrain). Dots indicate positions of stands in which actual records were obtained. Species of the community may be successfully planted in sites falling anywhere within the shaded pattern.*—After Boyko.[42]

Shade Tolerance. When, for any reason, individuals are eliminated from the canopy of a forest stand, their replacements will be from the seedlings and saplings already present below them at the time. If seedlings of the dominant species are shade tolerant, the floristic composition of the stand remains unchanged. If the dominants cannot reproduce in their own shade, but other species can, the latter will gradually take over dominance to form a new community. Such a succession of dominants is common under natural conditions, and

Fig. 36. *Trenched plot in a loblolly pine stand (40-50 yr.) four years after initiation (see Fig. 12). Contrast vegetation on trenched plot with floor of surrounding forest and control plot in foreground.*—Photo by C. F. Korstian.[219]

light is usually involved as a factor. The replacement of pine dominance by hardwoods or by spruce-fir is illustrative.

In forestry, when a forest stand is thinned or clear cut, the same situation arises. The reproducing species, which are the new replacements, may or may not be desirable, and the question of how to encourage or inhibit them, depending upon circumstances, has led to much study and theorizing on the causes of *shade tolerance.*

Since light is obviously reduced under a forest stand, it was once assumed rather generally that light is the controlling factor. Studies of "trenched plots" under forest stands gave results interpreted by many workers as indicating a greater significance for water since, within these plots, shade-intolerant species for a time grew well when root competition for water and nutrients was eliminated by cutting off the roots of the dominant trees.[219, 403] Extensive investigations of conifer reproduction in the Lake States indicate that, for each light intensity, growth could be increased by reducing root competition, and that at each level of root competition growth could be increased by increasing light.[372] Observations of the reproduction of certain southern pines[290] indicate that these shade-intolerant species may successfully meet extreme root competition if light is sufficient. It would seem that the successful growth of a seedling under a forest canopy may depend upon its ability to manufacture enough food with the light available to grow enough roots to meet the competition of the trees established there. Undoubtedly, shade tolerance cannot be explained on the basis of a single factor.

Physiological Responses. When the supply of food in an organism falls and remains below what is required for respiration and assimilation, the organism cannot continue to function normally and must eventually die. Since a green plant produces its carbohydrates through photosynthesis, the process must proceed at a rate sufficient at least to satisfy the immediate needs of the plant if growth is to be normal. Light, which provides the energy for photosynthesis, is sufficient during the growing season to supply plant needs anywhere on the earth. In fact, light intensities may be too high for some plants to grow in full sunlight, their seedlings being especially subject to injury. Such plants might well be restricted to habitats with partial shade; if their photosynthetic efficiency is insufficient to maintain them in forest shade, they might thrive in regions where light intensity is reduced by cloudiness or fog. The occurrence of some species in particular habitats might actually be determined by their high light requirement for germination. Probably the range of a species is rarely determined by light intensity alone, however; for it must be remembered that light effects are apparent in several processes and activities, which can rarely be considered independently. The production of chlorophyll, the opening and closing of stomata, and the formation of auxins are examples of light-conditioned phenomena with widely differing effects; but these activities must be considered in relation to each other when interpreting plant responses.

The production of chlorophyll, although (with a very few exceptions) accomplished only in the presence of light, is perhaps more apt to become limiting or significant in high than in low light intensities. Available evidence indicates a greater production of chlorophyll with decreasing light intensity and an ability of most plants to produce chlorophyll at light intensities considerably below those necessary for effective photosynthesis.

The opening and closing of stomata can usually be correlated with light, but there are enough exceptions to give warning against generalizations or interpretations based on the principle of alternate opening and closing with light and darkness. In some plants, stomata may open at night; in others, light seems not to be a controlling factor. Where stomatal movement seems directly responsive to light, other factors may at any time become more important and modify or counteract the effects of light, as when stomata close during the day if the water supply is insufficient. However, stomatal movement is usually correlated with light changes and, when other conditions are favorable, is apparently caused by turgidity changes in the guard cells resulting from metabolic activity, which varies with light. The opening and closing in turn may modify effects of light by varying gas exchange related to photosynthesis and rate of loss of water by transpiration.

The production of certain auxins or growth-controlling substances in plants is inhibited by light. As a result, through them, size, shape, movements, and orientation of parts may be influenced by light. A plant grown in complete darkness, since it produces a maximum of auxins, elongates excessively, with poorly differentiated tissues throughout and with almost no supporting structure. These characteristics in an intermediate condition are often recognizable in plants grown in heavy shade, as under a forest canopy or in close stands where plants shade each other. Such plants tend to be tall and spindly with widely spaced nodes and relatively few leaves. The better the light, the stouter and more compact the individual will be.

Should illumination be one-sided, the increased production of auxins on the shaded side usually stimulates sufficient extra elongation on that side to turn the growing portion of the stem toward the light. Some species—sunflower, for instance—are so sensitive to such differences of light that the floral portions shift from east to west with the sun daily as differential elongation in the stem progresses from one shaded side to the other.

The orientation of vegetative parts is such that every leaf receives a portion of the light available. Genetic differences determine whether the leaves are exposed in the form of a rosette or in a mosaic pattern, or whether they are supported by a spirelike central axis or several spreading branches, each of about equal size. The variations within such a general plan probably result from effects of auxins on growth of petioles and secondary branches.

Leaves normally become arranged with their broadest surface

Fig. 37. *Seedling of turkey oak* (Quercus laevis), *a sandhill species, whose leaves have already assumed the vertical position they maintain throughout life.*

exposed outward and upward on the side of the plant where they grow. This results in a maximum exposure to the available light at that point. However, plants growing under conditions of excessive light, especially where there is reflection from light-colored soil, not uncommonly have their leaves in a profile position, which, of course, reduces the light to which they are exposed. Turkey oak (*Quercus laevis*), which grows on sand dunes in the southeastern United States, regularly develops a twist in the petiole that turns every blade vertically. The leaves of wild lettuce (*Lactuca scariola*) are vertical when grown in full sunlight but maintain a horizontal position in the shade. Several so-called compass plants have leaves that are not only vertical but that also face east and west, exposing only their edges to the sun's rays at midday.

Plants growing in close stands characteristically lose leaves, and usually branches, from below when the light penetrates insufficiently to maintain necessary photosynthesis. Most monocots with grasslike leaves and underground stems are unaffected because their upright

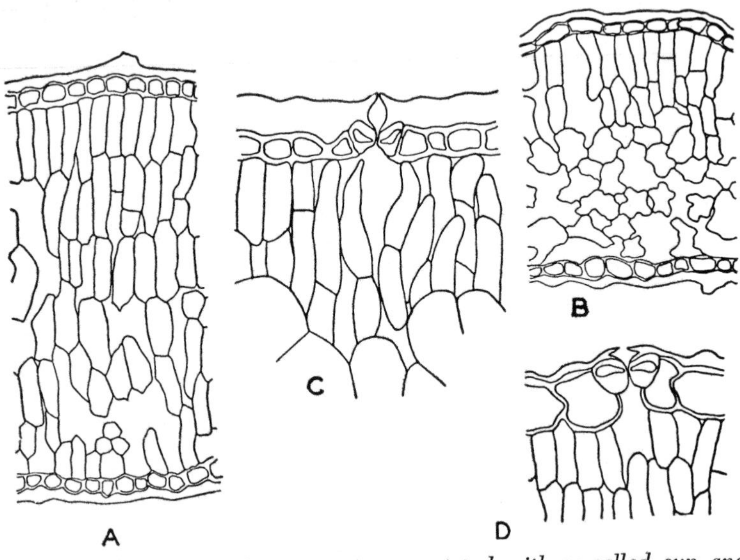

Fig. 38. *The anatomical characteristics associated with so-called sun and shade leaves of two chaparral species.* (*A*) Arctostaphylos tomentosa *from normal xeric habitat,* (*B*) *from mesic habitat.* (*C*) Adenostoma fasciculatum *from normal xeric habitat,* (*D*) *from stump sprout. Note differences in thickness of leaf and cuticle, and proportion of palisade to sponge tissue.*—From Cooper.[97]

linear leaves permit light to penetrate to their bases. In forest stands, this self-pruning may be economically important. Conifers that self-prune grow tall and straight with few knots and smooth grain. In contrast, those with dead branches down to their bases are difficult to handle and produce much less valuable wood when finally cut.

Leaves grown in full sunlight tend to be smaller, thicker, and tougher than leaves grown in the shade. This is particularly noticeable in plants of the same species and may also be observed on the same plant. A forest-grown tree may have sun leaves at the top and shade leaves near the base, or in the interior of its crown.

Certain structural differences are associated with the two types of leaves. Intense light results in elongated palisade cells and often the production of two or more layers of them. Conversely, weak illumination favors the production of sponge cells. A leaf that, with average illumination, has a single layer of palisade and several layers of sponge cells might have had, in intense light, two or three layers of palisade and a proportionate reduction in sponge tissue. In reduced light the sponge tissue is increased at the expense of the palisade. In extreme

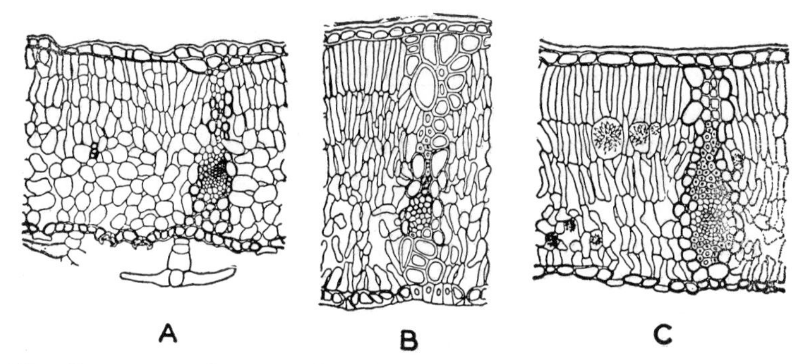

Fig. 39. *Structure of leaves of broad sclerophyll forest trees (A)* Castanopsis chrysophylla, *(B)* Quercus agrifolia, *(C)* Quercus durata. *Note compact structure, multiple layers of palisade, and tendency for all mesophyll to be palisade-like. Note also struts of mechanical tissue from epidermis to epidermis.*—From Cooper.[97]

cases there may be no palisade or no sponge tissue. The thickness of cutin and the amount of supporting tissue in the veins are likewise greater or less depending upon light intensity. These characters affect the relative toughness of the leaf.

What forces cause a developing cell to elongate at right angles to the leaf surface to form palisade or parallel to the surface to form sponge tissues cannot be stated with any certainty. The causes may not be entirely controlled by light, for unfavorable moisture conditions favor palisade production as does poor aeration. Sucker sprouts from stumps often produce leaves of the shade type in full sunlight, probably because of the favorable water balance maintained by the extensive root system of the tree. Certain advantages of shade leaf development are more obvious than the causes.

In strong light, cells elongate parallel to the light source. The more intense the light, the deeper its penetration into the leaf and the more layers of palisade there will be. Desert and alpine plants may have the mesophyll entirely made up of palisade cells. Leaves subject to reflected light from below commonly have palisade on the lower surface as well as the upper, and leaves growing vertically regularly have palisade on both sides.

When illumination is intense, chloroplasts arrange themselves along the side walls, and thus in palisade cells they receive a minimum of direct insolation. On the other hand, with weak light the chloroplasts tend to appear along the walls at right angles to the light source, and the form of sponge cells permits exposure of more chloroplasts to the greatest effectiveness of available light. There are added advantages in the thinness and greater area of the shade leaf, since both maximum exposure under conditions of reduced light and penetration of light to a high proportion of internal cells are thus assured.

Since reduced light favors elongation, vegetative growth, and delicacy of structure, it can readily be understood why several garden crops, cultivated primarily for their vegetative structures, are best grown in spring and fall or in regions with many cloudy days. A number of leaf crops are grown under artificial shade. The point is well illustrated by the production under artificial shading of the large leaves of tobacco needed for cigar wrappers.

Since intense light inhibits vegetative growth and favors, or is actually necessary for, flowering and fruiting, it is not surprising that centers of grain and fruit production characteristically have much clear, cloudless weather during the growing season. Here, too, is a partial explanation of the reduced size of alpine and arctic plants, which produce large and numerous flowers. Likewise it helps explain why trees in the open often fruit more prolifically than those in a closed stand, where overtopped individuals rarely produce a seed crop.

Photoperiod. A number of seasonal biological phenomena long have been accepted as such, without much concern as to causes. Violets, trilliums, bellworts, and many other wildflowers blossom in the spring, but asters, goldenrods, and chrysanthemums are expected to flower in late summer or fall. When a fruit tree occasionally blossoms in the fall, the occurrence is considered unusual. The controlling factor in such periodic phenomena was not recognized until Garner and

Allard [156] published results of their studies of *photoperiodism*, or responses of organisms to the relative length of day and night. Their investigations developed from difficulties experienced in growing new varieties of tobacco and soy beans in the vicinity of Washington, D. C. The tobacco grew vigorously and did not flower under field conditions, but in the greenhouse, during the winter months, it flowered and fruited abundantly. The soy beans flowered and set fruit at about the same date in late summer regardless of how long they had been in the vegetative condition, as determined by plantings spaced at wide intervals during the spring and early summer. When the length of

Fig. 40. *The effect of long day (15 hours), left, and short day (9 hours), right, on flowering of henbane* (Hyoscyamus niger), *a long-day plant. All plants received 9 hours of natural radiation. The supplemental light of the 15-hour lot was obtained from 100-watt incandescent lamps, which gave an intensity of only about 30 foot candles.*—Photo by courtesy of H. A. Borthwick, Bureau of Plant Industry, U. S. Dept. Agr.

daylight period was shortened for these plants by enclosing them in a dark chamber for a few hours each day, the tobacco flowered very soon and the formation of seeds in the soy beans was hastened materially.

Some Applications. It can readily be seen why garden plants grown for vegetative parts, if they are long-day species, develop best in spring and late fall and, if grown in summer, bolt to form flowering structures. The differences in photoperiodic response between varieties may be the sole reason for success or failure of a crop at a particular latitude and is an excellent reason for knowing one's seed stock and its potentialities. Flowering shrubs and herbs, too, if grown beyond their normal latitudinal range, may be pampered and kept alive but often fail to flower because the length of day is unsuitable, or may invariably flower too early in the spring or too late in the fall.

The cessation of growth and subsequent "hardening" of evergreen woody plants are initiated in response to length of day. If plants are put out within range of street lamps, some winter-killing may be anticipated. Street trees of several species retain their leaves on the side illuminated by street lamps long after dormancy and complete

Fig. 41. *An* Abelia *hedge in late fall that (left) ceased growth and hardened normally everywhere except section under boulevard light. Here, because of the extended photoperiod, the plants continued to grow and put out new shoots, which were killed by the first heavy frost (right).* —From Kramer.[220]

leaf fall on the opposite side, which does not have supplemental light.[261] On the Duke University Campus, lamp posts are regularly spaced in a long *Abelia* hedge, and every winter, frost injury results within a certain distance of each lamp because the plants here do not go into dormancy.[220]

Commercial greenhouses make use of supplementary lighting and controlled period of illumination to bring crops into flower for special days or to produce maximum vegetative growth. Growing a crop for its vegetative parts in one latitude for which seeds must be produced in another latitude is now a not uncommon practice.

Ecological Significance.[5] It is thus apparent why many plants in the tropics, where the light period is almost constantly twelve hours, flower throughout the year and, likewise, why so few plants in the United States, even in the South, have this characteristic. It is apparent, too, that arctic species must be long-day plants and why they rarely flower when brought farther south. Also, long-day species, which might grow perfectly well in the tropics, could not reproduce there and consequently, would not become established. Species requiring high temperatures and long days to mature are definitely limited in their northern range. The formation of abscission layers in leaves, preceding their fall, and the accompanying decline in physiological activity may be initiated in response to shortening days, regardless of reduction of temperature or age of the leaf.[278] Therefore, at or beyond the northern limits of their range, trees may be killed by frost because they are not yet sufficiently dormant to withstand low temperatures when they occur. But, within a species, there may be marked ecotypic variation in response. Clonal lines of *Populus*, grown under identical conditions, ceased growth inversely with their latitude of origin. Physiological variations within a species may therefore account for its success over a wide range of latitude.[296]

It should not be assumed that plant distribution is primarily determined by length of day. Many species are little affected by it. Also, temperatures, especially at night, have been shown to modify photoperiodic requirements and responses in several species. Again, consider the spring initiation of cambial activity, which occurs at about the same time every year in numerous species in an area. This strongly suggests a photoperiodic control, especially because correlation with temperatures was slight. However, the explanation probably lies in an interraction of light and temperature. Photoperiod is just another factor, which may operate with temperature, moisture, and light to determine the range and distribution of a species.

Temperature

General Plant Relationships. Each living thing is restricted to a definite temperature range, which may be quite dissimilar for different species and, depending largely upon the amount of water in the protoplasm, may vary for individuals of a species. The wide range of tolerance among species is illustrated, on the one hand, by subarctic conifer forests where −80° F. has been recorded and, on the other, by desert plants that withstand temperatures of 130-140° F. Dormant structures such as seeds and spores are practically without water and can, therefore, withstand the widest temperature variations and extremes.

Plant injuries from temperature changes are most often the result of freezing, which desiccates the tissues when the pure water on the cell walls crystallizes in the intercellular spaces and continues to crystallize as it is replaced from the vacuole and protoplasm. Injurious chemical changes, such as the precipitation of proteins, may accompany the desiccation. Some species, however—especially subtropical ones—are often killed before temperatures fall as low as freezing. Temperature injuries cannot always be explained in simple terms.

It is obvious that there must be seasonal physiological adjustments in some plants, which permit their survival as cold weather comes on. It is known, in this connection, that the concentration of the cell sap of most conifers increases in the fall. Gardeners make use of this characteristic, for young plants grown in greenhouses are "hardened" before they are set out and subjected to early spring temperature fluctuations. Such plants are most liable to injury when temperature changes are abrupt and extreme. On the other hand, many arctic and alpine species can grow, flower, and fruit during a period when they are subjected almost daily to alternate freezing and thawing.

Measurement of Temperature. Within their limitations, accurate standardized mercury-in-glass thermometers are useful for field studies even though more elaborate instruments are desirable for intensive and extended observations. Air temperatures are usually taken in the shade, with the thermometer exposed to the wind and away from the influence of one's body. Soil temperatures require a small well of some sort, or, when measurements are to be made periodically, a length of pipe may be permanently sunk to the desired depth. If the thermometer is suspended in the pipe by a string, it can be drawn up quickly and read before much change takes place. Soil temperatures at or very near the soil surface are difficult to obtain accurately with an ordinary thermometer because of the steep gradients from the surface downward, and upward into the air. The size of the thermometer bulb is sufficient to be affected by rather widely differing temperatures even when it is no thicker than 5 mm. Discrepancies have been observed as great as 11° C. between electrical (thermocouple) and ordinary thermometer readings at the surface. The errors are greatest in full sunlight and on dark soils.[121] It is under these conditions that the greatest care must be taken in placing the bulb. Precise temperatures in restricted space have long been obtained by use of a thermocouple for measuring electrical potential. A newly developed thermistor unit which measures resistance, is finding favor for this purpose[308] because of its small size and permanence.

Fig. 42. *Soil-air thermograph, which records the temperatures of soil and air continuously on a revolving drum. The cable at right is about six feet long and terminates in a sensitive bulb (not shown), which can be placed at any level in the soil.*—Courtesy Friez Instrument Division, Bendix Aviation Corporation.

Continuous temperature records are obtainable with thermographs. These usually consist of a bi-metallic or liquid-in-metal expansion element attached by levers to a pen, which records on a graduated sheet revolving on a drum or dial. Types with two and three sensitive elements and pens, attached by flexible cables, permit simultaneous recording of temperatures at different levels of soil, air, or both. Because environmental studies deal more and more with microclimate, they may require numerous simultaneous measurements at several locations. Under such conditions, the expensive recording instruments may have to be restricted to key points, with maximum and minimum thermometers installed elsewhere for comparisons. These, used in a pair, are read and reset at regular intervals to obtain the useful values of maximum and minimum temperatures for the period of exposure. The Six's type thermometer, which gives maximum and mimimum temperatures with a single instrument, has been used successfully in such studies when properly handled. It may well be added that recording instruments require shelters when used for extended observations. Circumstances may make such installations impractical but still permit use of the smaller instruments, which may be placed in or

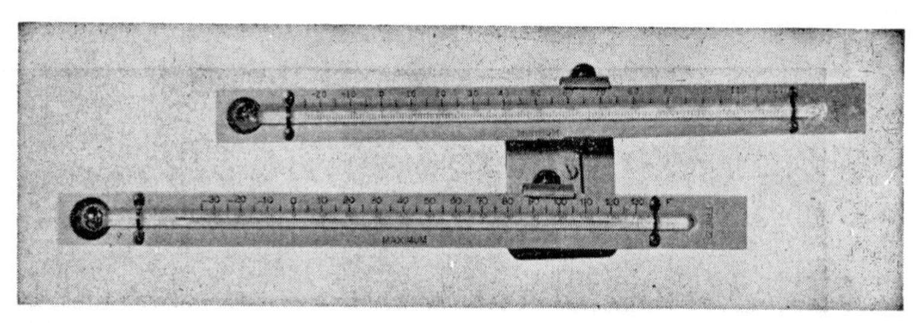

Fig. 43. *Maximum-minimum thermometers of a standard type for air temperatures. Installed in an instrument shelter, the holder permits whirling of the maximum thermometer for resetting.*—Courtesy Friez Instrument Division, Bendix Aviation Corporation.

on the ground, or equipped with an inconspicuous shield against insolation, in the air.

Temperature Records. Because temperature is so extremely variable, isolated or even numerous single determinations may be completely useless. A continuous record is most desirable, because it gives the duration of extremes and variations. Although extremes may be important in the reaction of a plant, their duration is apt to be what determines the plant's response. Therefore, a thermograph is desirable for thoroughly satisfactory work. The "mean temperature" as computed by the United States Weather Bureau[212] is usually the average of the maximum and minimum for the day. This is not accurate for, or truly indicative of, plant-temperature relations, because it ignores duration and is likely to run too high. A more appropriate mean is obtained by averaging the hourly temperatures for twenty-four-hour periods.

Annual mean temperatures are almost useless ecologically, for they do not indicate seasonal variation and duration. Temperate desert regions may have amazingly high annual mean temperatures and yet have winter frosts, which constitute an important limiting factor in the survival of certain species there. Subarctic areas may support forest vegetation because of the warm summers, yet mean temperatures may be so far below freezing that they suggest that little if any plant life would survive. It can be seen that mean monthly temperatures are desirable for evaluating ecological conditions, and this is equally true for monthly mean maximum and minimum values. Collectively, these indicate the extent of the growing season and the extremes to be expected during that time.

Temperature Variations. Since fluctuations of insolation result in fluctuations of temperature, seasonal and daily temperature changes, as with insolation, can be expected to follow a general pattern for any region. The pattern follows that of insolation but with temperature responses lagging behind changes in radiation. A daily maximum of atmospheric temperatures usually comes in midafternoon, and minimum temperatures occur just before sunrise. Soil temperatures below the surface lag even more, for their maxima may not occur until 8:00-11:00 P.M. and minima may not be reached until 8:00-10:00 A.M. This is, of course, due to the fact that soil is a poor conductor of heat. For the same reason, the soil surface, if unshaded, produces the highest temperatures for an area and likewise has the widest range of temperatures. It is the subsoil temperature that follows the trend indicated above. With increasing depth, daily fluctuations are reduced until at two or three feet they are not apparent. Seasonal air temperatures also lag as is indicated by the usual hot days of July and the cold of January, both extremes coming after the June and December solstice. Soil temperatures follow seasonal atmospheric trends with a further lag.[382]

Since the total insolation decreases with distance from the Equator, temperatures likewise decrease. Temperature zones, therefore, tend to run east and west, and the greater the latitude the lower are the temperatures to be expected.

There are, however, local and generalized exceptions. Large bodies of water are slower to warm up and slower to cool than land because of the higher specific heat of water. In addition, they reflect much of the insolation, and what heat is absorbed is distributed to much greater depths by water motion and convectional currents. As a result, temperature extremes are reduced around bodies of water as compared to those inland. The effect on plant distribution is particularly evident in the ranges of southeastern species, which often extend to the northern limits of the Atlantic coastal plain, where, undoubtedly, they are able to survive because of the maritime climate. The amelioration of temperatures is apparent about lakes as well as oceans,

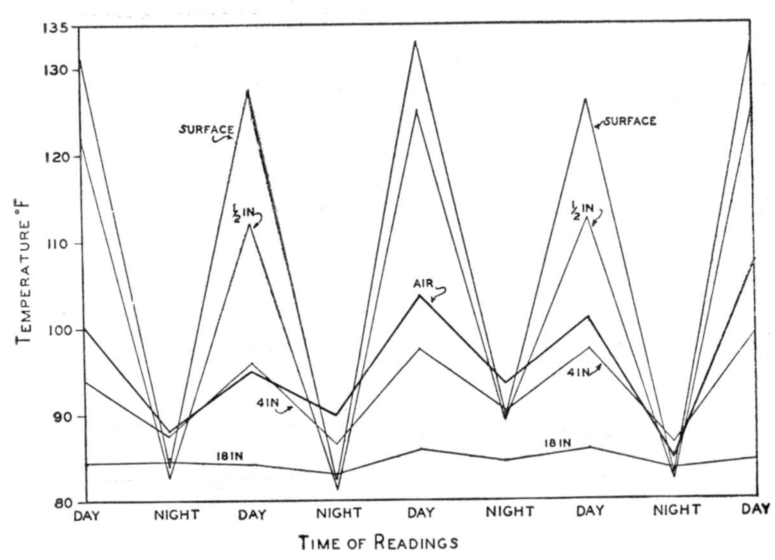

Fig. 44. *Maximum day and night soil temperatures taken on a sand dune at Beaufort, N. C. in August, 1947. Readings were made on successive days at 7:00* A.M. *and 7:00* P.M. *for night and day maxima, respectively. Temperatures were greatest at the soil surface and were successively less with increasing depth by day, but, at all depths at night, dropped as low, or lower than the maximum at eighteen inches. Minimum temperatures fluctuated within the range of 72-85° F. (difficult to show accurately on so small a scale). At eighteen inches the minimum was never more than one degree below the maximum, but the difference between minimum and maximum increased upward to the surface where one minimum was as low as 72° F.*

although to a lesser extent. The extremes of winter and summer temperature characteristic of the Dakotas are typical of the interiors of continents, but they are never experienced in lake-bounded Michigan, also in the interior and at about the same latitude.

The air near the earth's surface is warmed by absorption of insolation and reradiated heat from the earth. With increasing altitude the atmosphere becomes less dense and also contains less moisture and other heat-absorbing substances. Consequently, temperatures decline with altitude. Even the warm air rising from the earth is cooled by its expansion. Latitudinal temperature zones are, therefore, further disrupted by mountains where increasing altitude produces temperature reductions (approx. 3° F./1000 ft. of elevation) much more abruptly

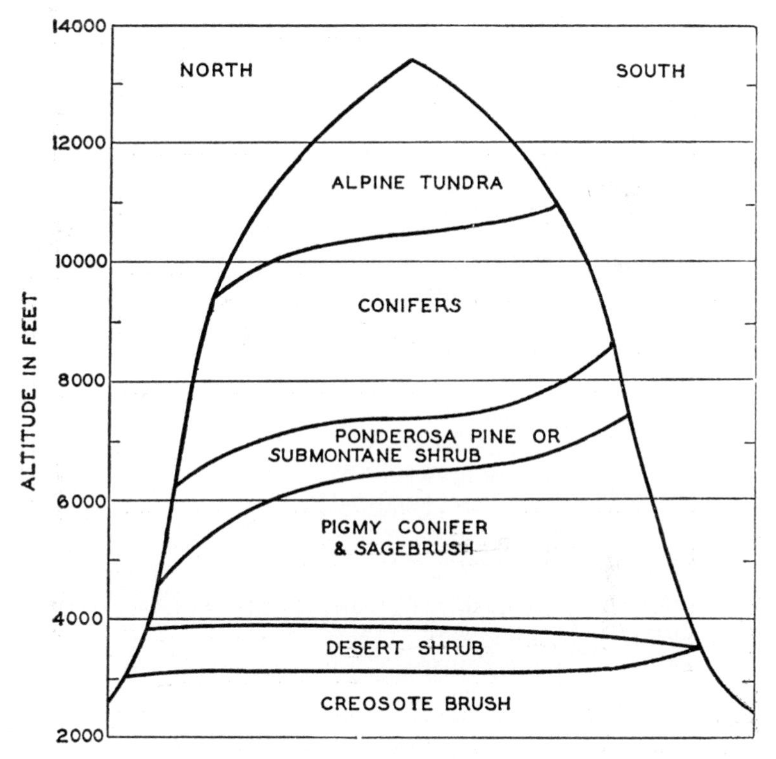

Fig. 45. *A generalized profile of altitudinal zones of vegetation in the mountains of Utah, which illustrates the effects of northern and southern exposures.*—Adapted from Woodbury.[434]

than increasing latitude. This is particularly noticeable on high mountains, where, because of the combined effects of temperature and moisture, one may see zones of vegetation altitudinally arranged, which at lower altitudes are latitudinally distributed over hundreds of miles.

Just as latitudinal temperature zones are irregular, so are the altitudinal zones not perfect. In some areas it is common for night temperatures in the lower layers of the atmosphere to be lower than those of upper strata. Such a temperature inversion results when rapid reradiation from the earth reduces its temperature below that of the neighboring air layer, which, in turn, loses heat to the earth by con-

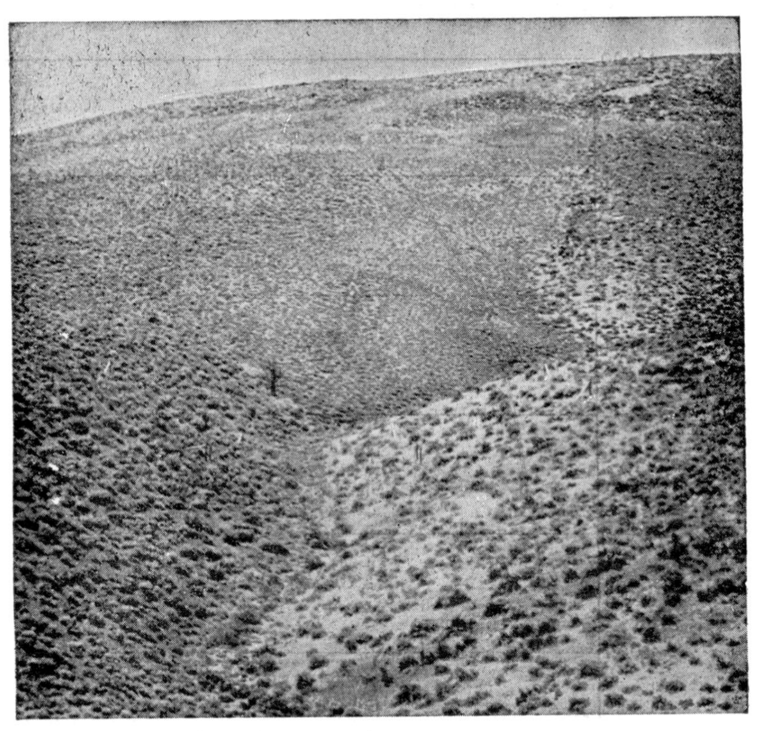

Fig. 46. *Effect of slope exposure is apparent in the desert, as elsewhere. Although species differences are not great, the south-facing slope at right supports a much sparser, more widely spaced stand of sagebrush than the opposite slope. Washoe County, Nev.*—Photo by W. D. Billings.

duction. A similar inversion is caused by cold air drainage which occurs at night when air from rapidly cooling ridges settles and flows down valleys under the warm air rising from them. It results in low night temperatures in the valleys when tablelands and upper slopes are much warmer.[194] The resulting thermal belts are sharply defined and may be distinctively marked by the vegetation they support. In mountainous country, orchards are frequently grown successfully at much higher altitudes on slopes than in valleys.[109] When cold air drainage is into a depression or an obstruction holds the cold air behind it, a frost pocket is formed, in which temperatures may be much below those of the general surroundings and, consequently, have marked effects on the vegetation.

Slope and exposure disrupt mountain temperature zones even more. Since the maximum effectiveness of insolation comes only when it strikes a surface at right angles, the greater the variation from a ninety-degree angle, the less radiant energy will strike a unit area. In the Northern Hemisphere, therefore, a south-facing slope receives more insolation per unit area than a flat surface, and a north-facing slope receives less (see Fig. 34). Thus the same temperature conditions found on a tableland may occur at a higher altitude on a near-by south-facing slope and at a lower altitude on a north slope. The distribution of vegetation being correlated with temperature and the consequent moisture differences, a particular community will be found above its ordinary altitudinal range on south slopes and below it on north slopes, and the extent of this irregularity in zonation is affected both by the angle of the slope and its exposure. In Wyoming, Douglas fir from the montane zone may come down to 7,500 feet on north-facing slopes while mountain mahogany from the lower woodland zone may be found extending upwards to better than 8,500 feet on south-facing slopes. In general, a vegetation zone extends higher on the south side of a mountain than on the north side.

Cover and Temperature. Anything that absorbs or reflects insolation before it reaches the earth will reduce both soil and atmospheric temperatures. Thus it is cooler in cloudy or foggy areas than in similar areas without clouds or fog, and any given area tends to be warmest on clear days. But, because heat radiated from the earth and clouds is held below a cloud blanket, the lowest temperatures also occur on clear days, and extremely low temperatures are not to be expected on cloudy days. Temperatures in and above bare soil, particularly dark soil, are higher than if that soil has some form of cover. Any type of vegetation absorbs some radiant energy and, consequently, reduces temperatures between itself and the soil, the reduction being proportionate to the closeness of the stand and how many strata compose it. Temperatures in forest stands in midsummer are commonly ten degrees lower by day than in the open and ten degrees higher at night. Soil temperatures under forest are lower than in the open during the growing season and usually higher in winter. However, soil temperatures under deciduous forest are subject to considerable winter variation.

Soil temperatures are further modified by dead or living cover on the surface. Any such cover reduces the range of extremes and the speed of variation. This amelioration of temperature may be important in the viability and germination of seeds and the survival of seedlings. Particularly affected are the physical and physiological processes involving water, its movement and availability in the soil, and its absorption and transpiration by the plant. Also, when soil is frozen, the runoff from heavy rains is much increased. Studies in Arizona[186]

Table 8. *The average day and night temperatures (°F.) in three upland forest communities in central Iowa. Air temperatures in contiguous prairie are higher than those in shrub by about 10° (day) and 4° (night).* From ([4]).

Community	Time	April	May	July	August
Shrub.	Day	58.8	65.1	76.7	73.9
	Night	45.3	52.8	64.0	61.1
Oak-hickory.	Day	57.5	63.8	80.1	77.7
	Night	42.2	51.9	70.4	68.4
Maple-basswood. . . .	Day	55.0	60.9	74.3	70.0
	Night	40.2	51.8	67.5	61.0

showed daily minimum soil temperatures to be five degrees higher under forest litter in the fall of the year than in bare ground and the daily maximum to be seven degrees lower. The average diurnal range was eighteen degrees in bare soil and only six degrees under litter. In North Carolina,[252] litter reduced the depth of frost penetration 40 percent, and, whereas the bare soil was frozen solidly, the soil under litter remained porous and loose, permitting deeper percolation during winter rains and thaws and causing very little runoff. The effects of snow as an insulator are much the same as are those of litter.

Temperature and Physiological Processes. For every species there is probably an average optimum temperature at which it grows most successfully, other factors being equal. Likewise there must be a maximum and a minimum temperature that it can withstand. These limits may result from the temperature tolerances of the protoplasm peculiar to the species, but they may likewise result from responses of one or more physiological processes, which vary from species to species.

The temperatures affecting germination might alone limit the range of a species. Among our cultivated crops, the minimum-maximum range of temperature for germination is 35°-82° F. for flax and 49°-115° F. for corn. The optimum for each, respectively, is 70° and 93°. That the center of production for flax is considerably north of the center for corn is therefore not at all surprising. Low temperatures may also be a necessity. Dormant seeds of some plants from cold areas require near-freezing temperatures under moist conditions to complete physiological changes, called after-ripening, before they will germinate. The conditions are well known for many species of economic significance. Flower buds may not form if temperatures do not fall sufficiently low. Again, flower buds may not break dormancy in spring unless they were sufficiently chilled for a long enough time during the winter. Peaches, which require at least 400 hours of temperatures below 7° C., thus have their southern economic range definitely restricted. Probably the ranges of numerous native species are affected by low temperature requirements as well as inhibitions.

Absorption of water is at a minimum when soil is frozen but increases, as do diffusion and capillary movement in the soil, with rising temperature. The optimum is surprisingly high as soil temperatures go, and the maximum approaches the boiling point in some instances. Absorption is reduced more at low temperatures for plants that grow normally in warm soil than for plants that grow, at least part of the year, in cold soil. For example, cotton absorbs only 20 percent as much water at 50° as at 77° F. while collards absorb 75 percent as much at 50° as at 77° F.[221]

Photosynthesis operates under a wide range of temperatures under natural conditions. Marine polar algae may live their entire lives at temperatures below 32° F. because the freezing point is depressed by the salts in the water. There is an often-quoted old report that spruce carries on photosynthesis at −22° F., but a recent study[149] using modern methods indicates that, although conifers do not lose their ability to carry on photosynthesis during midwinter, the species studied function only above 21° F. The process also goes on in desert plants at temperatures of 120° F. or more. The effective temperature range, however, is usually between 70° and 100° F. The rate of photosynthesis increases steadily as temperatures rise to the optimum. Beyond, in the narrow range to maximum temperatures, the rate declines abruptly. The rate of respiration also increases with temperature until at high temperatures the process becomes destructive of life. Van't Hoff's Law, which states that the speed of a chemical reaction doubles or more than doubles with each 18° F. rise in temperature, is applicable within limits to reactions in organisms. In photosynthesis it holds reasonably well between about 41° F. and 77° F. Beyond these limits there is much variation.

Growth, being a product of chemical and physiological processes, follows the same pattern and is favored by relatively high temperatures. At temperatures near or above the maximum, the water balance is apt to be thrown off by excessive transpiration. Reproduction follows the same rule regarding temperature, but flowering and fruiting have higher optima than vegetative processes in the same plant.

General References

H. A. ALLARD. Length of Day in Relation to the Natural and Artificial Distribution of Plants.

P. BURKHOLDER. The Role of Light in the Life of Plants.

R. F. DAUBENMIRE. *Plants and Environment.*

W. J. HUMPHREYS. *Ways of the Weather.*

H. L. SHIRLEY. Light as an Ecological Factor and Its Measurement.

U. S. DEPT. AGR. *Climate and Man.*

H. B. WARD and W. E. POWERS. *Weather and Climate.*

Chapter 6 Climatic Factors: The Air

Gases of the Atmosphere

The air surrounding the earth is made up of only a few gases in proportions that remain remarkably constant. The average volume percentages of dry air are: nitrogen, 78.09; oxygen, 20.95; carbon dioxide, 0.03; and argon, 0.93. In addition, there are minute but measurable quantities of several rare gases, which have no part in our discussion. Within the limits of the atmosphere that can affect plants directly, there is but slight variation in the proportions of these gases whether over the ocean or land, at sea level or on high mountains. Minor but rather consistent variations have been found over large industrial cities where quantities of carbon dioxide are constantly being produced.

Whenever an organism respires or a fire burns, oxygen is removed from, and carbon dioxide is added to, the atmosphere. Decomposition of organic matter also liberates carbon dioxide; photosynthetic activity of plants removes carbon dioxide and liberates oxygen. When these processes are not in balance, there may be local variations in the composition of the atmosphere, but air movement, combined with diffusion, is usually sufficient to eliminate gaseous differences quickly. Notable differences have been found in the first 8 cm. of atmosphere just above the soil, which at mid-day had as much as 0.101% of carbon dioxide under forest and somewhat less in grassland.[150]

Thus, regardless of its terrestrial environment, the organism is almost certain to be plentifully supplied with these gases that form a relatively constant part of the atmosphere; therefore, they need not be considered as important variables in the environment.

Gases of the Soil Atmosphere

Although normally there is never a shortage of oxygen in the air above ground, such a shortage sometimes occurs in the soil. Air space in the soil is limited and is partially, or sometimes wholly, occupied by water. Any change in the composition of the soil atmosphere is only slowly readjusted from the atmosphere above, for in soil, air movement and diffusion are relatively slow

Since all living structures in the soil respire, and this includes small animals and other microorganisms as well as all roots of plants, the supply of oxygen is constantly being reduced and carbon dioxide released. As a result, the soil atmosphere always contains less oxygen and more carbon dioxide than the air above. Oxygen tends to decrease with depth, and carbon dioxide to increase, but the greater the air capacity, as in granular or porous soils, the less the carbon dioxide and greater the oxygen. In the soil under closed stands of vegetation, carbon dioxide often equals 5 percent, and has been found in much higher concentrations. The constant use of oxygen and its extremely slow rate of diffusion when soils are saturated soon result in oxygen deficiency. Temporary saturation may not be serious, but, when prolonged, it results in death of the vegetation through inhibition of root growth and absorption. Under these conditions, several soil organisms may carry on anaerobic respiration for a time, but such activity results in chemical changes of several kinds, which may affect fertility of the soil or actually inhibit plant growth.

Available oxygen in an aquatic habitat probably is somewhat higher than in a saturated soil because of the movement of the water and because the oxygen is more readily replaced by solution from the atmosphere. If, however, the water is solidly frozen over, it is not uncommon for the oxygen supply to fall so low that many of the fish die. When such conditions develop in good fishing lakes, it is common practice to cut several holes through the ice and to pump air through the water until the depleted oxygen supply has been replaced. The mud at the bottom of a shallow pond probably has the least favorable oxygen concentration for plant roots. Most plants growing well in such places are of the emergent type, having at least part of their structure in the air and characterized by lacunar tissue, which permits gases to accumulate in, and move freely through the plant.

Water Content of the Atmosphere

In addition to the gases constituting the atmosphere, water is always present as vapor but in widely varying amounts. Since atmospheric moisture represents the indirect source of the plant's water and likewise controls the amount and rate of water loss by the plant, it is an environmental factor deserving considerable attention.

The capacity of air to hold water vapor increases as temperatures rise or pressure is reduced. The air is said to be saturated when it contains as much moisture as it can hold at a given temperature and pressure. If for any reason the temperature is raised or the pressure is decreased, the amount of water remaining constant, the air is no longer saturated. On the other hand, if the temperature of saturated air decreases, its capacity is reduced, and some of the vapor condenses as a liquid. Thus air that is not saturated will become so without change of vapor content if its temperature is lowered; and, when saturation is reached, the air is said to be at the "dew point." If the cooling continues, the vapor becomes a liquid, which may condense on objects near the surface of the earth as dew or frost or, if condensation takes place in the air, may result in precipitation.

Terminology of Atmospheric Moisture. Several expressions are used to describe the moisture content of the air. *Absolute humidity* is commonly interpreted as the amount of water vapor per unit volume of air and can be expressed as grams per cubic meter or any other units of mass and volume. In itself the absolute humidity has little bearing on the life of a plant, for it is not the total atmospheric moisture that determines evaporation and transpiration, but rather the difference between the amount of vapor present and the maximum amount the air could hold at the time. Thus the *relative humidity*, which is an expression of percentage of saturation, is more nearly related to the rate of water loss from a free water surface or from a plant. However, relative humidity depends upon temperature as well as the amount of moisture present, and, as a consequence, identical relative humidities do not indicate identical moisture conditions unless the temperatures are also the same. This means that every shift in temperature results in a change in relative humidity, regardless of moisture present, and a consequent change in rate of evaporation or transpiration.

Several authors have emphasized that, when considered independently of other factors, the actual amount of water vapor in the air has little if any influence upon evaporation. One illustration[7] especially serves to emphasize the ecological significance of this fact. Death Valley, Cailfornia, is probably the most arid region in the United States, yet its "dry" atmosphere contains on the average in July almost exactly the same amount of water vapor per unit volume as does the "moist" atmosphere of Duluth, Minnesota, at the same time of the year.

An atmosphere 70 percent saturated at 60° F. will contain much less water vapor than an atmosphere 70 percent saturated at 80° F., and the capacity to hold more water will be less in the first than the second case. Evaporation will, therefore, normally be more rapid at 80° F. even though the relative humidities are the same. It can be seen that a statement of relative humidity alone gives little indication of atmospheric moisture conditions, since a relative humidity of 80 percent may mean "dryness" if the temperature is high or "wetness" if the temperature is low.

It is desirable then to have a term indicating the amount of water that air can take up before it becomes saturated. Vapor pressure is a measure of the quantity of water vapor present, the temperature being constant, and is usually expressed in units of pressure (inches or mm. of Hg). Therefore, *vapor pressure deficit* is the difference between the amount of water vapor actually present and the amount that could be held without condensation at the same temperature. It is a direct indication of atmospheric moisture conditions, quite independent of temperature; and, therefore, compared to relative humidity, its values are much more indicative of the potential rate of evaporation.

When the relative humidity is 100 percent at 68° F., the vapor pressure is 17.54 mm. of mercury. If the relative humidity were 70 percent, the vapor pressure would equal 12.28 mm. (0.70×17.54), and the deficit would be 5.26 mm ($17.54 - 12.28$). If the relative humidity were the same (70%) at 59° F., the vapor pressure would be 8.95 mm. (0.70×12.79) and the deficit would be only 3.84 mm. ($12.79 - 8.95$). Tables of saturation pressures (vapor pressures) are usually available in handbooks of chemistry, and it is possible to transform relative humidities to vapor pressure deficits quickly when the temperature is also known. The relationships are shown in Table 9.

Greater general use of the vapor pressure deficit in ecological work seems desirable, for in spite of certain limitations, its accuracy is much

Table 9. *Vapor pressure deficit and vapor pressure as related to relative humidities at different temperatures.*

Temperature		Vapor pressure deficits (mm of Hg), reading down, at given relative humidities										
		Relative Humidities										
C°	F°	100%	90%	80%	70%	60%	50%	40%	30%	20%	10%	0%
50	122	0	9.25	18.50	27.75	37.00	46.26	55.51	64.76	74.01	83.26	92.51
45	113	0	7.19	14.38	21.56	28.75	35.94	43.13	50.32	57.50	64.69	71.88
40	104	0	5.53	11.06	16.60	22.13	27.66	33.19	38.72	44.25	49.79	55.32
35	95	0	4.22	8.44	12.65	16.87	21.09	25.31	29.53	33.74	37.96	42.18
30	86	0	3.18	6.36	9.55	12.73	15.91	19.09	22.27	25.46	28.64	31.82
25	77	0	2.38	4.75	7.13	9.50	11.88	14.26	16.63	19.01	21.38	23.76
20	68	0	1.75	3.51	5.26	7.02	8.77	10.52	12.28	14.03	15.79	17.54
15	59	0	1.28	2.56	3.84	5.12	6.40	7.67	8.95	10.23	11.51	12.79
10	50	0	0.92	1.84	2.76	3.68	4.60	5.53	6.45	7.37	8.29	9.21
5	41	0	0.65	1.31	1.96	2.62	3.27	3.92	4.58	5.23	5.89	6.54
0	32	0	0.46	0.92	1.37	1.83	2.29	2.75	3.21	3.66	4.12	4.58
		0%	10%	20%	30%	40%	50%	60%	70%	80%	90%	100%
		Vapor pressures (mm of Hg), reading up, at given relative humidities.										

greater than that of relative humidity. The potential rate of evaporation cannot be indicated with a single simple expression of atmospheric moisture, since the rate depends upon the vapor pressure gradient between evaporating surface and atmosphere. The gradient can be determined only when the temperature and vapor pressure of the liquid are known as well as those of the atmosphere. Vapor pressure deficit is directly related to evaporation only when the temperatures of the air and of the evaporating surface are equal.[399] Ecologists more often than not measure evaporation directly; but when evaporation is not known, in spite of the above, vapor pressure deficits could well be used instead of relative humidities.

Measurement of Atmospheric Moisture. The psychrometer and the hygrometer are the two instruments most useful to ecologists for this purpose. The former consists of two thermometers, one of which has

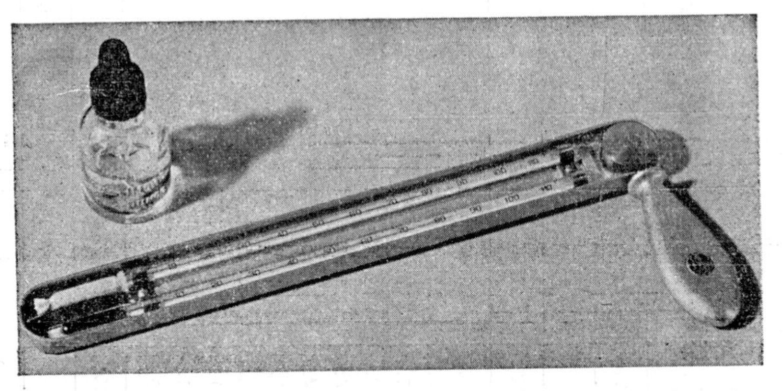

Fig. 47. *A sling type of psychrometer for determining relative humidity by the difference in temperature of the wet and dry bulb after whirling.*—Courtesy Friez Instrument Division, Bendix Aviation Corporation.

the bulb wrapped with a thin piece of wet cloth, and both of which are aerated in some fashion, usually by whirling. Evaporation from the wet cloth is controlled by the moisture in the atmosphere, and the bulb is cooled in proportion to the evaporation. The dry bulb gives the temperature of the atmosphere, and the difference between the dry and wet bulb readings gives the wet bulb depression. Knowing the barometric pressure and these two values, the relative humidity can be quickly determined from standard tables or from nomograms.[166]

The necessity for aeration of the thermometers, usually accomplished by rapid rotation, has led to the design of several "sling" type psychrometers. Because these must be whirled, they require considerable space for operation. The "cog" psychrometer, functioning like an egg beater, or a hand-aspirated type which forces air past the bulbs, can be used in much smaller spaces. Determinations can be made at

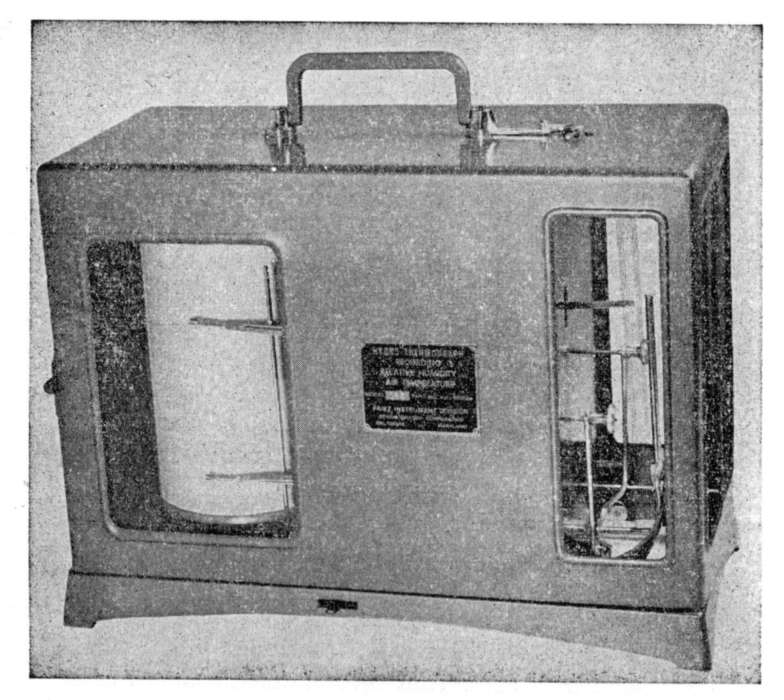

Fig. 48. *A hygrothermograph, which automatically gives a continuous record of relative humidity and temperature of the air.*—Courtesy Friez Instrument Division, Bendix Aviation Corporation.

freezing temperatures by coating the wet bulb with a thin layer of ice but the results are not very reliable.

The hygrograph is a continuously recording instrument in which an arm marks on a rotating drum the stretching and contracting of a strand of hairs, which respond to relative humidity. The drum is so calibrated that relative humidity is recorded directly. Often the device is equipped to record the temperature simultaneously and is then called a hygrothermograph. This automatic device is very convenient, particularly since it normally needs to be serviced but once a week. If, however, several stations are to be maintained, the necessary instruments may not be available, and the psychrometer is then the only solution.

With readings of the psychrometer and the hygrothermograph, the air temperature is also obtained, providing the means of calculating vapor pressure deficits with no extra determinations. A simple nomogram (Fig. 49) permits direct conversion from wet and dry bulb temperatures to vapor pressure deficit.

Evaporation and Transpiration. Measurement of transpiration under natural conditions is often practically out of the question. Although small plants may be potted or grown in cans and these may be weighed at regular intervals to determine water loss, only a limited number of plants can be used, and the labor involved can soon become prohibitive if a comprehensive study is to be made.

Relative rates of transpiration can be determined by the cobalt-chloride method, which is rapid and permits numerous determinations in a short time. Paper treated with cobalt-chloride is blue when dry and turns pink as it takes up moisture. Small squares of the dry paper can be attached to leaves between small glass plates by means of a wire clip. The time required for the paper to turn pink is taken as a basis of comparison. The method is simple but it has definite limitations. The close-fitting clips exclude all outside air and eliminate air movement as a factor, while at the same time diffusion into the air is practically stopped by the glass. Rarely will two leaves on a plant give identical readings, for their water loss varies with their positions on the plant and their ages. Thus several determinations must be made simultaneously to evaluate a single plant, while to compare this plant with others necessitates a considerable number of readings. Even then there is not a close conformity with gravimetric determinations, probably because the method measures only a component of the transpira-

---

Fig. 49. *Nomogram for the direct conversion of wet and dry bulb temperatures to vapor pressure deficit, at barometric pressures of 30, 29, 27, and 25 inches. To use, lay a straight edge across the appropriate temperature values on the wet and dry bulb scales and read off V.P.D. directly. Because of the reduction necessary for this reproduction, extreme accuracy is not possible in its use.*—By permission W. E. Gordon.[166]

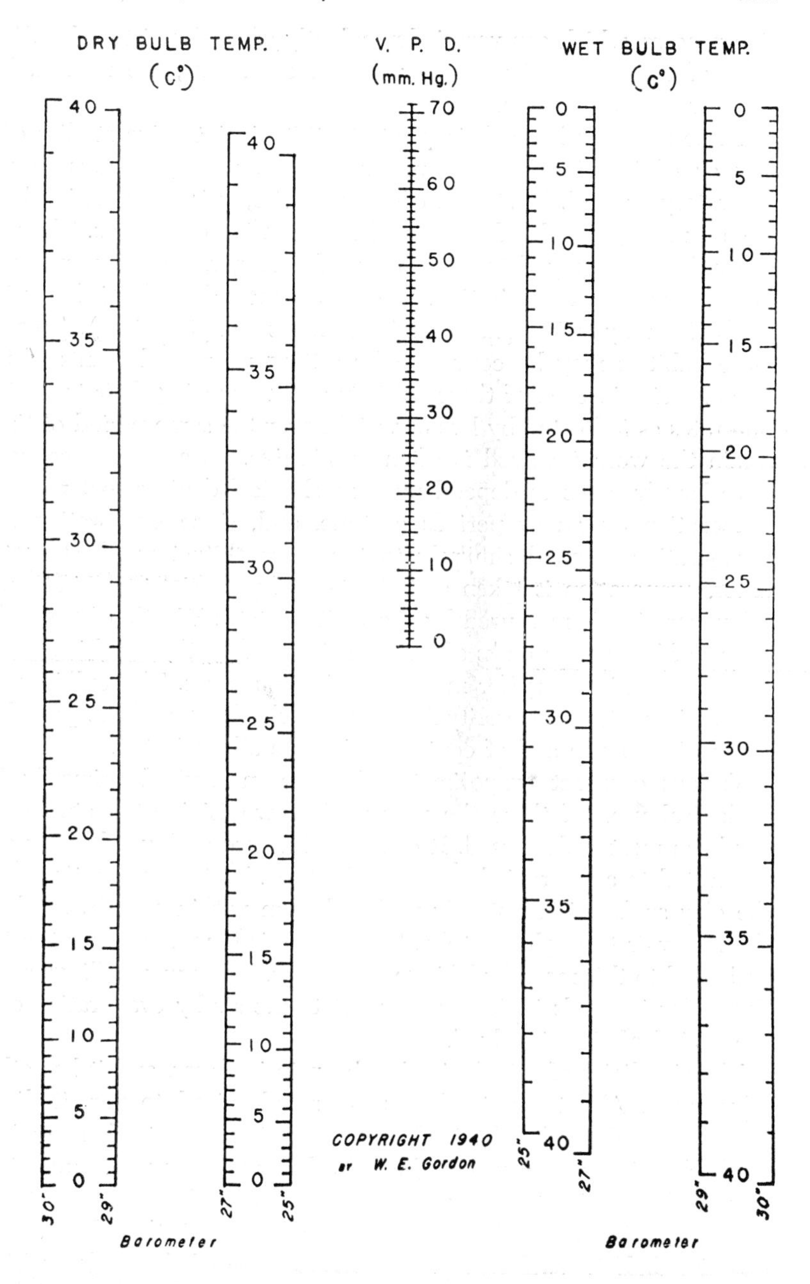
DRY BULB TEMP.
(C°)
V. P. D.
(mm. Hg.)
WET BULB TEMP.
(C°)
COPYRIGHT 1940
BY W. E. Gordon
Barometer
Barometer
30"
29"
27"
25"
25"
27"
29"
30"

tion process, which can vary independently of total moisture loss.[14] In spite of these limitations, under certain conditions, the method has been used to advantage.

These methods have their greatest utility in intensive studies of a few or of individual plants under experimental conditions in the laboratory or field. In studies of communities, it is often desirable to have a more generalized picture of transpiration conditions. The rate of evaporation may then be more useful than a limited number of measurements of transpiration. Perhaps the most desirable information is obtained by using plants as instruments (phytometers). Two or more habitats may be compared by setting up potted plants of the same species in each of these habitats and comparing their transpiration rates as indicated by loss of weight over the same period of time. Again the work involved is often prohibitive. As a result, ecologists have largely come to depend upon mechanical devices that measure evaporation over unit periods of time, and, since evaporation and transpiration respond similarly to the external factors affecting the latter, evaporation is taken as indicative of potential transpiration.

Evaporation is measured by the United States Weather Bureau by means of large open tanks of uniform size and depth, but this method is quite unsatisfactory for most ecological purposes. The bulkiness of the equipment, the necessity for frequent checking, and the probability of disturbance and of contamination are all against it.

Various compact evaporimeters have been devised primarily for ecological use. Of these the now well-known Livingston atmometer has been most widely used. It consists of a porous clay sphere or cup connected to a reservoir by means of a tube. Water evaporating from the clay surface is constantly replaced from within and drawn from the reservoir through the tube. The reservoir is marked near the top and filled to this mark by lifting the stopper. Subsequent fillings made at regular intervals indicate water lost to the air by evaporation over the period of time involved.

The simplicity of this device has been in its favor, and it has other advantages. All atmometer bulbs are standardized to permit direct comparison of results obtained with every instrument wherever it is used. If, as is frequently true, the bulbs become dirty or accumulate a film of algae or fungi, they must be restandardized.

The spherical form of the atmometer bulb gives it the advantage of exposing half its surface to the sun regardless of the sun's position. Other evaporimeters with different shapes have been less useful for this reason alone. Black bulbs can be used in combination with white and the increased evaporation resulting from their greater heat absorption may be used as a measure of relative light intensity in different habitats.

Since the bulb permits evaporation, it also will absorb water during rainy spells. For field work, therefore, it is necessary to install one of the various mercury traps designed to permit water to rise in the tube

Fig. 50. *Two atmometers set up and in use in a study of grassland environment. The improvised shelter was used for max-min thermometers.*—Photo by R. B. Livingston.

but not to let it return to the reservoir. A simple but effective trap consists of a drop of mercury in the lower end of the tube, held in position between two plugs of pyrex glass wool.

When temperatures fall below freezing, clay atmometers cannot be used because of the danger of breakage. A summary of the development, uses, and limitations of atmometers is given by Livingston.[240]

A good, relatively simple method for evaluating water losses to the atmosphere from specific communities would have numerous practical considerations as in watershed management, where water yield is of first concern, or in predicting success of introduced or managed vegetation types under any circumstances. The problem is obviously complex since all the factors affecting the individual must somehow be integrated for the community. Nevertheless, results of some recent studies[24, 304] are promising. By using the mathematical relationships between moisture gradients and temperature or radiation measurements, estimates have been made that show a high correlation with those based upon weighings of entire plants and parts of plants. This approach is related to the growing conception that, when soil moisture is not limiting, transpiration is controlled more by radiation than by any other atmospheric factors. That unlike vegetation types lose different amounts of water when soil moisture is low is interpreted in terms of differences of rooting depth and the supply of water there.

Condensation of Atmospheric Moisture. If air is sufficiently cooled, its relative humidity increases to 100 percent, or saturation. The slightest cooling beyond this point will result in condensation of part of the vapor to form a liquid. The temperature at which condensation occurs (dew point) will, of course, vary with the moisture content of the air.

Cooling of air masses is commonly caused by their expansion when air rises in convection currents or when moving air is forced to rise, as when it strikes a mountain slope. Cooling also occurs cyclonically; for then masses of warm and cool air may meet, and, depending upon which is least stable, warm air moves up over the cool (a warm front) or the cold air underruns the warm (a cold front). Of considerable local ecological significance is the contact cooling resulting when relatively warm air moves over a cooler surface or when cold air moves in over a body of warm water. Under these conditions, fogs or clouds may form, which not only may result in precipitation but may also modify the effects of solar radiation.

Dew. When air containing water vapor comes in contact with a cold surface, the moisture condenses as dew, or if the surface temperature is below freezing, as hoar frost. Dew probably affects most vegetation very little but in some desert areas without summer rain, extreme nightly cooling of surface soil may result in heavy dew formation that is significant for plant life. In Palestine there are numerous areas where annual dewfall equals 100 mm. per year and some where it exceeds 150 mm. It is estimated that the annual total of dew forming on plants is at least as great as the annual rainfall in the western Negeb.[12] Since humidity is high on dewy nights, evaporation is at a minimum and the combination certainly contributes to the survival of plants under extreme moisture conditions.

Fog. Any minute particles of matter in the atmosphere with hygroscopic properties may serve as condensation nuclei (there is disagreement as to their necessity) about which droplets of water form, the size of the droplets depending upon the speed of condensation. Contact cooling usually produces only small droplets, which remain in the air and are visible as fog. Coastal fogs are of this type when they are

Fig. 51. *Ocean fog pouring over crest of Coast Range, Oregon.*—Photo by W. S. Cooper.

the result of prevailing winds coming off the warm ocean and striking a cooler land mass. Such fogs are usually dissipated as the day progresses, evaporating as the temperature rises. Coastal fogs may also be caused by winds blowing from areas of warm water across cool currents. In summer, along the Pacific coast, warm air moves in from far offshore across the cool California current flowing from the north. Fog forms over the cold current and is blown inland, where it disappears if the land is warm but persists at night when the land is cooler. Because they affect light, temperature, and moisture conditions, fogs may be of extreme importance in determining types of coastal vegetation and the agricultural possibilities of an area. The distribution of coastal redwoods of our Pacific coast forms a striking example of the effects of fog. In a region almost without summer rainfall, the coastal redwood and several associated species are almost precisely limited to the humid fog belt along the coast. Fogs inland are usually over low ground, swamps, or small bodies of water, and are common in valleys where air movement is reduced and radiation cooling is effective.

Fig. 52. *Coastal redwood forest in California, showing the characteristic fog that is a factor in its survival.*—U. S. Forest Service.

Smog. Dust, pollen, and spores may occur in the atmosphere anywhere, even in the arctic and over oceans, but with little significance except to sufferers from hayfever. However, when an abundance of minute, solid particles accumulates in the air of valleys an inversion of temperature may result which causes the cool lower atmosphere to become stagnant and may keep it so for days. If fog forms in this stagnant layer it is much deeper than usual fogs, it is denser because of the numerous condensation nuclei, it clears more slowly, and light

Fig. 53. *Complete destruction of vegetation by smelter fumes with resulting erosion. Vicinity of Ducktown, Tenn., in 1935.*—U. S. Forest Service.

penetration to the ground is much reduced.[350] Such pollution, including particles, sulfur dioxide, carbon dioxide, and carbon monoxide, derived principally from industrial and home combustion products and the exhaust fumes from automobiles, has been termed "smog." To what degree smog affects vegetation is not known, although there is evidence of some toxic effects as well as the light reduction. In man, and undoubtedly other animals, bronchial afflictions are aggravated, and in some instances deaths have resulted. The release of industrial fumes, such as those from smelters, which destroyed vegetation, has been prohibited and it is likely that air pollutants of all kinds are due for much more attention.

Clouds. Clouds differ from fog only in their position. Both are made up of droplets of water suspended in the air because they are so minute that they do not settle out. Clouds are frequently formed when air is carried upward by convectional currents and is cooled to the dew point as it rises. Cooling and condensation with consequent cloud formation also result when air is forced upward over a mountain range and from cyclonic disturbances.

Clouds are classified on the basis of form and position, the terminology being derived from an early simple classification in which four types were recognized: cirrus (curly), cumulus (piled up), stratus (flat), and nimbus (rain or storm). Modern systems divide clouds into families, each with its own type of clouds distinguished by descriptive names that are combinations of the old terminology.[418] For details about clouds and cloud forms, an illustrated manual should be consulted.[411, 193]

Precipitation. Fogs and clouds reduce intensity of solar radiation that reaches the earth and may thus be of constant, though minor, ecological significance in certain areas. But, of more general importance, they are the source of precipitation when, because of rapid condensation, their tiny droplets increase in size sufficiently to respond to gravity and fall to the earth. Not all clouds produce rain because convection may not be rapid enough or persistent enough to produce drops of sufficient size.

Summer rains are usually convectional in origin, resulting from local, vertically ascending air currents of high velocity, and thus are frequently short and heavy. When moist air is forced upward mechanically as it rises over a mountain, the effect is the same (orographic precipitation). The highest precipitation records are usually obtained on windward mountain slopes and are produced by such forced ascents of air. Tropical rains, often short and very heavy, are usually convectional or orographic.

During cool seasons, in the middle latitudes, precipitation is most often cyclonic in origin. This refers to the great air masses circling counter-clockwise about a passing center of low pressure. The air in the forward half of an advancing low, being relatively warm and moist, ascends over the cooler local air in a long, gradual, inclined slope—a typical warm front. This extended inflow and ascent produces a large area of sustained precipitation. In the rear half of the cyclone, cold air undercuts warm, moist air to form a cold front that often results in heavy precipitation but over a narrow belt. The slower the low moves, the longer a front remains over an area and the more protracted the precipitation.

The relative stability of annual and seasonal precipitation in an area where the cyclonic type predominates is due, to a large extent, to the fact that the pattern of origin and path of movement of low and high pressure areas tends to be seasonally similar, year after year. Because these patterns repeat themselves, short-time predictions of the direction, and the precipitation and temperature effects of an advancing low can be made with considerable accuracy. Middle latitude vegetation must be adapted to the rythmic and often sudden march of cyclonic precipitation and associated temperature changes.

Other forms of precipitation include snow, which is formed like rain but at temperatures below freezing and under conditions that permit the crystals to fall before they melt. Sleet is rain that falls through air strata of low temperature and then reaches the earth as clear pellets of ice. If rain falls on a cold surface and freezes, it is called glaze. Hail, which falls almost exclusively in summer because of its dependence on convectional storms, starts with a snow or ice nucleus, which falls to a stratum of sufficiently high temperature to be partially melted. When carried upward again, the moisture on the surface freezes, and condensation adds to the size. If the process of falling and being carried up again is repeated several times, a concentrically layered mass of snow and ice is formed with sufficient size to fall to the earth as a hailstone.

Since hail is primarily a summer phenomenon occurring only under exceptional conditions, it is of little consequence to plants as a source of water. It may, however, do serious physical damage, often stripping foliage completely from woody plants and damaging herbaceous structures beyond recovery. Sleet and glaze are in the same category. Glaze may be so heavy as to cause great damage to forest trees through breakage. Conifers are particularly susceptible to such damage because of the load of ice that can accumulate on their many needles. In young stands, the trees may be broken down so that they die, or they may be so bent and twisted that, should they grow to maturity, they form badly distorted trees.

Snow is an important source of soil moisture and, in addition, may serve to modify the effects of low temperatures. Roughly ten inches of snow are equivalent to an inch of rain although the moisture content of snow is highly variable. Under average temperature conditions, water derived from melting snow might make up from 5 to 25 percent or more of the total precipitation, but its importance is not determined entirely by amount. Since conditions in the spring may be such that a heavy blanket of snow disappears in a few hours, the water may run off rapidly, especially if the soil is frozen, and be of no more significance than that of an extremely heavy rain of short duration. That same

Fig. 54. *Northern hardwood stand of birch, hard maple, elm, and ash after a glaze storm in New York. Scarcely a tree escaped damage.*—U. S. Forest Service.

amount of snow, if it melts over a period of weeks, can release water so slowly that practically all of it will soak into the soil, to become a part of a reservoir to be drawn upon during dry periods weeks or months later. Again, under semidesert conditions where the vapor pressure deficit is high, this may not be true, because, if the snow remains for long periods, much of it may be lost by evaporation or sublimation.

The reserve of ground water derived partly from snow becomes of greatest importance where the total precipitation is relatively low. The grasslands of our Middle West are much more dependent upon the reserve of ground water than are forested regions where the total precipitation is greater and is more evenly distributed throughout the year. The success of agriculture, especially wheat production in the mixed prairie region of the Dakotas, Nebraska, Colorado, and Wyoming, is to a great extent, dependent upon the reserve of soil water derived from snow. To be sure, where snowfall comprises a high percentage of the total precipitation, it must be of relatively greater importance than elsewhere. Subalpine forests are often almost com-

Fig. 55. *Average snow pack as it appears in March in the Sierra Nevada. Echo Summit, Calif.*—Courtesy of W. D. Billings.

pletely dependent upon soil water derived from snow. The red fir forest along the crest of the Sierra Nevada receives practically no rain throughout the growing season. However, the cool summer days at these high altitudes do not create high water losses, and since snowfalls of sixty feet have been recorded there, the resulting water is adequate to maintain the forest and to provide, as it runs off, an excess usable for agriculture at lower altitudes.

It is apparently possible to increase the amount of snow accumulated in such forests as well as to reduce the rate with which it melts. Since the greatest storage and the slowest losses both occur in stands with numerous small glades or openings,[426] forest management in this direction may increase the water supply and prolong the streamflow.[215]

Snow water is of prime importance in those arctic and alpine regions where there is practically no rain. Here the plants are shallowly rooted, not uncommonly limited to the surface soil by perpetual frost a few inches below. Surface water must then be supplied continuously to maintain plant life. This is provided by the melting snow,

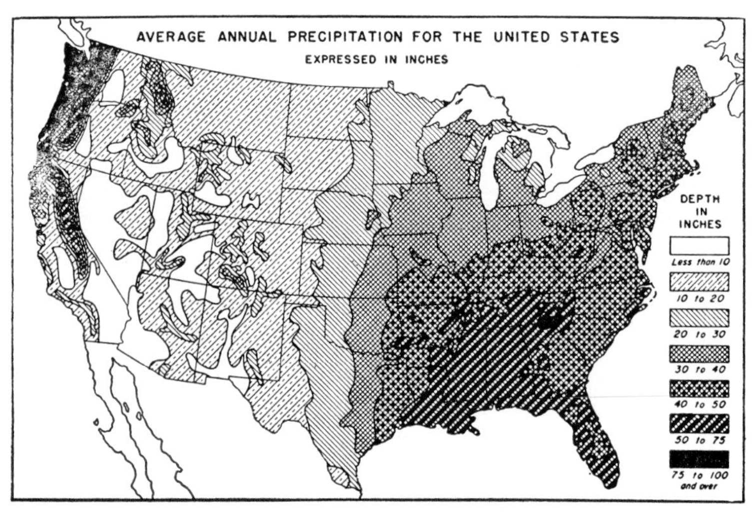

Fig. 56. *Average annual precipitation for the United States.*—By permission, from Bernard,[25] in *Hydrology,* copyrighted 1942, McGraw-Hill Book Company.

some of which, in depressions or other protected places, may remain throughout the growing season. The richest flora in best condition will usually be found at the margins of snow patches and in drainage lines below them. Ridges and raised ground are the first to be exposed at the beginning of the growing season, and there growth begins almost immediately. As the season progresses, more ground is exposed by melting snow, and plants there begin growth. Thus, at distances of a relatively few feet, may be found plants of the same species, that have flowered, fruited, and dried up, and, in the moist soil beside the snow, plants which have just begun their growth.

The total annual precipitation of an area is only a rough indication of moisture conditions for plant growth. A light rain of 0.15 inches usually does not affect soil moisture, for most of it will be intercepted by vegetation and will evaporate quickly. That which reaches the soil

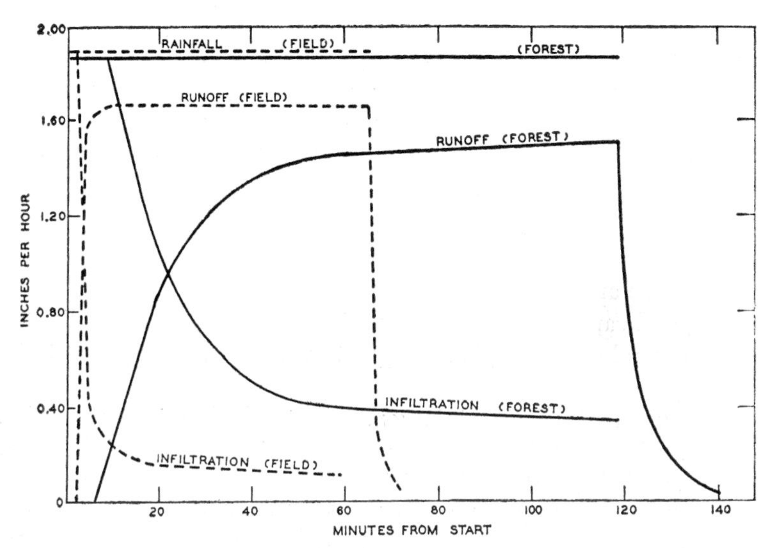

Fig. 57. *A comparison of surface runoff and infiltration on forested pineland (55 yr.) and on bare, abandoned land in Mississippi when precipitation was at essentially the same rate for both areas.*—Adapted from Sherman and Musgrave.[369]

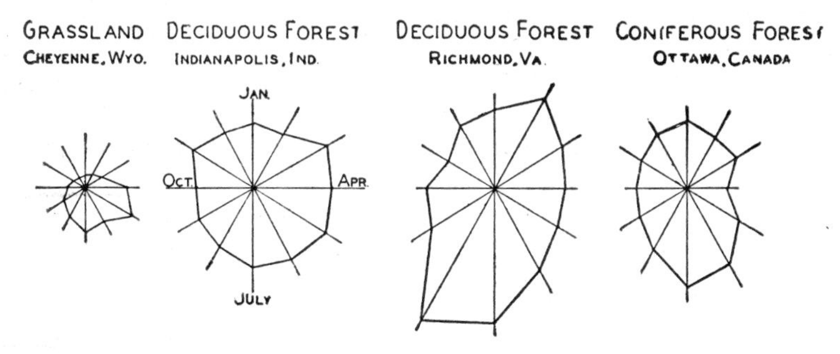

Fig. 58. *Annual precipitation patterns (based on averages) for several stations, which illustrate the relative amounts and distribution of precipitation throughout the year for areas supporting grassland, deciduous forest, and coniferous forest.*—Adapted from Transeau.[406]

will wet only the surface and likewise be lost to the air. Several inches of the total rainfall may, therefore, be of no significance whatever except to raise the humidity temporarily and reduce transpiration for a short time. If rain falls heavily for short periods, say 2 or 3 inches in the same number of hours, much of it will be lost by runoff, the amount varying with steepness of slope, nature of the soil, and amount and kind of cover. Again, the seasonal distribution of rainfall may be of much more importance than the total amount. If rainfall is uniformly distributed throughout the growing season, moisture conditions may be far more favorable with 25 to 30 inches than they would be with 40 to 45 inches if the growing season is interrupted by one or more protracted dry spells. If precipitation is regularly seasonal, the type of vegetation may be definitely limited. For instance, grasslands characterize those areas where rainfall is rather light and concentrated in the spring and early summer. Winter rains with dry summers, characteristic of several coastal regions, support evergreen shrubby vegetation.

Measurement of Precipitation. A standard rain gauge (U. S. Weather Bureau) is a cylinder 8 inches in diameter and 20 inches high, which has a funnel built into the upper end that permits the water it catches to run into an inner cylinder with exactly one-tenth the cross-sectional area. The ratio of the outer to the inner cylinder being 10:1, the measurement of water collected in the tube must be divided by 10 if taken directly, or it can be measured with a standard graduated rod. The 10:1 ratio makes accurate readings possible to 0.01 inch. Exceptionally heavy rains may overflow the tube, and the water in the large cylinder must then be poured over into the emptied tube for measurement. Several types of recording gauges have been devised, primarily to indicate time and rate of fall.[266] Those most often used either (1) register increments of fall as a small bucket fills, tips, and records, or (2) weigh and record accumulative precipitation as it falls.[25]

For generalized field studies, the precipitation records from the nearest weather station are often useful. However, there may be wide local variations, especially if the topography is irregular; and, in mountainous regions, only local measurements have real significance. A study of this variability on a 67 acre tract showed that 31 gauges would have been necessary to get a statistically acceptable annual mean, and many more for monthly means.[343] In addition, under forest stands, the precipitation reaching the soil will vary from stand to stand because of variation in interception. Thus, for intensive work it is desirable to maintain rain gauges at each sight of study. Although standard gauges are desirable, it is possible to obtain satisfactory

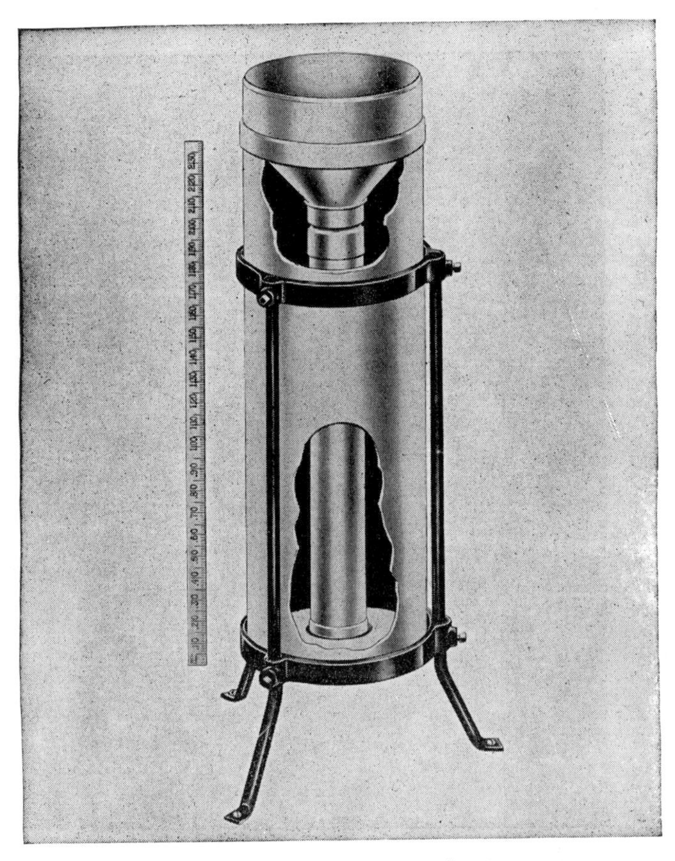

Fig. 59. *A standard rain gauge and measuring stick. Cutaway view to show funnel and inner tube.*—Courtesy Friez Instrument Division, Bendix Aviation Corporation.

records for comparison by using straight-walled jars or cans with a known and equal diameter. Size of the gauge does not apparently affect measurements materially. More important is that the rims be precisely horizontal, at the same height above level ground, preferably clear of obstructions. On steep slopes the most accurate measurements are obtained with tilted gauges having the rims parallel to the ground surface.[182]

Measurements with regular rain gauges of precipitation falling as snow are not dependable and rarely comparable. Because air move-

Fig. 60. *Snow sampler being weighed to determine water content of the snow core it contains after being forced vertically downward into snow.*—Soil Conservation Service.

ment about a gauge modifies the amount of snow intercepted, shields are necessary to control the wind. Shielding of gauges has not been standardized. Usually, records are obtained by direct measurement of the snow that has fallen. These measurements are made at a point where the wind has not caused drifting or disturbance, and the equivalent value in rain is computed from samples of the snow. Depending upon the density of the snow, the ratio may range from 5:1 to 50:1, but 10:1 is fairly average. Careful records of snowfall and water equivalents have not been generally kept until recently. In the western mountains, where melting snow may be the only source of water for distant low country, observations made year after year on numerous "snow courses" (commonly mountain meadows), make possible forecasting of floods and, more particularly, the supply of water available for irrigation.[76] The sampling is done with an aluminum tube of easily coupled sections, which is forced vertically into the snow to cut a

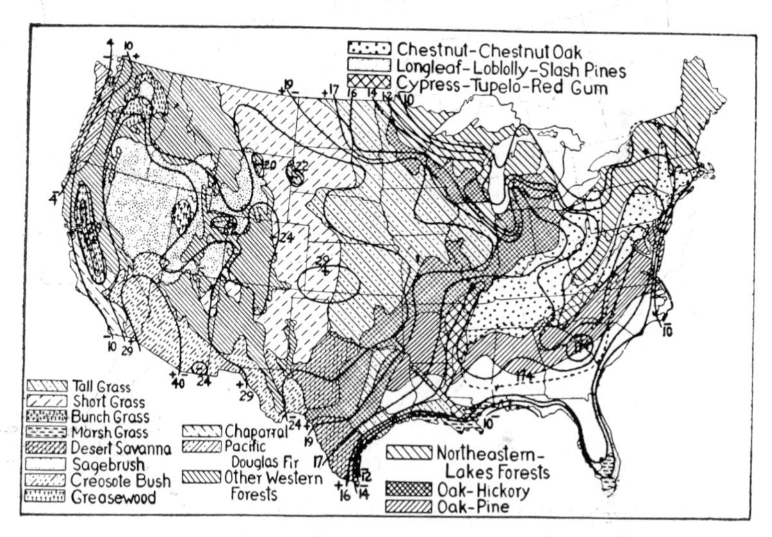

Fig. 61. *Isoclimatic lines of vapor pressure deficits and vegetation areas of the United States.*—From Huffaker.[190]

core. The depth of snow is indicated by markings on the tube and water content is determined by weighing tube and core on a simple scale. The design is such that one ounce of core equals one inch of water depth.[436]

Atmospheric Moisture and Vegetation. It should be clear that any single atmospheric factor is insufficient in itself to explain the distribution and survival of species or plant communities. Precipitation records are only suggestive, for they must be interpreted in terms of seasonal distribution, and they are not at all indicative of soil moisture conditions or of the evaporating power of the air to which a plant must be adjusted if it is to survive. The variation in the seasonal pattern of precipitation from place to place becomes particularly apparent when illustrated with twelve-point polygonal diagrams[406] (Fig. 58), which make possible easy comparison of amount and time of precipitation by months. Evaporation alone is a poor criterion of ecological conditions since it does not take into account the amount of water supplied by the soil. When points of equal vapor pressure deficits are connected by lines on a vegetation map,[190] the zones come nearer to matching

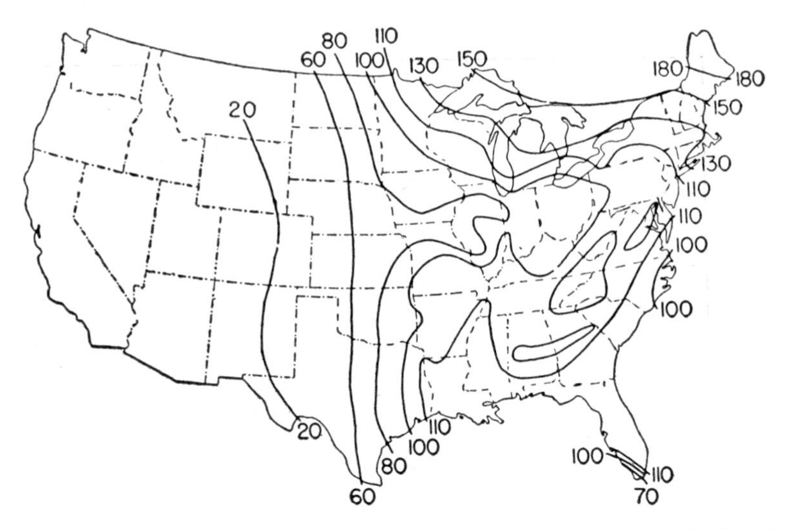

Fig. 62. *Precipitation-evaporation ratios for the United States calculated according to Transeau.*[405] By permission from Jenny, *Factors of Soil Formation,* copyrighted 1941, McGraw-Hill Book Co.

the distribution of vegetation types on a regional basis than similar ones based on evaporation.

Seeking a single comprehensive value that would include several factors operative in plant distribution, Transeau[405] used the ratio of precipitation to evaporation (P/E) for plotting climatic zones. These zones match remarkably well the vegetation types of eastern North America, where the method was initially applied; but wide applications are limited by the paucity of adequate evaporation data.

For the same reason the numerous attempts to formulate bioclimatic expressions indicative of vegetation have all had to depend primarily on temperature and precipitation data.[233] As a result, no matter how

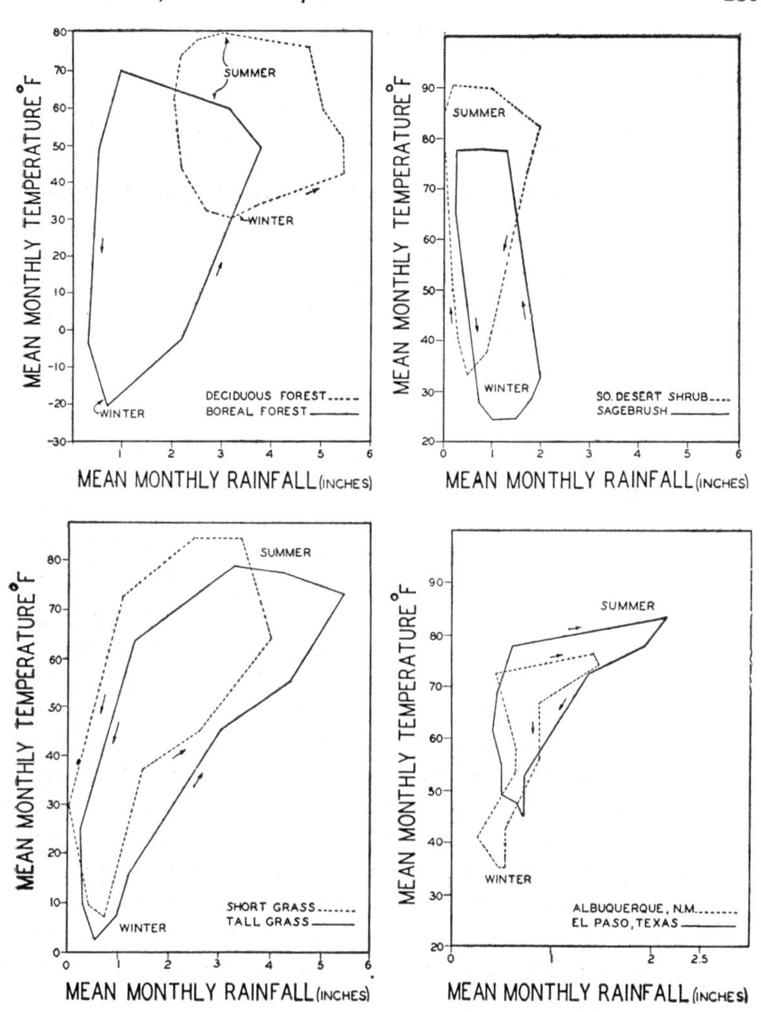

Fig. 63. *Three sets of composite climographs, which permit comparison of forest, desert, and grassland climates, as well as differences within these general types of vegetation. The fourth set, which shows the similarity of climates at stations in New Mexico and Texas, has been used as support for classifying El Paso grassland as short grass, of which Albuquerque is representative, and not desert grassland as some have done. Monthly means of temperature and rainfall have been plotted against each other and the points connected by lines. Those with less than 12 angles have points along straight lines.*—From Smith.[383]

much the formulae are complicated by attempts to correct for evaporation, soil moisture, or seasonal variations and extremes, the resulting isoclimatic lines have a great similarity. They conform roughly to broad vegetation types but have little utility for distinguishing their variations.

One of the more recent attempts[400] to provide a useful formula for evaluating water balance in terms of vegetation has attracted much interest in spite of its complicated nature. Because adequate presentation of the method is impractical here, only an outline will be given. Monthly potential evapotranspiration is determined from a calculated monthly heat index related to mean monthly temperature, and corrected for latitude. Summation gives the annual water need. Interest in evapotranspiration is in part related to the possible application of this estimate of water need to specific communities and habitats. Regional application involves another step. The amount that mean monthly precipitation is greater or less than monthly potential evapotranspiration is determined for each month and summed for the year. Then, the difference between annual surplus and deficit is divided by annual evapotranspiration to give a ratio, or annual index, which like other such indices, has been used for mapping climatic types.

A graphic method for distinguishing differences and similarities in atmospheric conditions is the climograph, in which mean temperature is plotted against mean relative humidity by months, and the points are connected to form highly distinctive twelve-pointed figures. Introduced by Ball [15] for indicating climate of geographic areas, it has been variously used for comparing climates in studies of the distribution, migration, and success of populations of man, birds, and insects. The system is subject to modification and has been used also as a graph of temperature-precipitation (sometimes called a hythergraph). The latter method has been used [383] for characterizing climates of widely differing climax types in different parts of the world and for distinguishing grassland climates in North America.[71] Applications of the method have not been exhausted.

Wind

Air moves from a region of high pressure to one of low pressure, and the differences in pressure are largely the result of unequal heating of the atmosphere. The equatorial regions receive more heat than regions to the north or south; consequently. low pressures normally exist in the lower latitudes. The tendency, then, is for air to move from the poles toward the Equator, there to rise and return toward the poles. This pattern, although true in general, is modified by the de-

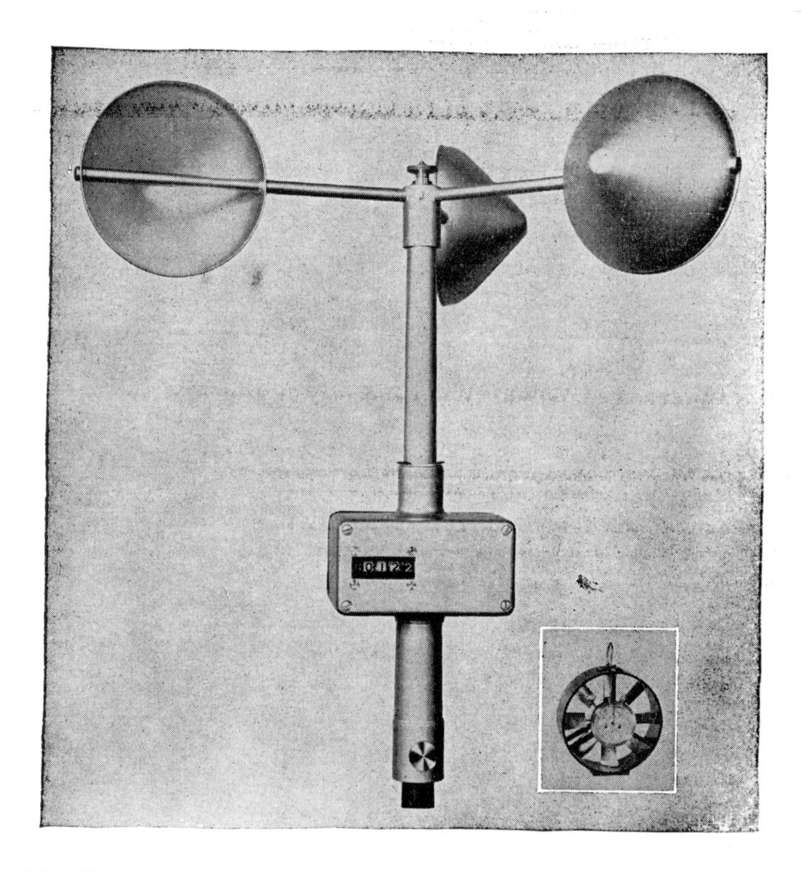

Fig. 64. *Cup anemometer, Weather Bureau type, for relatively permanent operation, and a Biram type anemometer, convenient for short-time measurements.*—Courtesy Friez Instrument Division, Bendix Aviation Corporation.

flecting action of the earth's rotation and by differences in temperature resulting from oceans and land masses.

Continents in temperate zones tend to become very hot in summer, and the resulting low pressures produce winds that blow inland. The cold of winter reverses the pressure, and winds tend to be outblowing. In mountainous areas or along sea coasts these seasonal trends may have daily variations again produced by temperature-pressure differences. Mountain valleys and slopes, which are often warmed rapidly during the day, produce valley breezes blowing upward. At night, the rapid cooling of bare high ridges results in a flow of cold air down the valleys. The contrast between day and night temperatures of land and water results in an offshore breeze at night, as the land cools rapidly and higher pressures result. During the day, the land again heats up rapidly above the temperature of the sea, and an inshore breeze develops that may be noticeable for several miles inland. This brief outline of factors producing wind, although greatly over-simplified, should serve to emphasize that air is almost constantly in motion and should suggest that, within limits, the general plan of motion is predictable for seasons and parts of the earth.

Measurement of Wind. Wind speed is measured with some form of *anemometer.* The cup anemometer used by the United States Weather Bureau has three or four hemispherical or conical cups, each attached to horizontal arms that rotate on a vertical axis and thus drive a gear system, which turns indicator dials. These are readable in miles per unit of time, usually expressed as miles per hour. More elaborate instruments are equipped with automatic recording devices. As with other environmental measurements, position of the instrument must be considered in terms of objectives. Results will vary with height above the ground and nearness of obstructions such as trees, buildings, or topographic irregularities.

The cup anemometer is inconvenient to carry and operate in the field. It is designed for relatively permanent installation. In the Biram portable anemometer, a small fan drives the dial indicating air movement. The device is useful in small spaces and for short readings. Since it has no vane, it must be set to face the wind.

Physiological-Anatomical Effects of Wind. The movement of air being in general characteristic of all environments, its direct effects on plants are negligible under average conditions. In certain situations, however, wind may be an extremely important factor. Plants growing in habitats exposed to continuous winds of moderate velocity transpire more rapidly than unexposed individuals. If the prevailing winds are from one direction, the side of a plant toward the wind may be so desiccated that new growth is killed before it is well begun. Lateral buds taking over the growth may or may not survive, and a scrubby, matted growth develops on the windward side. To leeward, the new shoots are protected by the rest of the plant, and growth goes on there, resulting, over a period of years, in asymmetric growth forms of amazing shape. Such one-sided growth is commonly found in exposed places at high altitudes in the mountains where otherwise upright plants may be prostrate and form mats fitting into hollows or behind protecting rocks. Not uncommonly a forest stand on the protected side of a ridge or in a ravine may appear as though every tree had had its tip sheared to an exact height limit. Again, this is due to the desiccating effect of the prevailing wind.

Fig. 65. *Prostrate and matted, wind-sheared trees* (Pinus albicaulis, Tsuga mertensiana) *on a leeward slope near timber line, Mt. Hood, Ore. The twisted form is commonly termed* Krummholz.—U. S. Forest Service.

Asymmetric growth, matted vegetation, and sheared tops as seen along the coast are likewise produced by wind to some extent, but here an added factor plays a part. The wind picks up spray as it comes in over the breaking waves. The spray may be carried several miles inland, especially in severe storms, but its major effects are most noticeable near the coast. The spray that strikes any obstacle is dropped there; and, of course, the salt from the spray accumulates on that object. Few dune and coastal plants are completely tolerant to salt spray, but fortunately, most strong winds are accompanied by rain, which minimizes the effects by dilution and washing. If a severe windstorm is not accompanied by or soon followed by rain, much vegetation will be injured or killed by salt spray even for some distance inland.

Those plants growing near the beach are sprayed lightly almost daily[41] and, as might be expected, show different degrees of tolerance. This results in zonation of vegetation associated with exposure to the wind.[286] Undoubtedly salt spray is one of the strong factors in determining the make-up and distribution of all plant communities on coastal dunes.

When trees grow on one side only, they may become so heavy as to uproot themselves, but usually the eccentric growth is slow enough to permit compensating anatomical changes, particularly in the trunk. Secondary growth may cease completely on the windward side

Fig. 66. *Asymmetric growth of a live oak* (Quercus virginiana) *exposed to ocean wind and salt spray from the right. North Carolina coast.*—U. S. Forest Service.

of the trunk and increase proportionately on the leeward side, thus forming a brace under the added top. An extreme illustration is a section of trunk, now at the University of Minnesota, taken from a Monterey cypress that grew on Cypress Point, just north of Carmel Bay, California. It is 74 inches in the diameter that grew parallel to the prevailing wind but is only 9 inches in the opposite diameter. Only 50 growth layers were formed on the windward portion of the section, but the leeward portion (71 in.) has 304 layers. This section was taken 24 feet above the ground.

That diameter growth of the lower trunk is greater in free swaying trees than in those held rigidly or in closed stands has been demonstrated for *Pinus radiata.*[196] Experimental trees were held rigidly with guy wires and after ten years the controls had notably greater trunk diameters, somewhat eccentric in line with prevailing winds, to a height of 25 or 30 feet. Their roots were also enlarged near the trunk. Height growth was unaffected. The results suggest why diameter growth in heavily thinned stands sometimes appears to be greater than might be expected because of increased space.

Fig. 67. *A white pine stand in New Hampshire after the storm of 1938. Such damage was prevalent over much of New England at the time.* —U. S. Forest Service.

Other physiological effects might be mentioned, but they are largely brought about within the plants themselves through adaptations that serve to reduce the rate of transpiration through their effects on stomata. In the drier sections of the country, such as the plains and desert, the almost continuous dry winds increase transpiration rates materially and serve to accentuate the effects of low water supply.

Physical Effects on Plants. Most people have seen the effects of a strong wind (25-38 miles per hour) upon vegetation. It is not uncommon for dead branches to be broken from trees; an occasional tree, especially if overmature and diseased, may be blown down. Closed forest stands usually suffer no major damage because the trees give support to each other. With greater velocities the wind becomes increasingly destructive. At gale velocities (39-54 m.p.h.) live branches are broken, and a full gale uproots trees with ease.

Fig. 68. *Wind throw often results because trees are uprooted, especially if on shallow or wet soil. Here is shown a giant Douglas fir in Washington whose torn-up root system had a spread of fifty feet.*—U. S. Forest Service.

Many of the destructive storms along the Gulf coast approach hurricane speeds, and it is fortunate that they infrequently reach the mainland. Several destructive hurricanes in recent years have moved northward along the Atlantic coast and struck inland at speeds in excess of 70 miles per hour. The destruction in their paths was extreme. From the Carolinas to New England whole forests fell before the wind, the trees uprooted or broken off. An added factor in the destructiveness of some of these storms was the saturated soil, produced by a preceding period of heavy rain, which contributed to the ease and amount of uprooting and wind throw. Storms of such force and destructiveness are rare in North America, but lesser winds may cause considerable damage. When closed forest stands are thinned or selectively cut, the remaining trees are subject to wind throw for a number of years even though wind does not blow with great velocity.

In addition to physiologically-produced flag forms of woody vegetation, there are those resulting from purely physical effects of wind. A study of asymmetric trees in the Columbia River Gorge[229] showed that, when branches are continually bent in one direction by prevailing winds, the branches become "wind trained" and hold their positions permanently. Some grew completely around the trunk from the windward to the leeward side. Still another cause of asymmetry was found here. Severe winter storms, coming largely from one direction, cause much breakage, especially when accompanied by sleet, and almost complete pruning of branches on the windward side often results.

Transportation by Wind. We have already indicated how important to precipitation are the vapor-laden winds moving inland from large bodies of water and how transporting salt spray may be of local significance. Wind plays a more direct role in transporting pollen and in dissemination.

Wind-borne Pollen. Many pollen grains are light and small or, as in conifers, have bladder-like wings, which increase their buoyancy. As a result, they may be carried for great distances by the wind. The chances that an individual pollen grain will accomplish its function must be extremely small. This uncertainty is compensated for in quantity of pollen produced. Here is the pollen production for several wind-pollinated European species:[143] *Rumex acetosa* produces 30,000 grains per stamen, *Acer platanoides,* 1,000; pollen output per staminate cone of gymnosperms may be judged by *Pinus nigra,* 1,480,000; *Picea excelsa,* 590,000; and *Juniperus communis,* 400,000; production per flower of angiosperms ranges from *Rumex acetosa,* 180,000, through *Tilia cordata,* 43,500, to *Acer platanoides,* 8,000. Such figures for single stamens and flowers serve to explain the continuous and enormous rain of conspicuous pollen that may fall in season, especially from conifers. Sidewalks, porches, floors, tables—everything in the vicinity of a coniferous forest—may be dusted with pollen.

Not all noticeable pollen is locally produced, and a great deal of

Fig. 69. *Flag-form trees in the Columbia River Gorge, Ore. (1-2) Storm-pruned Douglas fir, deformed by breakage and killing due to glaze storms and strong east winds of midwinter. (3-4) Wind-trained Douglas fir shaped by long-continued pressure of strong west winds of late spring and summer.*—From Lawrence.[229]

evidence has been accumulated to show irregular and normal distributions. There is a story that, in the early days of the city of St. Louis, it was at one time continuously showered with a yellow dust, which gave residents some concern until botanists identified it as pollen of *Pinus palustris* transported from the coastal plain far to the south. Some quirk of pressure and wind was depositing the pollen upon St. Louis.

There are numerous records of pollen being transported long distances.[143] Spruce, pine, and birch pollen was collected on lightships in the Gulf of Bothnia about 19 and 35 miles off the coast. Spruce pollen is carried from southern to northern Sweden. Peat samples taken in Greenland contained pollen of *Picea mariana* and *Pinus banksiana*, which must have originated on Labrador or southwestward. One of the most interesting studies of pollen transport was made on a crossing of the Atlantic from Gothenburg to New York. Using a vacuum cleaner equipped with filters, a more or less continuous quantitative record of pollen in the air was obtained on the entire trip. Numbers of grains decreased with distance from land, but at no time did sampling fail to show some pollen. The evidence is to the effect that birch, pine, oak, willow, sedge, and grass pollen are carried in quantities for more than 600 miles over the ocean.

The amount of pollen in the air and the distance it is transported is of significance to some plants but more so to many people who suffer from hay fever. Recently the kinds and numbers of pollen grains in the air in many sections of the country are determined daily and publicly announced.

In general, wind-pollinated plants grow in the open or in exposed places. Even in a forest it is the trees of the upper strata that are characteristically wind pollinated; the flowers are small and inconspicuous, with simple or reduced structure. The corolla is often lacking, and there is an absence of bright colors, odor, and nectar. Stamens and pistils are commonly borne in different flowers, the stigmas are usually feathery, and the stamens are long and pendant. In spite of its apparent wastefulness, the system produces satisfactory results.

Dissemination. Plants migrate from one point to another by means of spores, seeds, fruits, fragments of plants, or entire plants. The agent of transport may be water, animals, or wind, depending upon the various adaptations of the *disseminules,* which facilitate the movement.

Dissemination by spores is characteristic of all plants except spermatophytes. Wind-disseminated spores, like pollen, are small and dry and may be transported great distances. Everywhere that pollen is carried, spores are found too. Their transportation over long distances can be of great ecological and economic importance. A spore carried by a freak wind into distant territory may establish a species where it has never grown before, thus extending the range of the species and possibly necessitating adjustments within the community in which it develops.[433] The economic considerations are obvious when it is remembered that fungi that produce diseases of both plants and animals are all propagated by spores. The fight against wheat rust is a case in point. Whenever a resistant strain of wheat is developed, it is immediately subject to attack by mutating strains of the rust, whether these strains are of local origin or not. There is evidence that strains of rust appearing in the Dakotas have come from wind-borne spores produced as far away as Mexico.[388]

Seeds, fruits, and fragments of plants are effective as disseminules in proportion to the devices that facilitate their transport. Wind dissemination is increased by the presence of winged structures, bladder-like protrusions, or plumose extensions of the surface (see Fig. 99). Seeds, because of their small size, are apt to be carried farther than fruits, but for all, the kind of adaptation is an important factor in transport. The effectiveness of a parachute-like pappus is illustrated by the ubiquitous dandelion and related composites of field and roadside. Many winged fruits do not travel far because of their size, but often the wings (ash, elm, maple, basswood) are sufficient to assure transport beyond the shading and competitive effects of the parent tree.

The transport of entire plants is well illustrated by the tumbleweeds (*Salsola, Cycloloma*). These have but a single main root, which, when broken at the ground surface, releases the spherical plant to roll before the wind until caught, perhaps in some fence corner. As it rolls, its seeds are gradually shed, sometimes miles from the place of growth.

The pioneers in a new habitat usually have effective means of dissemination and an abundance of seed. The same is true of weeds of cultivated fields and waste ground. The more common and widespread a species is, the more efficient are the mechanisms that facilitate its dispersal, regardless of whether the agent be wind, animals, water, ice, or gravity.

Fig. 70. *Approaching dust storm near Springfield, Colo. (1937), which was typical of conditions in the "dust bowl" during the drought of the 1930's.*—U. S. Forest Service.

Fig. 71. *Soil blowing out of a Kansas wheat field in the 1930's and piling up on highway, where fences and trees partially checked its movement.*—U. S. Forest Service.

Wind and Soil. The slightest air movement shifts dust particles from place to place, and increasing velocity results in the transport of larger particles of soil in increasing amounts. Although fine materials are everywhere being shifted by wind, its greatest effects are noticeable in dry climates where there is a prevailing wind and a minimum of vegetation. During extended droughts, the cultivated, semiarid regions of our Midwest and Southwest have at times become shifting seas of drifting soil, and the clouds of fine materials carried about in the air have given rise to the term, "dust bowl."

Fig. 72. *Road cut through a deep deposit of loess in Missouri. The almost vertical banks have stood for eighteen years without eroding.*—U. S. Forest Service.

Over an extended period of time great quantities of materials may be transported and deposited by wind, as is demonstrated by the enormous deposits of *loess* in various parts of the world. This fine-grained fertile soil occurs in deposits from a few to fifty feet deep or more over thousands of square miles in the central Mississippi Valley region. Our richest farm lands in Iowa, Nebraska, and Kansas are on loess soils. The deposits occur along the Rhine in Europe and in the pampas of Argentina, and reach their greatest extent in Asia, particularly in north-central China. Loess probably originated during the glacial period as dust was swept up from the barren flood plains of glacial rivers and carried high into the air, from which it settled more or less uniformly over wide areas.

Sand beaches and some desert regions are dry, free of vegetation, and swept by prevailing winds, which carry the soil along near the earth's surface. Any obstacle that checks the velocity of the wind causes some of its load to be deposited and starts a mound or ridge called a *dune*. Some dunes grow, by the deposit of more sand, to a height of several hundred feet, but usually they are much smaller.

Fig. 73. *Extensive active sand dunes on the coast of Oregon showing transverse ridges that have typical form with gradual slope to windward and an abrupt drop to leeward.*—Photo by W. S. Cooper.

Most of the sand is deposited near the crest or on the lee slope; this results in a characteristic gentle windward slope and a sharp drop on the lee slope, the steepness of which is determined by the angle of rest of the sand. Because wind frequently changes direction, dunes are rarely stable for long and present a constantly shifting pattern. Along sea coasts they tend to move inland as sand is carried from the windward side and dropped down the lee side.

A dune is never completely stable unless covered with a continuous mat of vegetation. Should this mat be broken for any reason, a "blowout" results, which may enlarge and start again the shifting of the entire dune. Many cottage owners have learned this to their sorrow when—as has happened on Lake Michigan dunes—they have returned after a single year to find their summer homes almost completely buried under a shifting dune that had been stable for years. The encroachment of dunes on forest areas is not uncommon. Whole forest stands may be buried and subsequently, with shifting winds, be uncovered again to expose "graveyard" forests of dead trunks and branches.

Fig. 74. *Blowout in Oregon coastal dune that was once completely stabilized by vegetation. This is a compound blowout as indicated by the partially stabilized surface of an earlier blowout (lower right), which was again excavated to a lower level by a later blowout.*—Photo by W. S. Cooper.

The extensive dunes on the banks along the South-Atlantic coast[283] have in recent years become increasingly active because their cover was broken or reduced by overgrazing and other disturbances by man. Acres of maritime forest have been buried, buildings have been destroyed, and channels in the waterways have been blocked. Here, as in the dust bowl, are problems that require drastic measures for solution, but such measures must take into consideration the ecological factors involved. Cover crops, strip cropping, mulching, and other modified methods of cultivation have been used with demonstrated effectiveness in the dust bowl area. Long ago many European coastal dunes were planted with forests and effectively stabilized. The Carolina dunes, though, occupy thousands of acres with almost bare sand on which forests cannot be planted until some stability is attained. Kill Devil Hill, the dune from which the Wright brothers made

Fig. 75. *"Graveyard forest" near Florence, Oregon. Once a closed stand (probably mostly* Pinus contorta*) growing on the soil layer which is broken through in the foreground, the forest was completely buried by the dune now appearing in the rear which subsequently moved on to uncover it again. The view is to leeward.*—Photo by W. S. Cooper.

their historic first aeroplane flight, was stabilized with grasses by Army engineers, after much effort and considerable cost, through use of sodding, seeding, and watering. Such methods are impractical on thousands of acres. The efforts of the Civilan Conservation Corps in the early 1930's were at least partially successful. Taking only the native dune grasses, they transplanted them according to several spacing systems and with some regard to habitat variation over several hundreds of acres. Combined with plantings, brush fences were installed at regular intervals across the largest blowouts. A considerable part of their work was initially effective. Had it been given a minimum of later attention it might have brought many acres of shifting dunes under control.

Fig. 76. *Coastal sand dune moving inland and encroaching on evergreen maritime forest near Kitty Hawk, N. C. Grasses in foreground have been planted.*—Photo by C. F. Korstian.

Microclimate and Microenvironment

Whereas the general climate, characteristic of a broad region and including the entire depth of the atmosphere, has a controlling influence on the nature of vegetation, the immediate environment of the specific habitat determines what species will be there. This restricted portion of environment that affects plants directly is commonly called the microenvironment. The term is sometimes applied to special, limited habitats such as caves or even crevices in rocks, but more often it refers to the portion of general environment that is immediately adjacent to the earth's surface. The concept of microenvironments has led, logically, to consideration of microclimates as they occur in small spaces, or, more frequently, as very small parts of large spaces.

In practice, microenvironmental studies most often deal with near-surface layers of the atmosphere below the 4.5-foot height at which standard weather instruments are usually installed in shelters. The variability of climate below this height was what determined this instrument position initially and also explains the need for intensive study of this layer to interpret plant relationships. This is, of course,

Fig. 77. *Planted grasses and brush fences set up on shifting sand as part of a dune-stabilization program developed by the Civilian Conservation Corps. The attempt was partially successful but was not followed up with later work, which would have added to its success.*—Photo by C. F. Korstian.

the critical environment in which all seedlings must become established and in which most shrubs and herbs live their entire lives.

Throughout the discussion of various climatic factors, the important implications of their variations near the ground have been adequately stressed and need not, therefore, be repeated. The intent of this brief section is to point out that there is a considerable knowledge of microenvironment in general,[158] what kinds of measurements are desirable, and how they should be made. Furthermore, the broader this knowledge, the more apparent the need for obtaining data on the specific habitats for which interpretation of vegetation is attempted.

General References

R. F. Daubenmire. *Plants and Environment.*

R. Geiger. *The Climate Near the Ground.*

W. J. Humphreys. *Fogs and Clouds.*

H. Lundegardh. *Environment and Plant Development.*

W. E. K. Middleton and A. F. Spilhaus. *Meteorological Instruments.*

C. W. Thornthwaite. Atmospheric Moisture in Relation to Ecological Problems.

G. T. Trewartha. *An Introduction to Weather and Climate.*

C. O. Wisler and E. W. Brater. *Hydrology.*

H. B. Ward and W. E. Powers. *Weather and Climate.*

Chapter 7 Physiographic Factors: Pedogenesis

Although physiography includes the physical relations of earth, air, and water, physiographic factors in the environment of plants will here be restricted to those natural phenomena of the earth's surface as evidenced in soil and topography. If related to the origin and nature of soil, they are *pedologic* (from *pedos*—soil or earth) but when considered specifically in their soil-plant relationships, they are *edaphic* (from *edaphos*—referring to soil as a foothold for plants). The discussion is divided accordingly. This section, headed "pedogenesis," deals with the origin and development of soils and some biological implications. The next chapter is on "edaphology" or the relationships of specific soil characteristics to plants.

Soil

Land masses of the earth are covered by an unconsolidated surface mantle of mineral particles derived from parent rock by processes collectively called *weathering*. The depth of the mantle is variable depending upon disturbances and time, while its physical and chemical properties depend upon the nature of the parent rock and the weathering agencies that may have affected it. This inorganic material may be termed *soil* but is usually not so considered until organic materials have accumulated from organisms that have lived in or upon it.

Soil Formation. Weathering may result in purely physical change, as when rock masses are broken into smaller and smaller sizes, or may be of a chemical nature, producing changes in composition of the material. The two processes function together normally. Disintegration is largely accomplished by physical agents, such as water, wind, ice, and gravity, and by expansion and contraction resulting from temperature changes. The first four agents are functional through the erosive action of the load of cutting material they transport and are, therefore, effective in proportion to speed of movement or to force and pressure.

Fig. 78. *Wind-swept subalpine habitat in Utah with typical coarse, angular rock particles and little organic material.* Krummholz *at left is of* Picea engelmannii *and* Pinus flexilis (*see also Fig. 31*).—U. S. Forest Service.

The effects of temperature are the most widespread although not always conspicuous. Differential expansion and contraction of rock materials result in cracking, which is especially marked when temperature changes are abrupt. The widest temperature fluctuations occur in arid regions and at high altitudes where their effectiveness is indicated by consistently coarse and angular soil particles. To a lesser extent the process goes on everywhere. Prying action of plant roots and excavating or burrowing by animals may contribute to disintegration, but these activities are certainly of greater importance in their facilitating of chemical processes. Openings in the soil increase aeration and the percolation of water. Shifting the soil about exposes new particles to chemical action and likewise helps to incorporate organic matter.

The chemical or decomposing processes all tend to result in increased solubility of soil materials, which, in solution, may then be available for the use of plants but are also subject to leaching, or washing out, of the surface layers by rain water. Both oxidation and hydration, the addition of oxygen or water to a compound, are common and result in softening of rock. Carbonation, or the taking up of carbon dioxide, produces carbonic acid merely by union with water, and the acid is an effective solvent of many rocks. Water itself is a weak solvent, and, with the addition of carbonic acid, which is always present, its action is much increased. Decaying vegetation, when present, also contributes acids that facilitate solution. In solution, salts ionize and the relative effective concentrations of the basic and the acid radicals thus formed determine whether the soil solution will be alkaline or acid in reaction.

These and other chemical processes operating more or less continuously, together with physical processes, constitute weathering, which produces soil material that retains few characteristics of parent rock. However, soil is not a product of these processes alone, for biological activity also contributes to its formation. Organic material is an essential part of soil, and its decomposition and incorporation are accomplished largely by microorganisms, whose numbers and activities increase as more complex organisms, particularly higher plants, gradually occupy the surface.

Soil Profile. Processes resulting in the formation of soil material also contribute to soil *development.* As weathering proceeds, fine materials in suspension and solution are carried downward by percolating water to a lower level, where they gradually accumulate. As a soil develops, therefore, a rough stratification becomes apparent in which the horizons characteristically have different physical and chemical properties. These horizons, collectively called the *soil profile,* are designated and recognized as follows:

A Horizon. The upper layer of soil material from which substances have been removed by percolating water.

B Horizon. The layer below the A Horizon in which these materials have been deposited. Layer of accumulation.

C Horizon. The underlying parent material, relatively unweathered and not affected as above.

Litter accumulated on the surface of the mineral soil may be termed the A_0 Horizon. It is often convenient to subdivide the major horizons as A_1 and A_2, B_1 and B_2, etc. A_1 is a particularly useful subdivision, for it is applied to the portion of the A horizon, distinguish-

Fig. 79. *A soil well that illustrates a soil profile (White Store sandy loam) in which the A_0 horizon is very thin, the sandy gray-white A horizon is sharply distinguished from the plastic red clay of the B horizon, and the rocky C horizon shows in the bottom.*—Photo by C. F. Korstian.

able by its darker color, in which organic material has become incorporated.

Soil profiles may be observed in any fresh road cut. When studied in connection with vegetation, a rectangular pit is usually dug some

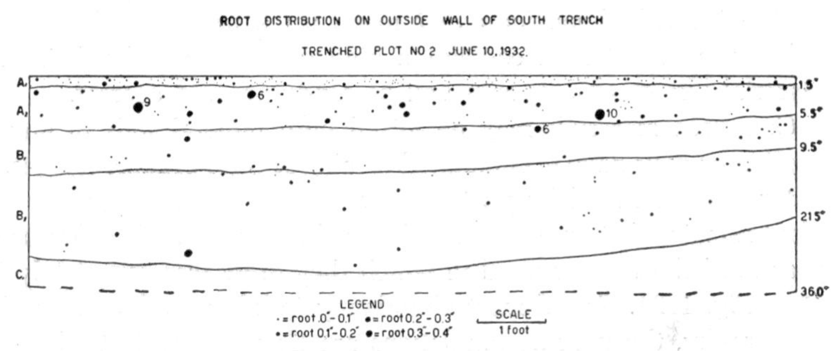

Fig. 80. *An illustration of root distribution in soil horizons and of a method for mapping roots in the wall of a soil well.*—From Korstian and Coile.[219]

four to six feet long, and wide enough to stand in comfortably. One face is kept vertical and cut cleanly to observe the horizons—and possibly the root distribution. Depth of the pit is, of course, determined by local conditions and position of the parent material.

Local Soil Variations. Size of soil particles (*soil texture*) and shape of particles, which determines how they fit together (*soil structure*), may vary markedly within short distances. Texture and structure primarily affect the plant through their influences on air and water in the soil. Organic materials, in addition to modifying soil structure, are the source of plant nutrients that may be quite unavailable from mineral sources.

These variables are a product of the manner in which the soil originated and the time involved in its development. Great areas of the earth are covered with soils that overlie the parent rock from which they were formed. These are *sedentary* soils, whose materials are termed *residual,* if of mineral origin, or *cumulose,* when deposited as organic matter. If soil material has been brought to its present location by some agency such as wind, water, gravity, or ice, it is said to be *transported* and will accordingly have distinguishing characteristics.

Soils Formed in Place. Residual materials are most weathered at the surface and become progressively more like the parent rock with increasing depth. Where parent rocks differ in hardness or solubility, the resulting soils will differ. Fine-textured clayey soils may represent the leached residue of easily soluble rock, such as limestone, or may be the individual particles that made up a fine-grained hard rock. When the parent rock contains a high proportion of hard, insoluble material like quartz, its soils will be sandy or even coarser.

Cumulose materials may be mixed with mineral soils in any proportion or may have accumulated as almost pure organic masses. The latter are illustrated by peat bogs, which are commonly made up of plant remains that only partially decayed and were added to year after year until the lake or pond in which they grew was completely filled (See Fig. 112). Found most abundantly in lakes produced by glacial topography, the peat accumulations are likewise greatest where temperatures are low enough to limit the activities of organisms that produce decay.

Transported Soils. On the great part of the earth's surface covered with residual soil, the effects of transporting agents are commonly noticeable only locally. To the ecologist, however, these localities are of interest because the soil conditions are usually different enough to cause vegetational differences too.

Except for loess, discussed elsewhere (p. 152), soils of aeolian origin are usually sandy deposits, which wind picked up from wide exposed beaches of lakes or oceans. Normally occurring as dunes, they usually form unfavorable habitats because of the low water-holding ability of sand, its relative sterility, and because the dunes are subject to blowouts should the surface cover of vegetation be incomplete (see Figs. 72, 73). In contrast, stabilized dunes of arid or semiarid regions form relatively favorable habitats, because almost all the water that falls upon them is available for plant use.

Alluvial soils have been deposited by streams, which, as transporting agents, are effective in proportion to their velocity and the size of particles involved. Since currents are rarely constant, the size of transported particles varies, and deposits are always noticeably stratified. Alluvial soils are characteristic of lowlands that formed as deltas in or at the mouths of streams or as flood plains along streams that periodically overflow their banks. The greater the distance from the main channel of the stream, the finer the texture of the soil materials deposited. Alluvial deposits usually make desirable agricultural land if properly drained; and, because of favorable moisture conditions, they usually support the richest natural flora of a region.

Colluvial materials are transported by gravity. Except in regions of rugged topography or in mountainous areas, they are rarely extensive. Generally they occur as talus slopes at the bases of cliffs from which the material has fallen. They are usually potentially good soils because they are mixtures of coarse and fine materials, often originating from several kinds of rocks, and organic matter is likewise

Fig. 81. *A wide flood plain in an old river valley whose alluvial soils constitute the best farming land in the region. Hiawassee River, Tenn.* —U. S. Forest Service.

mixed with the mineral components. The favorableness of the habitat is primarily determined by the moisture supply, which is strongly variable, depending upon exposure.

Glacial ice plucks and gouges quantities of soil material from whatever surface it traverses. Carried in the ice, these materials are ground, pulverized, and mixed until they are deposited as moraines at the limit of advance or dropped as the ice recedes. The glacial debris is heterogeneous in composition and texture, and the depth of its deposit is highly variable. Drainage is imperfect, but melt water from the receding ice is plentiful. Its early, rapid, and haphazard flow results in the transporting and assorting of a large amount of soil material, which, as drainage lines become established, is deposited to form topographic and soil features associated with glacio-fluvial activity. The water-assorted soils deposited in the valleys of glacial streams or carried from terminal moraines to form outwash plains are characteristic.

Although glacial deposits may include weathered rock and some organic material, these are usually not abundant in the beginning. Weathering and the establishment of vegetation at first proceed slowly

Fig. 82. *Colluvial cones, still in formation in Colorado. Only in such rugged mountain topography is gravity of direct significance in soil transport.*—U. S. Forest Service.

on glacial soil, but as they progress, a generally good, productive soil is formed. The soils of the northeastern United States and most of Canada are almost entirely of glacial origin.

Regional Soil Variations. Climate, which varies with latitude and longitude, includes the important factors in soil formation, especially temperature and rainfall. Within a climatic area, differences in parent material and topographic position often are reflected in soil variations, which may be chemical or physical. Such variations are most pronounced where parent rock is newly exposed or where soil materials have weathered but slightly, as below a receding glacier. After longer exposure the developing soils become much more alike, and the longer the time involved, the less noticeable will be differences related to local conditions. Evidence is sufficient to indicate that, within a climatic area, soil development progresses toward a particular kind of soil and profile regardless of the origin or nature of the materials; likewise, that the ultimate soil group for similar climatic regions will be similar.

Since climatic conditions determine the activities and kinds of organisms of a region and these organisms in turn contribute to soil development, it is not surprising that vegetation types and soil types are closely related. The development of a soil is paralleled by vegetational changes, the vegetation contributing to soil maturation and the soil controlling a progression of plant communities. Ultimately, the majority of the soils in a region and the communities they support attain a state of equilibrium with each other under the existing climate. The soils will then have similar mature profiles and the vegetation will be in climax condition (see Chap. 10). Likewise, the general area in which mature soil profiles are similar will coincide with the general range of a climax vegetation type.

The recognition of climatic soil types originated in Russia. The approach is well illustrated by Glinka's (1927) grouping of the great soil groups of the world primarily on a climatic basis. Acceptance of the idea has become rather general although sometimes in modified form. The use of specific climatic factors, such as the relationship between precipitation and evaporation, for delimiting effective climate produces regions that correspond closely to the major soil groups.[198] In the United States,[258, 259] soils are usually grouped on the basis of mature profiles into zones of considerable extent. Since only the mature profile is considered, it is a recognition of the same basic approach used by those determining regional limits through climate, although it requires that the profile must exist in reality, not as a potentiality.

Profile Development. Three major processes of soil development are involved in the production of the profiles characteristic of different climatic conditions.

Podsolization occurs typically in humid, cold temperate regions where rainfall exceeds evaporation and where vegetation produces acid humus. The acid decomposition products from the litter increase the solvent power of the plentiful percolating water so that soluble materials and colloids are almost completely removed from the surface soil, which is, therefore, of single grain structure at maturity. Although podsolization occurs under hardwood and pine forests, its strongest development takes place where spruce, fir, or hemlock are dominant. The process is partially a product of the vegetation, for the content of bases in the needles of these trees is notably low, and decom-

Fig. 83. *The layer of calcium accumulation in a pedocal soil under sagebrush desert as shown in a road cut in Nevada.*—Photo by W. D. Billings.

position products of the litter they produce always give an acid reaction. The Russian term "podsol" refers to the ash gray color of the A horizon.

Laterization is characteristic of tropical conditions with high temperatures and abundant rainfall. It is essentially the leaching of silica from the surface soil. The low acidity produced by decomposition of tropical litter promotes the solution of silica as well as alkaline materials. After laterization, the surface soil is high in iron and aluminum, which are not removed by the process. True laterites (Latin—brick) are red throughout.

Calcification may occur anywhere but is most important in regions with low rainfall unevenly distributed throughout the year and with temperatures producing a relatively high rate of evaporation. Under these conditions, a permanently dry stratum may develop in the profile below the depth to which rainwater penetrates. Carbonates produced by carbonation in the surface layers, as well as those that may be present in the original soil material, are carried downward in solution toward this dry layer. When the water is removed by plants or evaporation, the carbonates are left behind, at or above the dry layer, depending upon the depth of penetration of the moisture at the time.

Climatic Soil Types. On the basis of absence or presence of a lime carbonate layer formed by calcification, the mature profiles of all soils

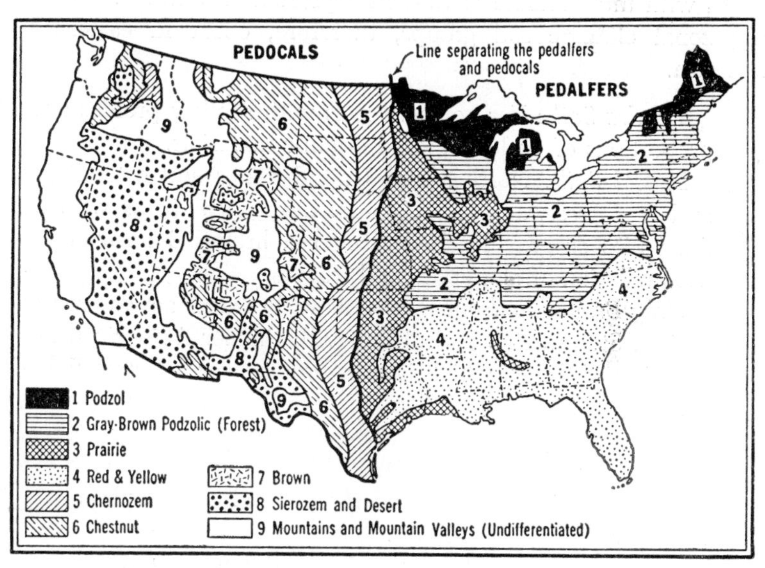

Fig. 84. *General distribution of the important zonal soil groups of the United States.*—After Kellogg[207] and Klages.[216]

of North America fall into two groups: *pedalfers,* without the layer; *pedocals,* with the carbonate layer. The two conditions occur regardless of the nature of parent material or its geological origin, and their distribution is obviously controlled by climate. Soils of eastern North

America are all pedalfers, for the unfavorable balance between rainfall and evaporation necessary to carbonate deposition does not occur here. West of a line (about the 99th meridian) running from the eastern Dakotas to the southern tip of Texas, where annual precipitation is normally less than twenty inches a year, mature profiles almost invariably show pedocal characteristics except where climatic conditions are variable, notably in the mountains and in parts of California. Climate, vegetation, and soil have corresponding distributions. The pedalfers occur principally in association with forest regions, while the pedocals do not support forests but are typically covered with grassland or desert.

Pedalfers. Although mature soils lying east of the line marking the western boundary of the prairie are usually of this type, they vary considerably. The range of temperatures within the area is so great that podsolization is characteristic in the north and laterization in the south with intermediate conditions represented between. The following zonal climatic soil groups, therefore, occur in eastern North America.

Tundra Soils: Far northern soils with shallow profiles and high proportions of undecomposed organic materials.

Podsol Soils: Northeastern United States and extending north and northwestward into Canada. Distinct horizons with a thick A_0, white or gray leached A over a brown B horizon with its accumulation of aluminum and iron.

Gray-Brown Podsolic Soils: A wide band across east-central United States. Like podsol but with thinner A_0 horizon and less leaching of the A, which is gray-brown over a brown B horizon.

Red and Yellow Soils: Southeastern United States where humid, warm climate produces both podsolization and laterization. Colors bright, low in organic matter, high in clay, strongly leached. Yellow soils in the sandy, poorly drained coastal plain; red soils in the well-drained uplands.

Prairie Soils: Western margin of the pedalfers. Intermediate between forest and grassland soils. Black or dark brown with brown subsoils that differ little in texture from the surface.

Lateritic and Laterite Soils: Subtropical and tropical. Represent extreme in mineral weathering. Leached of silica.

Pedocals. Zonation of these soils from north to south has not been recognized as for pedalfers. Moisture being more effective than temperature in producing variation in pedocals, the conspicuous zones lie in a north-south position. Their location and brief characterization follow:

> *Chernozem Soils:* A broad band extending from Canada into Mexico just west of the Prairie Soils. Rich in organic matter. Black soils with brown or reddish calcareous subsoils. Strong carbonate horizon but normal horizons indistinct.
>
> *Brown Soils (also known as Chestnut Soils):* Bordering Chernozems to the west and developed under successively drier conditions, they contain successively less organic matter westward and southward and become lighter in color, as indicated by their division into Dark Brown and Brown Soils. Occupy mainly the area usually called the Great Plains.
>
> *Gray Soils:* Desert and semidesert soils largely in the Great Basin and southward. Gray with yellowish to reddish calcareous subsoils. Negligible organic content. Weathering largely physical.

Within these climatically determined soil regions, are local variations that, because of time, topography, or parent material, bear no resemblance to the mature *zonal* soil type. The immature soils of swamps, islands, and flood plains or of steep slopes are *azonal.* There may simply not have been enough time for soil development to produce the regional profile, or erosive forces may prohibit a mature profile from ever forming. Again, poorly drained flats or depressions usually have distinctive soils with little resemblance to that of surrounding uplands. In areas of low rainfall and high temperatures the depressions are characteristically saline from the accumulated salts that remain after drainage water is evaporated. The profile is indefinite and remains so unless the drainage pattern and leaching become more effective. Undrained depressions in moist climates fill with organic material to form bog soils, and poorly drained flats may have highly organic soils like those of wet tundra, or, if surface-weathered, a plastic clay B horizon. Finally, parent materials may indefinitely prevent formation of the zonal profile because of chemical or physical characteristics. Obviously, a typical pedocal cannot form if calcium is lacking, and a sand dune will never have the same profile as that produced by parent material of diverse mineral elements. Often occurring locally, or as islands scattered through a zone, outcrops of various rocks retain distinctive soil characteristics. Altered andesite in the sagebrush desert,[28] serpentine, wherever it occurs,[416]

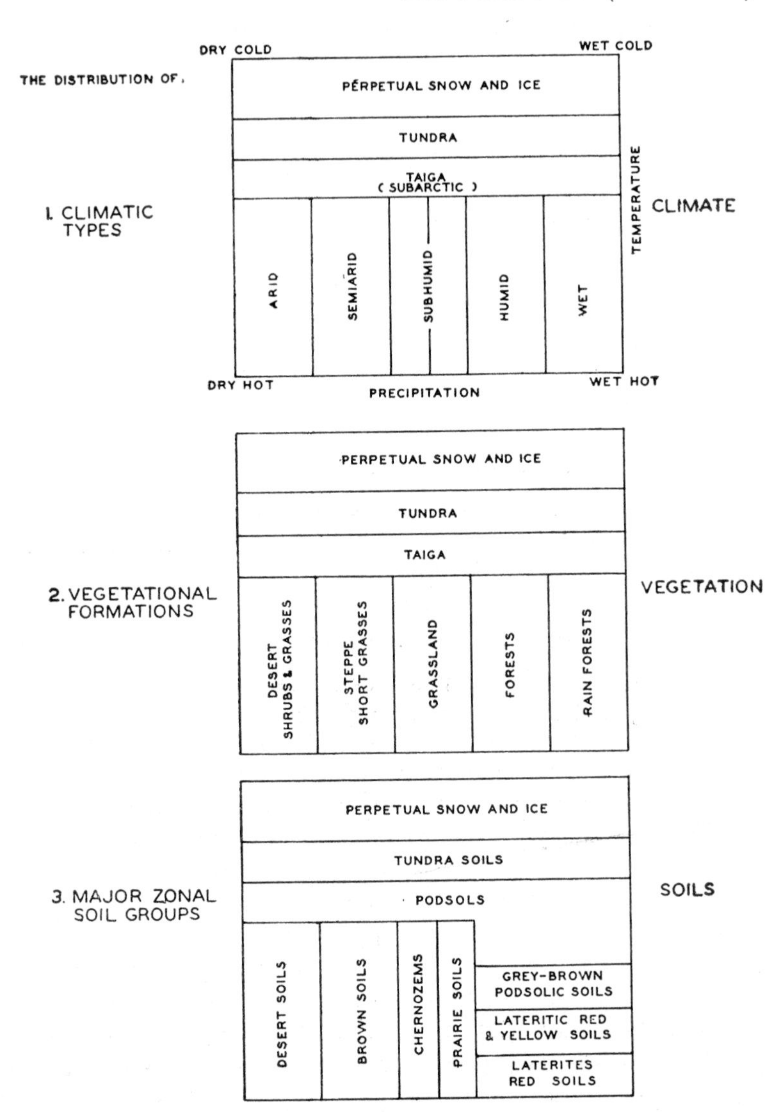

Fig. 85. *Schematic representation to show the interrelated distribution of climatic types, vegetational formations, and major zonal soil groups.* —After Blumenstock and Thornthwaite.[32]

and various shale "barrens"[307] are illustrative. These special circumstances may be *interzonal* in that they may occur in more than one zone.

The climatic classification of soils is useful because it makes possible broad considerations of regional problems. It is logical because it bases the major categories upon mature conditions, which remain stable with the climate, and yet permits explanation of local variations. Best of all, it has world-wide application. Enough investigations have now been made to show that the same general soil types are repeated in those parts of the world where climatic conditions are duplicated. Thus, it has been feasible to devise several schematic representations of the relationship of temperature and moisture to soil formation that are reasonably applicable anywhere. A relatively simple climatic system[398] is shown below, in which temperature-evaporation and precipitation-evaporation relationships are used as criteria of climatic control. It serves to emphasize the importance of moisture in pedocal development and grassland areas but shows that temperature is more effective where pedalfers develop with the forests they support.

Vegetation and Soil Development. The close similarity between the distribution of major vegetation types and zonal soil types has been mentioned. It has also been suggested that the characteristics of a mature profile are partially produced by the vegetation or that they are possible only because of the kind of vegetation supported by that soil in the given climate. The soil and its processes do not constitute an independent system, but rather are a part of the larger ecosystem which includes vegetation and all of its environment.[110] This point should be further emphasized. Newly formed soil material has no profile and bears no resemblance to the mature soil of the region. It cannot support the vegetation that grows on a mature soil, but the plants that can grow upon it contribute to its development, probably most effectively through their decomposition products, and so, in time, the resulting soil changes permit other plants to grow. There results, sometimes over a long period, a succession of edaphically controlled vegetation types, leading ultimately to a climatically controlled community. Paralleling the plant succession are changes in the soil—called *soil development*—which are primarily possible because of the plants and which lead to the mature profile, also controlled by climate. Soil development and vegetational development are intimately related and together are controlled by climate.

General References

K. D. Glinka. *The Great Soil Groups of the World and Their Development.*

H. Jenny. *Factors of Soil Formation.*

C. E. Kellogg. *Development and Significance of the Great Soil Groups of the United States.*

C. F. Marbut. Soils of the United States, in *Atlas of American Agriculture.*

U. S. Dept. Agr. *Soils and Men.*

Chapter 8 Physiographic Factors: Edaphology

The earlier discussion of physiographic factors dealt with the origin, development, and general nature of soils. There remain to be considered the characteristics of soils which, directly or indirectly, affect plants. These are the *edaphic* factors. Some were given incidental attention under pedogenesis, but a systematic consideration of their variations and effects is necessary to an appreciation of the complex of soil-plant relationships.

Soil must provide plants with anchorage, a supply of water, mineral nutrients, and aeration of their roots. Not all plants require these essentials to the same degree, but unless all are present to some extent the average plant cannot become established. On this basis, soil has four major components: (1) mineral material derived from parent rock, (2) organic substances added by plants and animals, (3) water, and (4) soil air. These components vary in amount and proportion from place to place, and the variation may be a significant factor in determining the occurrence of species and vegetation types.

Mineral Components and Physical Character

Soil Texture. One of the most useful bases for classifying soils is that of size of particles. The local soil variations previously discussed are all reflected in soil texture, which in turn has much to do with soil moisture, aeration, and productivity.

The standard classification in the United States is that of the United States Department of Agriculture, which recognizes the following sizes of soil particles by name:

Name	Diameter, mm.
Fine gravel	2.00 —1.60
Coarse sand	1.00 —0.50
Medium sand	0.50 —0.25
Fine sand	0.25 —0.10
Very fine sand	0.10 —0.05
Silt	0.05 —0.002
Clay	<0.002

The percentage weight of these size classes in a soil sample is determined by mechanical analysis. The larger classes may be separated satisfactorily by means of sieves, but the fractions of small size are determined by the pipette method [279] or, better still, the use of a hydrometer.[39] Both methods are based upon the differential rate of settling of particles in water.

After mechanical analysis, accurate textural description is possible by using the names for the fractions singly or in combination. The soil classes are named primarily for the predominating size fraction,[128] but when many sizes are present, the term, *loam,* is introduced. Thus a soil may be termed *gravel* or *clay* if either of these sizes is present almost exclusively, but if gravel or clay merely predominates and is mixed with several other size classes, the soil is called *gravelly loam* or *clay loam.*

A knowledge of the textural grade of a soil suggests numerous other characteristics of that soil. With experience, even a rough estimate determined by "feel" is useful, for texture indicates other physical properties, particularly those affecting moisture, aeration, and workability.

Soil Structure. The arrangement of soil particles becomes especially important when small size classes are involved. Sands have single-grain structure; but silts, and more particularly clays, tend to have particles aggregated in clumps. Aggregation is largely caused by the colloidal portion, less than 0.001 mm., of the clay. Just as clay soils with their tremendous internal surface swell when wet, they also contract as they dry. The minute particles are drawn together by cohesive forces in large or small aggregates whose size and shape affect drainage, percolation, erosion, and aeration (Fig. 86).

If the granular structure is lacking or destroyed by mismanagement, as when trampled by livestock or worked too wet, the soil puddles or

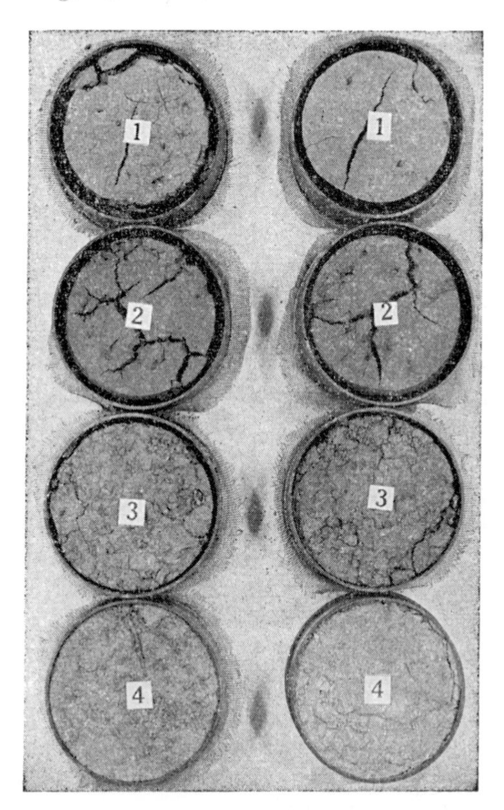

Fig. 86. *Shrinkage upon drying as illustrated by some Piedmont soils. Samples obtained in place (see Fig. 87), then initially saturated with water and oven-dried. B horizon clays—(1) Orange, (2) White Store, (3) Tirzah; A horizon sandy loam—(4) White Store. Such shrinking and swelling in the B horizon affects soil aeration and water movement.*—From Coile.[86]

bakes into hard solid masses, and shrinkage results in the formation of deep cracks. In a loam soil or one with a high organic content, these undesirable features are reduced, while the desirable characteristics produced by colloids are retained.

Organic Content

The amount of organic material in soil may greatly modify its physical characteristics as determined by the mineral components. In addition, organic material is the major source of certain plant nutrients, especially nitrogen, so that fertility and productiveness are usually correlated with it.

Under natural conditions, organic matter in soil is derived from remains of plants and animals. Mostly these remains accumulate on the surface of the mineral soil to form a layer of litter, which, if sufficiently thick, may reduce the effects of insolation, check erosion, and prevent compacting resulting from precipitation. When decomposition of litter does not exceed accumulation, the A_0 horizon has a surface stratum of undecomposed twigs and leaves, which is termed the L layer. Beneath this is a stratum of decomposing but still identifiable plant remains, which is marked by fungal hyphae in abundance and is called the F or fermentation layer. In contact with the mineral soil there may be an H or humus layer if the climate is sufficiently cool and moist. The term *humus* is applied to material decomposed beyond obvious recognition. Soil animals and percolating water carry the humus into the soil where, through further decomposition, its chemical constituents are slowly released for use by succeeding generations of organisms.

When a distinct layer of humus (H layer) is present with a rather abrupt transition to mineral soil, the humus type may be designated as *mor*. If there is no distinct layer of humus but rather it is mixed with the surface mineral soil, the humus type is *mull*.[183]

Local variations in amount, nature, and rate of decomposition of humus are to be expected. Evergreen leaves do not decompose as readily as deciduous ones, nor do they have the same chemical composition.[415] Even the leaves of deciduous species do not all yield the same decomposition products. Organisms causing decomposition may be active and abundant in one habitat but quite incapable of living in another because of such factors as temperature, moisture, and aeration. Consequently, humus may be unequally effective in different habitats, and soils of similar origin may have quite different productive qualities.

Soil Water*

Soil water probably affects plant growth much more commonly than any other soil factor. It follows, therefore, that a basic understanding of what causes differences in amounts and availability of soil moisture and what such differences may mean to a plant is knowledge necessary for an ecologist.

Classification of Soil Water. A simple, arbitrary system of classification that divides soil water into four general categories is sufficient for most ecological purposes.

* Much of this section is adapted from a review by Kramer,[222] which includes an extensive bibliography.

1. *Gravitational water* occupies the larger pores of the soil and drains away under the influence of gravity. For a short time after a heavy rain or irrigation, the soil may be completely saturated with water, the air in it having been displaced from the noncapillary pore spaces between the particles. Under the influence of gravity, the free water soon percolates downward through the soil toward the water table unless prevented by some barrier, such as a hardpan or other impermeable layer. Within two or three days after a rain, all the gravitational water usually drains out of at least the upper horizons of the soil, and the pore spaces refill with air accordingly. If the soil remains saturated with gravitational water for several days, serious injury to root systems may result from lack of oxygen and accumulation of excess carbon dioxide. Hence gravitational water is of little direct value to most plants and even may be detrimental.
2. *Capillary water* is held by surface forces as films around the particles, in angles between them and in capillary pores. Immediately after gravitational water has drained away the capillary water is at its peak, and a soil is then said to be at its *field capacity*. Much of this film water is held rather loosely and is readily available to plants, but some of it, which is held by colloidal material and which is in the smallest pores, is relatively unavailable. It is in this connection that the size of particles becomes important. A cubical sand grain one millimeter on the edge has a surface of only 6 square millimeters, but if it were divided into cubes of colloidal size, 0.1 micron on the edge, the total surface resulting would be 60,000 square millimeters. The increase in surface and angles between particles would thus increase tremendously the total capacity for holding capillary water. However, the water available to plants does not increase proportionally, for the greater curvature of the films and the sharper angles sufficiently increase the force with which water is held to materially increase unavailable water.
3. *Hygroscopic water* is held in a very thin film on the surface of particles by surface forces and moves only in the form of vapor. The moisture remaining in air-dry soil is usually considered as hygroscopic and is, in general, unavailable to plants. Distinction between this and capillary moisture is difficult, for exposure

of soil to increasingly moist atmospheres may increase the water content even to saturation.

4. *Water vapor* occurs in the soil atmosphere and moves along vapor pressure gradients. It is probably not used directly by plants.

Origin of Soil Water. Precipitation in the form of rain, hail, or snow is the ultimate source of water found in the soil, but not all precipitation becomes soil water. The steeper a slope, the more water will run off from its surface before it can enter the soil. Excessive precipitation in a short period of time results in greater runoff than that following a gentle rain, since infiltration cannot keep pace with the rate of fall. If soil becomes saturated and precipitation continues, little, if any, will enter the soil. A larger proportion of water from slowly melting snow is apt to enter the soil than from an equal amount of rain. Infiltration into a fine-textured, clayey soil is slower than into a coarse-textured, sandy soil; and a compact mineral soil absorbs water more slowly than a loose soil or one with a high organic content or heavy litter. The particles of a bare mineral soil tend to pack at the surface when rained upon for only a few minutes and thus reduce the rate of infiltration (see Fig. 57). Variation of local conditions may, therefore, modify the effectiveness of a given amount of precipitation.

Movement of Soil Water. Water moves downward in quantity during and immediately after rain or irrigation. Later it may move upward or laterally to some extent when evaporation and use by plants reduces the amount near the surface. Its principal movement occurs as a liquid in capillary films or through noncapillary pores, but some movement also occurs in the form of vapor. Gravity, hydrostatic pressure, and capillary action are the forces involved; and movement may be the result of interaction of all three.

The rate at which infiltration takes place is at first determined by surface conditions. When they are favorable, practically all of a light rain is absorbed. Within a half hour or less, however, absorption declines and is then controlled by conditions in the lower horizons, where percolation may be very slow. Movement of gravitational water through the soil is controlled by the number, size, and continuity of the noncapillary pores through which it percolates. Drainage is rapid in coarse-textured soils, but in clays movement is slow since the pores are small and may be blocked by the swelling of colloidal gels or by trapped air. Channels left by earthworms or other animals and those left by dead roots greatly facilitate downward movement. If there is no impermeable hardpan layer and if the water table has not been raised too near the surface, all gravitational water drains from surface strata within two or three days after a rain leaving the soil water content at *field capacity.*

A simple explanation of the movement of capillary water may be entirely adequate for most ecological purposes. Since capillary water forms a continuous, thin film around soil particles and in the small spaces and angles between them, it is obvious that surface tension of the water creates inward pressure in the film and that water, therefore, tends to move from regions with thicker films to regions with thinner films. An explanation with broader applications considers the difference in attraction for water between two portions of soil having different moisture contents, and expresses this attraction or force as *capillary potential*—that is, the force required to move a unit mass of water from a unit mass of soil. Various methods of measuring this force indicate that the potential is directly related to the water content and that there is no change in the state of water, as moisture content is reduced from field capacity to an oven-dry condition, but merely an increase in energy required to move it. On this basis, the boundaries between gravitational, capillary, and hygroscopic water are too indistinct to be recognized. That these boundaries are indistinct is, in fact, true regardless of the point of view. Such relatively simple considerations seem entirely satisfactory for an adequate understanding of plant-water relationships, although recent studies of soil moisture by soil physicists have become increasingly technical.

Movement of capillary water is closely related to soil texture. In wet soils, it is rapid in sand and slow in clay, but the rate is reversed as soils dry out. Capillary rise, or the distance that capillary force will move water, is much greater in clay than in sand, although the rate of movement is less in clay. The rate is surprisingly slow at all times and probably is quite insufficient to maintain an adequate film on the soil particles from which a root is removing water. The water coming to a root by capillary action does not at all equal the amount made available in new films that the root contacts because of its elongation and production of new root hairs. When soil water is below field capacity, capillary movement is probably insufficient to replace the film on particles from which roots of an actively transpiring plant are removing water. The continuous elongation of these roots with the production of new root hairs brings them in contact with new films and helps to keep up the supply of necessary available water.

Movement of water vapor is along vapor pressure gradients, which are affected by temperatures and vapor pressures of the air and the different soil horizons. There must, therefore, be some movement in all soils, but its effects are most noticeable in semi-arid regions where there is no connection between the water table and capillary water near the surface. In winter or in any cool period, water vapor moves upward from the warmer subsoil and cools and condenses in the surfaces layers. When temperatures rise at the surface, evaporation takes place into the air, and the total ground water is reduced. Usually the surface soil is warmest in summer and results in downward movement of vapor with condensation at lower levels. If the surface soil is cooler than the air above it, water vapor may move into the soil and condense there in quantities sufficient to be of significance under semiarid conditions.

Water Lost to the Atmosphere. The loss of water from soil to the air by evaporation varies with the factors affecting the steepness of the vapor pressure gradient. Temperature, humidity, and movement of the air, as well as temperature and moisture content of the soil, are factors which in turn are modified by exposure, cover, and color of the soil. Probably the loss of water by evaporation is much less than is commonly supposed, for numerous studies indicate that there is little capillary rise to replace water lost by evaporation unless the water table is within a few feet of the surface. In those areas where water lost by evaporation might be critical, the water table lies so deep that precipitation rarely wets the soil down to it and, consequently, the upward rise is of no consequence. In general, the loss of water by evaporation seems mostly to be from the top foot of soil. Under natural conditions, this probably affects few species and is rarely of significance.

In agriculture, water lost by evaporation has been the subject of much argument, particularly with regard to the effects of cultivation. Evaporation from a dry soil surface is much less than from a moist one, because diffusion through soil is very slow. Since, if no rain falls, a dry soil surface can be moistened only by an upward capillary movement of water, it has been maintained that cultivation of the surface must reduce loss by evaporation, because it prevents capillary movement. It is now known that, unless the water table is very near the surface, capillary rise is negligible under any circumstances. This being true, the dust mulch, or cultivated surface, has little to support it. In fact, if the surface capillary water is not connected with the water table, as is frequently true under irrigation, cultivation for a mulch probably increases the loss of water. Organic mulches seem to be more effective in reducing water loss, probably because they shade the soil and reduce its temperatures, increase the distance of diffusion from soil to air, and protect the soil from the drying effects of wind.

Water lost to the atmosphere through transpiration far exceeds that lost by evaporation. Whereas evaporation seems to be effective only in the surface soil, plants remove water from considerable depths. Studies of orchard soils in different parts of the country indicate that all readily available water may be removed to a depth of 3 to 6 feet in three to six weeks, depending upon atmospheric conditions and the kind of soil. Sandy soils, of course, are exhausted more quickly than clayey soils. The relative losses by evaporation and transpiration are illustrated by experiments,[412] in which water was lost from a bare soil surface in a tank at the rate of 4.7 pounds per square foot during one growing season, while a four-year-old prune tree removed water from a similar tank at the rate of 416 pounds per square foot of soil surface. An acre of deciduous fruit trees near Davis, California, used 8 acre-inches of water in six weeks in midsummer. Corn grown in Kansas requires some 54 gallons of water per plant to mature. If this were applied at one time, as by irrigation, it would cover a cornfield to a depth of about 12 to 15 inches. Plants growing naturally have similar requirements. The knowledge that transpiration is the chief means of reducing capillary water in the soil has led to a consideration[213] of what kinds of plants on watersheds will least reduce the supply of water by transpiration and still prevent erosion.

Soil Moisture Constants. To compare the moisture characteristics of soils or to discuss them with respect to plants, quantitative expressions of hydro-physical properties are a necessity. These properties, determined under fixed conditions, are called *constants.*

The *hygroscopic coefficient* is the moisture content, expressed as a percentage of the dry weight, of a soil in equilibrium with an atmosphere that is near saturation. Accurate determination is difficult. Although indicative of unavailable water, its usefulness to plant scientists is limited.

Maximum water holding capacity is the water held by a saturated soil. It may be determined by weighing a unit volume of soil before and after it has been immersed in water for 24 to 48 hours.

Field capacity is the amount of water a soil retains after all gravitational water is drained away. Soils in the field attain this condition within 1 to 5 days after a rain except when the water table is near the surface or saturation extends to a depth of many feet. After prolonged rain, soil may be assumed to be at field capacity if samples taken at 8- to 12-hour intervals have essentially the same moisture content.

It is now common practice to express most soil moisture values on a volume basis. In addition, it is desirable that most of these values should apply to the soil as it lies in the field. It is, therefore, advisable to obtain undisturbed samples of a certain volume and to make as many determinations as possible without modifying the structure of the samples. Such samples may be obtained with metal cylinders,[85] which, when forced into the soil, cut a sample of exact volume, which is then enclosed with airtight lids. Rocky soils may make it impossible to obtain undisturbed samples. It then becomes necessary to use special techniques, which, although they give much the same results, require more time and pains than are ordinarily necessary.[249] Some investigators obtain all their samples only when the soil is at field capacity. This system has several advantages, such as eliminating the problems relating to swelling on wetting, simplifying sampling, and giving a value for field capacity that is strictly determined by field conditions. When soils are dry, it is often possible to soak them, in place, and permit them to come to field capacity before sampling.

Capillary capacity (Water holding capacity) is the water retained against the pull of gravity. Although this appears to be essentially what is meant by field capacity, it is a value determined under laboratory conditions and may run slightly higher than field capacity. The saturated samples of undisturbed soil used for determining maximum water holding capacity are permited to drain over sand for a fixed time, usually two hours, and the weight of water retained, expressed as a percentage of the volume of the sample, is termed the capillary capacity. Physical measurements of greater precision have largely replaced this determination in recent years.

When the maximum water holding capacity, the field capacity, or capillary capacity, and the dry weight of an undisturbed sample are known, it is relatively simple to calculate pore volume, air capacity, volume weight, and specific gravity of soil material.[248]

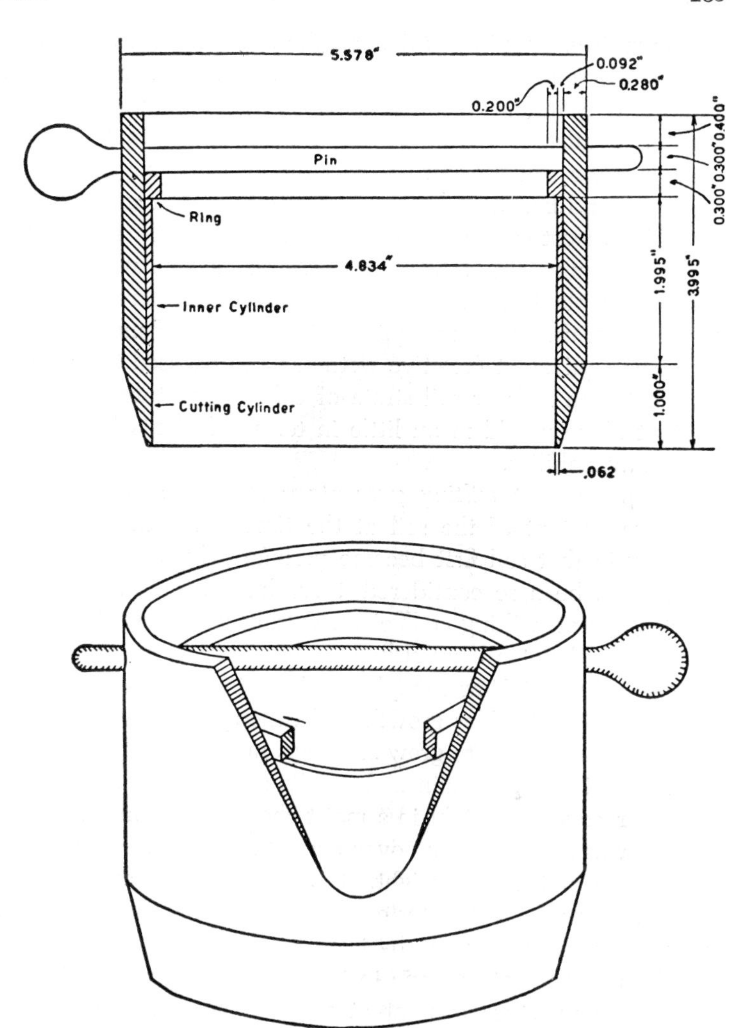

Fig. 87. *Sampler for obtaining undisturbed soil for determining volume-weight, air space, and water holding capacity. A counter-sunk steel plate or a block of wood placed on the cylinder prevents it from being battered when driven into the soil with a sledge hammer. The inner cylinder (see Fig. 86) is removed with the sample (600 cc.) and is covered with tightly fitted lids for transportation. The design is adaptable to circumstance, e.g.—a standard, sealable tin can has been used for the inner cylinder.*—After Coile.[85]

The *moisture equivalent* denotes the water content of soil after it has been subjected, usually for 30 minutes, to a centrifugal force of 1000 times gravity in a soil centrifuge. Its determination is simple if equipment is available. Within limits, it bears a constant relationship to certain other soil moisture values or, at least, suggests what these values should be. Its ratio to field capacity is near unity, but the relationship is least constant with coarse-textured soils. In many soils the moisture equivalent is 1.84 times as great as the water left in those soils when plants wilt. Unavailable water can, therefore, be approximated from the moisture equivalent. The ratio of moisture content to moisture equivalent (relative wetness) can be used to make comparisons between soils or soil strata of different textures where moisture content alone would mean little in terms of plants because of variation in availability.

The *permanent wilting percentage* should be considered as the moisture content of the soil at the time when the leaves of plants growing in that soil first become permanently wilted. Because it has not always been so considered, there have been various other terms (wilting point, wilting coefficient, wilting percentage) applied to the concept, and not all investigations have produced the same results. Briggs and Shantz[50] first emphasized the importance of this soil moisture condition to plant growth and called it the "wilting coefficient." Their procedure was to grow seedlings in glass tumblers of soil sealed with a mixture of paraffin and vaseline. When the leaves wilted and did not recover overnight in a moist chamber, the moisture content of the soil was determined by oven drying at 105° C. and calculated as a percentage of the dry weight. It is generally agreed that permanent wilting marks the soil water content at which absorption becomes too slow to replace water lost by transpiration.

Briggs and Shantz[50] came to the conclusion that soil texture alone determines moisture content at which plants wilt permanently, regardless of the species, their condition, or the environmental conditions. This conclusion was not immediately acceptable to everyone, and numerous studies were made to check its validity with different kinds of plants of different ages under a variety of conditions. It is now generally agreed that permanent wilting of any species occurs at the same water content of a soil of a certain texture regardless of the age of the plant or environmental conditions under which it grew. Uniformity of results is assured if noncutinized herbaceous plants are used and if permanent wilting of the lowest pair of leaves is used as an end point. This eliminates the problem of recognizing the onset of permanent wilting and variations related to the ability of some plants to live much longer than others after the onset of wilting.

Briggs and Shantz also concluded that it was possible to calculate the wilting point from the moisture equivalent because the following relationship held in their soils:

$$\text{wilting coefficient} = \frac{\text{moisture equivalent}}{1.84 \pm 0.013}$$

Although this often holds true, it does not apply to all soils. Studies in different parts of the country indicate that the ratio ranges at least from 1.4 to 5.65. Attempts to relate the moisture content at the time of wilting to other variables have been equally unsatisfactory, and it appears, therefore, that its determination is most reliable when observed

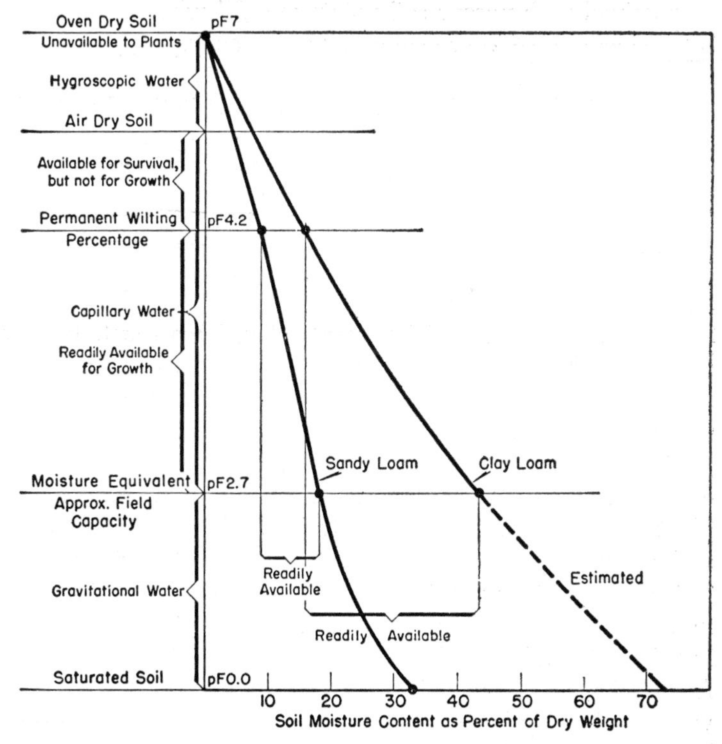

Fig. 88. *Relative availability of soil water in sand and clay loam with respect to soil moisture constants and* pF. From Kramer.[224]

directly. Because the expression, "wilting coefficient," has been so often associated with calculated values, it is logical to restrict it to that usage and to apply the term, "permanent wilting percentage," to determinations made by direct observation.

Readily available water is that which can be used by plants for growth and is, therefore, the moisture present in excess of the permanent wilting percentage. This may include gravitational water, but its rapid drainage makes it of little consequence. The remaining usable water is in the range from field capacity or moisture equivalent down to the permanent wilting percentage. This range is narrow in sand and wide in clay. Obviously, the wider the range, the longer plants can resist drought; and, in cultivation, the less frequently irrigation is necessary. The rate at which water moves from soil to an absorbing surface is strongly indicative of plant-soil moisture relationships at the time. An indication of the availability of soil water to the plant may be obtained with porous soil point cones,[241] whose rate of absorption is taken as the basis for evaluating *water supplying power* of the soil.

Availability of Soil Moisture to Plants. Gravitational water is readily available to plants only when present in a saturated soil, a condition that rarely continues long enough to be of importance. Normally, then, readily available water is that capillary water in the range between field capacity and the permanent wilting percentage. It is usually lowest in sand, and highest in clay. The following values for readily available water occur in some North Carolina soils:[222] sand, 2 percent, sandy loam, 14 percent, clay, 19 percent. However, this generalization does not always hold, for some clays may have high field capacities but also have high wilting percentages. A California clay with a moisture equivalent of 31 percent was found to have a wilting percentage of 25 percent, and, therefore, it could contain only 6 percent of available water. Such a soil would store less water for plant use than many sandy soils, and plants growing in it would suffer from drought much sooner than its soil texture would indicate. This also explains why, in contrast to the usual situation, sand dunes in deserts have more favorable moisture conditions than the surrounding clay soils. When both are at or near the wilting percentage, as they frequently are, a typically light rain provides little or no available water in the clay but does provide some in sand, in addition to penetrating more deeply, because of the lower wilting percentage of sand.

Whether or not all available water is equally available to plants is not entirely agreed upon. The evidence from a variety of sources seems to favor a decreasing availability as the supply is reduced toward the permanent wilting percentage and particularly in the lower half of the range of available water. Another factor affecting the availability of soil water is the concentration of the soil solution, which, if high, may have a toxic effect on plants and also modify their osmotic activity. Soil temperature, too, may be effective. Water supplying power may be reduced by half when soil temperature is lowered from 77° F. to 32° F. Probably the increase in viscosity of water at low temperatures reduces the rate of movement from soil to absorbing surface.

***p*F and Soil Moisture.** It should be apparent that soil water is held with varying degrees of tension depending upon the gravitational, hydrostatic, and surface forces involved at any time. With high moisture content, the tension is low; and as water is removed, the tension of remaining water increases. The force required to remove water held at a given tension can be measured by applying pressure to a thin layer of soil supported on a porous membrane. On an energy basis, the force may be expressed in terms of the height in centimeters of a column of water whose weight (or pressure) is equal to the particular force. Because the moisture content of a soil can be determined directly with such a pressure-membrane apparatus for any pressure up to 25 atmospheres,[333] moisture content can be plotted against moisture tension to form curves that accurately show the entire range of availability of water to plants in a given soil. At the moisture equivalent, tension equals about one third atmosphere, and at permanent wilting it is near 15 atmospheres. These constants being of greatest importance in interpreting plant responses, determinations are most frequently made at these two pressures.

Soil-moisture tensions, when expressed as a water column, have so great a range that they cannot be used on a compact scale. This is overcome if logarithms of the values are used. These are termed *p*F values (*p* indicates a logarithmic expression of the energy or force, F, involved). Ovendry soil then has a *p*F of 7.0; wilting percentage, 4.2; and moisture equivalent, 2.7. Note that the greater the *p*F, the less water there is and the greater the tension, and that the values of *p*F units, being logarithmic, represent pressures that increase by multiples of ten.

The form of *p*F curves is similar for different soils but not identical.

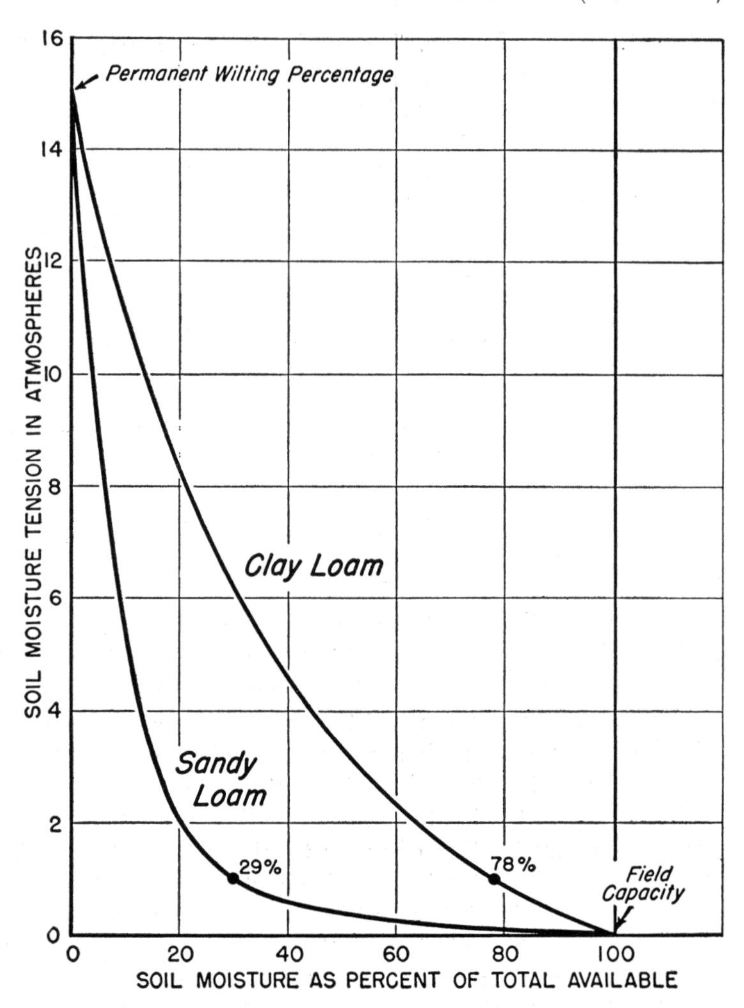

Fig. 89. *Soil moisture-tension in relation to available moisture.* From Kramer.[224]

This explains why soil moisture constants do not have exactly the same relationship to each other in all soils. If the pF curves all paralleled each other, determination of one constant would permit the calculation of others by a conversion factor.[123]

Measurement of Soil Moisture. For ecological purposes, it is of prime importance to know how much soil water is available for plant use and often to be able to follow its variations from day to day throughout a growing season. Because of soil variation, it is usually desirable to have determinations from numerous places in a stand and usually from more than one stratum in the soil. It is undesirable to use sampling methods that disturb any considerable amount of the soil or injure roots in the experimental area; and, again, any expression of soil moisture should preferably refer to a unit volume of sample obtained in an undisturbed condition. The last qualification is advisable because interest is in the volume of water available to roots occupying a given volume of soil, rather than weight of water in a given weight of soil.

It should be clear from our previous discussion that, to interpret soil moisture conditions, several soil moisture constants are necessary and that some physical analyses of the soil may be desirable. A single collection of samples from each local area of study may suffice for these purposes. Thereafter, some method must be fixed upon, which, within the time available to the worker, will give as adequate a notion as possible of the variations in soil moisture content of the experimental areas. Finally, it must be possible to express the soil moisture data in terms of what is available to plants.

Methods currently in general use are of two types: (1) determining the actual content of water, (2) measuring the forces with which water is held or the rate at which it is supplied to an absorbing surface.

The actual content of water is determined by weighing samples before and after drying to constant weight in an oven at 105° C. The loss in weight, representing the water content, is expressed as a percentage of the dry weight or, if the samples are undisturbed, on a volume basis. The disadvantages of the method are numerous. Sampling takes time and disturbs the soil, the samples must be transported, weighing and drying are time-consuming, and a continuous record is impossible. However, the method has its uses, and, where only a few determinations are wanted, it is undoubtedly the procedure to use. Note, too, that it requires no equipment that is not ordinarily available.

Several electrometric methods have been adapted to the measurement of soil moisture.[293] All require calibration[187] in terms of the wilting percentage of the soil involved but thereafter permit rapid determinations at short or long intervals and direct translation of measurements into available water. The methods most in favor at present measure the resistance between two electrodes, either imbedded in gypsum blocks,[40] or separated by fiberglas in a monel metal case,[89] and buried in the soil. The resistance varies inversely with amount of soil water and also with soil temperature. The fiberglas units contain a thermistor for temperature readings at the time of moisture determinations. Other methods measure dielectric constant, or electrical or thermal capacity of the soil, values that vary with changes in soil moisture.

Two physical measurements, making use of (1) tensiometers[334] and (2) soil point cones,[241] have been used successfully.

A tensiometer measures the tension existing between the soil and the soil water. It consists of a porous cup set in the soil and connected to a manometer by a tube of small diameter. Water in the instrument makes connection through the porous cup with the soil water, from which equilibrium tension is transmitted to the mercury of the manometer. Since the height of the mercury column indicates the tension in the soil, the manometer can be calibrated for a wide range of soil moisture values, and readings can be taken at any time and translated directly into values for available water. The instrument is accurate for values ranging from zero tension to approximately 0.85 atmosphere of tension, or from saturated soil to a reduction of 80 or 90 percent of available water.[349] Unfortunately, approaching the wilting percentage, in the moisture range most critical to plants, its values cannot be wholly trusted.

Soil points are small, hollow cones of porous porcelain, which can be inserted into the soil with a minimum of disturbance so that each has an equal area of surface in contact with the soil. The amount of water absorbed by the cone is determined by weighing and is taken as a measure of the water supplying power of the soil. In some types of studies, this value alone is sufficient to make comparisons between soils without any further analyses being necessary. It is also indicative of moisture conditions, for, at the wilting percentage, it approximates 0.085 g. in two hours. A variation of the cone is the soil pencil, graduated for reading height of rise of wetted area, and permitting comparative estimates of soil water.[132]

Soil Atmosphere

Organisms and Soil Atmosphere. It was pointed out earlier that air is a component of soil (p. 175). Both the amount and composition of this air are of importance to plants. Most plants require a well-aerated soil for growth and even for survival. Many seeds will not germinate unless well aerated even though temperature and moisture are favorable. Healthy roots must carry on respiration continuously, which means that oxygen must be present in the soil. At the same time, their activity produces carbon dioxide and carbonic acid, which tend to accumulate. To some plants, the increase in the proportion of carbon dioxide is more injurious than the decrease of oxygen. Since all aerobic microorganisms present are likewise using oxygen and releasing carbon dioxide, the balance of the two cannot be maintained unless there is a free exchange of gases with the air above the soil. If aeration is good, this may be accomplished by diffusion from the air. However, in any soil the proportion of oxygen decreases and that of carbon dioxide increases with depth, and the proportion of oxygen is not as great in soil as in air even when conditions are most favorable. (Table 10.)

Table 10. *Percentage composition of oxygen and carbon dioxide in soil air extracted at different depths in a silty loam soil.*[43] *Note that the percentage of O_2 decreases and of CO_2 increases with depth both winter and summer but that subsoil aeration is far better when the soil is dry in summer than when it is wet in winter.*

Depth (feet)	Oxygen		Carbon dioxide	
	Winter	Summer	Winter	Summer
1	19.4	19.0	1.2	2.4
2	11.6	17.4	2.4	3.7
3	3.5	16.7	6.6	5.0
4	0.7	15.25	9.6	8.55
5	2.4	12.95	10.4	11.85
6	0.2	11.85	15.5	11.9

Relation to Growth and Distribution of Roots. Since aeration becomes poorer and oxygen decreases with depth of soil, these conditions may limit the depths to which roots can grow. The deepest root penetration is in well-aerated soils. Species growing in wet lowlands are invariably shallow-rooted, for here aeration is poorest because the soil is periodically or continuously saturated and the only available oxygen may then be in solution. These shallow-rooted species will usually also grow well in uplands, but, if the naturally deep-rooted species are moved to lowlands, they do not do well or may actually die. Thus aeration may determine the rate of growth, an element of importance in forest stands, and may be the factor controlling the type of vegetation.

Soil Aeration and Plant Adaptations. Well-aerated soils may have an air capacity of 60 to 70 percent by volume, a condition determined primarily by structure and scarcely affected by texture. The amount of

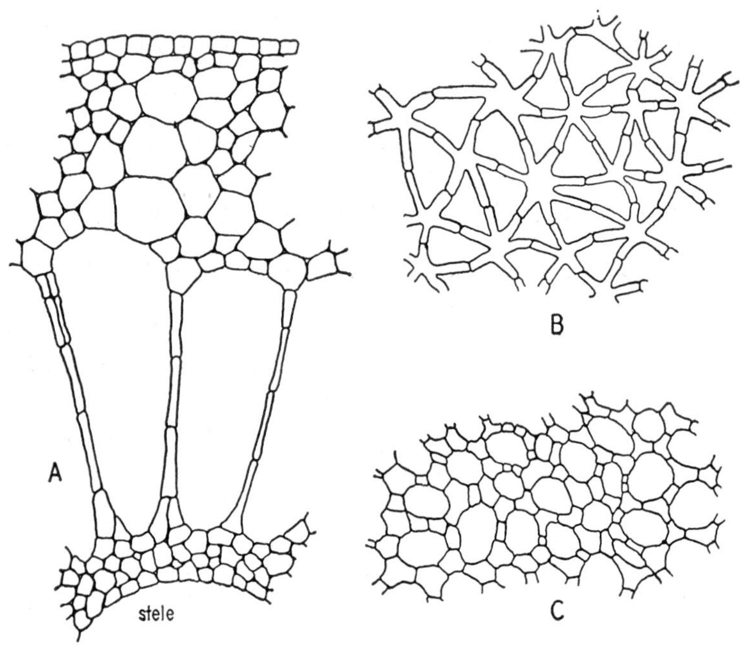

Fig. 90. *Some types of lacunar tissue found in stems of emergent and other aquatic vascular plants.* (*A*) *Cortex of water milfoil* (Myriophyllum). (*B*) *Ground parenchyma throughout stem of a rush* (Juncus). (*C*) *Same for a sedge* (Cyperus).

air varies, of course, with the water content of the soil, for air is forced from the spaces in the soil that become occupied by water.

Thus continuously saturated soil is poorly aerated, and the mud under a pond probably has the poorest aeration of any plant habitat. Most species growing in such habitats have adaptations that serve to counteract poor aeration. Many have large, continuous spaces—lacunar tissue—in their stems and roots permitting storage and free movement of gases within the plant. In emergent and floating-leaved species, these spaces are connected directly with the atmosphere through the stomata. Submerged leaves of aquatics are invariably finely dissected or extremely delicate, conditions that bring a majority of the cells in contact with the water, from which they must obtain oxygen in dissolved form. A few submerged species produce pneumatophores, or special branches that extend above water and give

Fig. 91. *Cypress swamp* (Taxodium distichum) *in the coastal plain of South Carolina. Note buttressed, somewhat planked bases of trees and an abundance of cypress knees, whose uniform height marks average high-water level.*—U. S. Forest Service.

direct connection with the air through lacunar tissues. In addition to shallow root systems, a number of swamp trees have other characteristics in common. Enlarged or buttressed bases and plank roots are frequent, especially in southern swamps, and the "knees" of cypress are in the same category. That these structures facilitate aeration has not been demonstrated,[225] but their formation seems to be in response to alternate inundation and exposure to air.[227]

Determination of Volume and Composition. *Total pore space* or *pore volume* (in c.c.) is equivalent to the weight of water (in g.) in the soil at saturation, for water then is assumed to occupy all the space in the soil. Actually not all air can be replaced by water, and the small amount of air remaining at saturation represents what is available to roots regardless of circumstances. *Air capacity* is the amount of air in soil that has been drained of all gravitational water. It is, therefore, equal to the difference between pore volume and the weight of water at field capacity. Since total water holding capacity and field capacity are constants, it follows that pore volume and air capacity are soil-air constants. The actual air content is not at all constant, for it varies inversely with the water content. Soils with a high air capacity are in general well aerated, but, after prolonged rain or flooding, they may for a time be poorly aerated because water fills so much of their internal space that actual air content is low.

Table 11. *Porosity, field capacity, and air capacity of some soils with different textures. After[18] from Kopecky.*

Character of soil	Particles smaller than 0.01 mm.	Total pore volume	Field capacity	Air capacity
Compact heavy clay........	86.7%	48.0%	47.6%	0.4%
Clay loam................	67.2	46.1	41.1	5.0
Compact loam.............	46.5	41.1	34.9	6.2
Very fine sand............	48.4	49.3	39.3	10.0
Friable loam..............	42.6	49.3	37.1	12.2
Friable fine sandy loam......	39.6	49.5	34.6	14.9

The composition of soil air may be determined with a portable gas-analysis apparatus.[173] The sample of air can be pumped from the soil through a sampling tube[342] or some similar device,[43] or it can be withdrawn from a unit volume of undisturbed soil. The total percentage of oxygen and carbon dioxide in the soil is usually very nearly that found in air, and, in general, an increase of one results in a proportional decrease of the other (see Table 10).

Some Chemical Factors

Soil Acidity.[380] Regardless of the nature of their parent material, soils tend to become acid in reaction if precipitation is sufficient to cause downward percolation of water during much of the year. This is largely the result of leaching of soluble basic salts. To illustrate, calcium carbonate is relatively insoluble in water but reacts with carbonic acid, ever-present in soil water, to form readily soluble calcium bicarbonate. This, of course, is leached from the surface soil by percolating water. Although the leaching bicarbonates may be re-precipitated at any time the soil dries out, they, nevertheless, tend always to move downward. Thus the surface horizons tend to be low in basic materials and may have a highly acid reaction because of the acids produced by chemical and biological activity in progress there. The surface strata have the largest accumulation of organic matter, which yields acid products upon decomposition, the greatest numbers of soil organisms whose activities may produce acids, and the most active chemical changes in the mineral components, also contributing to acidity. Consequently, acidity is normally greatest at the surface and decreases in the lower horizons of the soil.

A solution is acid when the concentration of H^+ ions exceeds that of OH^- ions, alkaline if the reverse is true, and neutral when they are equal. The concentration of H^+ ions expresses the degree of acidity or alkalinity because that at neutrality is known (0.0000001 or 10^{-7}) but because of the cumbersome numbers, the negative logarithm (*p*H) is used. Neutrality is *p*H 7.0 and a lower value represents acidity, a higher value, alkalinity. Each successively lower *p*H value represents an H^+ ion concentration, or acidity, ten times greater than the last. The *p*H of most soils will normally fall between 3.0 and 9.0, and, in humid regions, the range to be expected is considerably less, perhaps no greater than *p*H 4.0 to 7.5.

Under ordinary conditions, the hydrogen ions themselves probably have little direct effect upon plants, but degree of acidity of the soil may have a regulatory effect upon chemical processes that do influence growth. Increased acidity may reduce availability of nutrients, as when phosphorus combines with aluminum and iron to form insoluble phosphates. High acidity may, apparently, produce toxic effects, but these are not caused by H ions. It is more likely that they result from soluble compounds of aluminum and iron, which form in increasing amounts as the H ion concentration rises. Since calcium is a required nutrient, its characteristically low content in acid soils may be of more importance than the degree of acidity. Numerous soil organisms are sensitive to changes of acidity, and, if their activities are inhibited, decomposition of organic matter may be retarded, nutrients may not be released, and nitrification and nitrogen-fixation may be checked.

With such a variety of things that may be affected by soil acidity, it should be suspected that a simple relationship between *p*H and plant responses does not exist. Studies of soil *p*H and plant distribution bear this out, for, in general, if the environment is favorable and necessary nutrients are available, most species can tolerate a rather wide range of *p*H. At the same time, many of these species reach their best development or are most abundant within a restricted portion of that range of *p*H. It should be clear that, even under such conditions, *p*H alone cannot be the limiting factor.

Determinations of *p*H may be made colorimetrically by the use of indicator solutions or electrometrically with a potentiometer and a glass electrode.[320] A very useful approximation may be made with a universal indicator, which, when placed in the soil solution, takes on a color corresponding to a particular *p*H value. This is handy in the field since it requires no more than a small bottle of indicator and a pocket-size device such as a porcelain plate permitting comparison with color standards. More accurate colorimetric determinations require a series of indicators whose colors correspond to overlapping *p*H ranges. When electrometric equipment is available, it is preferable because of its accuracy.

Exchangeable Bases. Ecologists have given relatively little attention to the ways in which the mineral nutrients of the soil affect plant distribution and growth of wild species. An important part of the mineral nutrition of native and cultivated vegetation is derived from the *exchangeable bases* or cations adsorbed on the surfaces of the soil colloids. When these vary considerably in amount or kind, there may be marked differences in the type of vegetation or at least in rate of growth. For example, it has been shown that, in soils derived from hydrothermally altered rocks in the Great Basin, sagebrush and its associated species fail to grow because of the very low percentage of exchangeable bases as compared with the normal brown soils of the sagebrush zone.[28]

The colloidal portion of the soil is composed primarily of alumino-silicates. These colloidal particles are almost always negatively charged, and upon their surfaces are adsorbed great numbers of cations. These cations are principally H^+, Ca^{++}, Mg^{++}, K^+, and Na^+, named in the decreasing order of tenacity with which the cations are held. The hydrogen ion is held more tightly than calcium and replaces calcium more readily than calcium will replace hydrogen.

This same relationship holds between calcium and magnesium, and so on down the series. The displaced cation usually enters the soil solution. This phenomenon, in which one cation may replace another on the colloidal particle, is called *base exchange*.

Plants are almost entirely dependent on this proces of base exchange for their supply of calcium, magnesium, and potassium. Of the anions, only PO_4^{---} is held to any extent by colloidal adsorption, the other anions, such as No_3^-, being readily soluble in the soil solution and therefore, readily leached. One source of the H ions that can displace the bases and make them available is the carbonic acid formed when carbon dioxide from root respiration is released into the soil solution. This was shown experimentally for the calcium ion.[199] Another common source is the organic acids derived from humus.

Soils differ widely in their ability to supply cations because of the effects of climate, parent material, and vegetation. The maximum amount of exchangeable cations a soil can hold is called the *base exchange capacity* of the soil. Obviously, a soil high in colloids will have a high capacity as compared with one low in colloids, as, for example, a sand. Even the kind of clay may make a great difference in the base exchange capacity of a soil. For example, kaolinite has a very low capacity compared to clays of the montmorillonite group, which have relatively high capacities.

Since soils are constantly losing some of their adsorbed bases due to replacement by H ions, the soil is rarely, if ever, saturated with bases to its capacity. The degree of saturation at any given time is known as the percentage of base saturation of the soil. The base exchange capacity of a soil minus the percentage of base saturation is theoretically equivalent to the percentage hydrogen saturation of the soil, since hydrogen is the replacing ion.

Both climate and vegetation have great effects upon the amounts of exchangeable bases present in soils. On soils derived from the same parent material, sugar maple-beech-yellow birch forest maintains a soil at a higher percentage of base saturation than that under a red spruce forest.[72] This seems to be due largely to the ability of the hardwoods to absorb calcium from the subsoil and to add it to the surface soil by leaf fall.

Many investigators have shown the relation between precipitation, percentage base saturation, and *p*H. In brief, it may be stated that, in regions of high precipitation, the bases are readily replaced by hydrogen ions and then leached from the soil. The excess of hydrogen ions results in lowering the *p*H and creating an acid soil. Such conditions prevail in the cool, moist, coniferous forests of the north. Just the opposite conditions prevail in the soils of arid regions where low precipitation and scanty vegetation combine to allow the bases to remain on the colloids, thus maintaining a high percentage saturation and *p*H. These relationships are represented schematically thus:

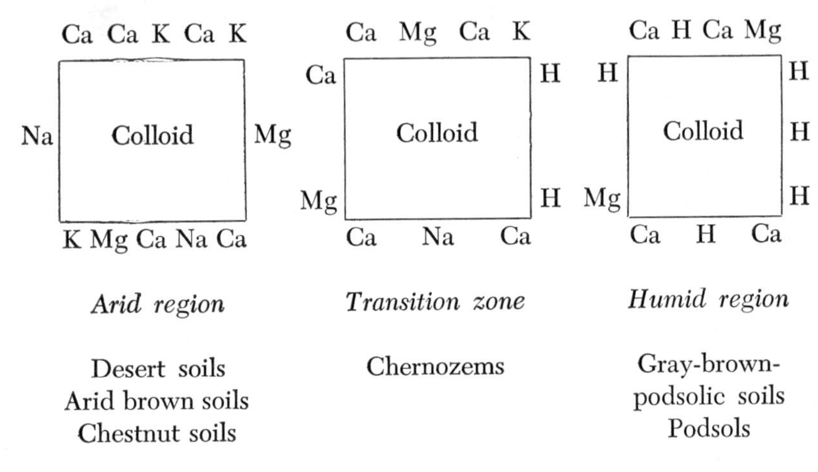

Arid region	*Transition zone*	*Humid region*
Desert soils Arid brown soils Chestnut soils	Chernozems	Gray-brown-podsolic soils Podsols

Inhibition of Growth by Plant Products. That certain plants produce soil conditions inhibiting the growth of other plants is probably true.[427] Over a hundred years ago it was argued that crop rotation was necessary for this reason and that fallowing of land favored the next crop because it permitted the leaching of harmful excretions or by-products of decomposition resulting from the previous crops. Today we cannot entirely ignore this line of reasoning, for explanations of the benefits of rotation and fallowing based upon nutrient deficiencies are not always adequate. Likewise, there is some evidence that toxic substances are released in the soil as excretions,[347] or when external root cells are sloughed and decompose,[338] or when the plants disintegrate after death.

A number of grasses inhibit growth of other plants. In lawns, certain strains of bluegrass almost completely check the growth of white clover.[1] Walnut inhibits the growth of a number of herbs. Fairy rings of both fungi and higher plants may be the result of toxic products produced by the plants, for other explanations do not always suffice. If water, supplied in excess to flats of experimental plants, is permitted to percolate through the soil and is then used as the water supply for other plants, the latter are frequently inhibited in growth even under the most favorable conditions.[22] Extracts from decomposing plant remains have produced similar results. Apparently toxic or growth-inhibiting substances are produced by a number of plants, which may affect germination of seeds and growth of seedlings, or even of mature plants of the same or other species. Some species are affected, others are not. Whether higher plants are affected directly is not always clear. Perhaps effects upon soil microorganisms and their activity in turn affect the higher plants.

The entire subject is controversial, and some evidence is conflicting.[272] The limited information that is available is often derived from observation of agricultural soils and cultivated plants. Cultivation, probably because of better aeration, reduces the effectiveness of inhibiting substances, and the problem is practically eliminated by crop rotation and the compensating effect of fertilizer. It is, therefore, not surprising that investigators have turned to other things. In natural soil, however, these artificial modifications are absent, and, consequently, in view of the possible implications in interpreting associations of species[151] or the causes of succession,[203] it is surprising that the subject has not been given more attention.

Alkalinity. Soils with an alkaline reaction have usually originated from limestone, dolomite, or marble in which calcium carbonate is the basic mineral. The $CaCO_3$ tends to neutralize acids that appear in the soil, and the degree of alkalinity is proportional to the solubility of the limestone. Dolomite contains more $MgCO_3$ than $CaCO_3$, and gypsum is largely $CaSO_4$, but the soils they form contain $CaCO_3$, and their floras are essentially similar to that of limestone. In our arid West, soils are often alkaline in reaction because of the sodium ions, which accumulate as sodium hydroxide (NaOH).

Neutral or alkaline soils favor the activities of most soil organisms and the availability of nutrients for higher plants. At the same time, the tendency of soil colloids to aggregate and produce crumb structure in the presence of lime results in soil structure with water, air, and temperature conditions favorable to plant growth. Thus most cultivated crops do best on soils with a pH ranging close to neutrality. Native plants, in general, respond similarly, but there are exceptions, which require, on the one hand, high concentrations of $CaCO_3$ or, on the other, extremely acid conditions regardless of other factors.

Not all species found growing in calcareous soils are *calciphiles.*

The distribution and occurrence of many show no correlation with alkalinity of the soil. A considerable number of these widely distributed species may, however, grow more luxuriantly when on calcareous soil. Some, although not restricted to the habitat, will be found there characteristically. These are true calciphiles. There are, in addition, obligate calciphiles, which grow only in calcareous habitats.

The exceptional vigor on calcareous soils of otherwise widespread species may result simply from the improved aeration, moisture, or nutrient conditions produced by lime. Calciphiles may grow on other than alkaline soils if competition from non-calciphiles is not too great. The less favorable are the general conditions for growth, the more the calciphiles are restricted to their alkaline habitat, and, as a result, at or near the limits of their ranges they often appear as obligate calciphiles.

Salinity. Under conditions of poor drainage and high temperature, much of the water deposited in low places evaporates and leaves behind the salts it has carried from the soil of surrounding slopes. If precipitation is seasonal and alternates with extreme drought, there is insufficient leaching to prevent accumulation of these soluble salts, which then form *alkali soils,* so called regardless of the salt involved. Alkali soils of various kinds occur in all parts of the world and are common in the arid portions of western North America. Lowlands bordering the oceans are subject to periodic inundation with sea water and, consequently, contain relatively high concentrations of salts.

Plants that can tolerate the concentrations of salts found in saline soils are termed *halophytes.* How they survive where ordinary plants have little chance has been the subject of much debate. If not actually dry, these saline habitats may be termed "physiologically dry" because of the high concentrations of salts, which would limit osmotic activity and, consequently, absorption of water by the ordinary plant. The morphological and anatomical characteristics usually appearing in plants of arid regions are common in plants of saline habitats. Succulence is particularly general. Yet these xeromorphic characters have been shown to be relatively ineffectual in maintaining low transpiration rates in halophytes. They must then be able to absorb water in spite of the high salt concentrations, and this is possible because of their own high salt contents.

Not all species are equally tolerant, and, therefore, they will often be found in zones adjusted to the concentrations of salts in the soil and the plant. Flat areas with uniform salt concentration may support a constant group of species over their entire extent. The number of species tolerant to salinity is not great and many of the same genera are found in all parts of the world where similar conditions occur (e.g., several Chenopodiaceae). Because certain species in alkali areas are tolerant to definite ranges of salt concentration and, in addition,

Fig. 92. *Margin of a saline flat in the Smoke Creek Desert, Nev. The shrub at the margin is the relatively salt-tolerant greasewood* (Sarcobatus vermiculatus). *Extending farther into the playa is salt grass* (Distichlis stricta), *which is more tolerant but soon also fades out until nothing grows over most of the area.*—Photo by W. D. Billings.

to particular salts, they may be rather positive indicators of soil conditions. There are other species that are not so limited. In some, the concentration of the cell sap adjusts itself to changes in the soil and permits growth under a variety of conditions. Some can tolerate only small amounts of salt and do better in its absence, while a few others absolutely require salt to survive, some even requiring a fairly high concentration. The extreme in salinity is illustrated by portions of the Great Salt Lake area in Utah where salt concentrations are so great that no vascular plants can grow.

Topography

Although topography affects vegetation indirectly by modifying other factors of the environment, it has nevertheless a significant influence upon all plant communities. If an area is so level that topographic variations are negligible, then, other factors being equal, uniform vegetation may be anticipated throughout. Normally, however, such areas of any extent are rare, and slopes, bluffs, and ridges with different exposures, lowlands, drainage lines, and depressions are present.

Such irregularities in topography produce light, temperature, and moisture conditions that differ greatly between north and south slopes or ridges and depressions. The effect of exposure on these individual factors having been previously discussed (pp. 90, 109), it is necessary here only to emphasize that vegetation on slopes is the resultant of interaction of light, temperature, and moisture differences. In the northern hemisphere, south-facing slopes receive more light, have higher temperatures, and are drier than the average site in the area, while north-facing slopes receive less light, are cooler and moister than the average. Of course, these differences vary with degree and extent of slope, but, in general, the environment of north and south slopes differs sufficiently to maintain distinctive vegetative types.

Apart from the interaction of the factors mentioned above, slopes affect runoff and the amount of soil water and, likewise, the degree of erosion. *Normal* or *geologic erosion* is universal. If there is any slope at all, and regardless of vegetational cover, surface material moves down a slope. However, this movement is so slow that soil processes reduce to a minimum the nutrient value of the moving particles at the same time that depth of profile is maintained by extension into the subsoil at a rate paralleling the loss of surface materials. Incorporation of new unweathered material in the profile from below contributes to retention of natural fertility, and makes normal erosion highly desirable.

Accelerated erosion exceeds the rate of profile maintenance and is, therefore, destructive. Evidence occurs on many hilltops but is most apparent, and often extreme, where man has disturbed natural processes. An opposite extreme is the local area so level that natural erosion is insufficient to carry off depleted surface minerals. Their accumulation, together with poor drainage, produces a characteristic interzonal soil with a leached A horizon and a thick clay hardpan for a B horizon.

Since water always moves toward depressions, these are invariably moister than uplands and usually support distinctive vegetation. If topography is immature, as in glaciated northeastern North America, drainage is relatively poor and depressions contain ponds or lakes supporting aquatic vegetation. Some lakes fill with sediment, marl, and organic materials to form bogs, which likewise have their characteristic species. With more mature topography, depressions are connected by streams, which make drainage far more effective. Even so, the streams are usually bordered by flood plains supporting vegetation requiring more favorable moisture conditions than obtain upon the uplands.

The greatest differences in vegetation associated with local variations in topography can usually be correlated with moisture, either in respect to an excess or to a deficiency. If the latter, adaptations that facilitate absorption or restrict transpiration are likely to characterize the plants. In a region where moisture is rarely a critical factor, slope and exposure produce scarcely noticeable differences in vegetation. This occurs only under conditions where a combination of fog, clouds, or rain maintains a humid atmosphere, low transpiration rates, and a plentiful supply of water.

In addition to local topographic effects are those of a regional nature associated with mountains. The decrease in temperature and the usual increase in precipitation with increasing altitude result in vegetational zonation. Within these zones, the local effects of topography again become apparent so that zones lie at higher altitudes on a south than on a north slope and the species of a particular zone will be found extending downward in ravines and upward on ridges.

A mountain may affect conditions for growth at some distance from itself. Some mountains are centers over which rain clouds form and from which they often move to provide moisture for surrounding lowlands. At the same time, streams starting in mountains and fed by precipitation there, flow down into valleys below and the plains beyond. Other mountains act as barriers when they lie at right angles to the prevailing winds, for so much moisture may fall upon the mountain that little is left for the area beyond. This explains the lack of moisture in the Great Basin. The prevailing winds coming from the

Pacific lose their moisture over the Coast Ranges and the Sierra Nevada.

Finally, it is probable that mountains act as barriers to the natural migration of some species that are unable to compete with the flora upon the mountain or to withstand the successive changes of environment associated with increasing altitude.

General References

L. D. BAVER. *Soil Physics.*

P. J. KRAMER. *Plant and Soil Water Relationships.*

H. J. LUTZ and R. F. CHANDLER. *Forest Soils.*

F. J. RUSSELL. *Soil Conditions and Plant Growth.*

U. S. DEPT. AGR. *Soils and Men.*

Chapter 9 Biological Factors

Associated organisms having mutual relationships to each other and to their environment are recognized as a community. Many, if not all, of the organisms in a community are thus not only a part of the community but also a part of the environment of every other organism there. The dominants obviously compete with each other and with subordinate individuals. At the same time, they provide conditions that permit the survival of lesser organisms, which, though quite inconspicuous, may yet markedly affect the permanence of the community as a whole. Both plants and animals are factors of the environment of any community, and man is not the least of these factors.

Plants as Factors

Competition. It has been shown that, within a community, competition occurs between individuals of the same species, or between different species, whenever some requirement of the organisms is available in amounts insufficient to supply all demands adequately. Each organism involved in competition is a factor in the environment of all other organisms so involved. The effects of competing organisms upon each other are more apt to result from their influence upon physical or physiological conditions of the environment (such as available water or nutrients, light, temperature, humidity, and air movement) than they are from direct action. An extreme example of direct competition as a factor is that of the strangling fig, an epiphyte of tropical forests, whose roots grow down, envelop the tree trunks and eventually kill the trees as pressure is increased. When the tree falls, it may pull down numerous others over which the fig has sprawled. The community, however, is only locally disturbed and soon readjusts itself, for the forest is climax and these giant epiphytes are a part of it.

The introduction of new species into a community, by man or other agents, usually results in failure because the plant cannot meet the competition of the normal species, which are adapted to each other and their environment. However, an occasional species reverses the rule, establishes itself as a part of the community, and often produces community changes. Japanese honeysuckle (*Lonicera japonica*) was introduced in the southeastern states many years ago and has spread widely. In lowland woods particularly, it sprawls over all the low

Fig. 93. *Japanese honeysuckle in bottomland hardwood forest. When the vine is as dense as this, few tree seedlings come up through it. If they do, they are soon pulled over and the honeysuckle forms mounds upon them, as at the left.*—Photo by L. E. Anderson.

vegetation and climbs well up into the trees. Under favorable conditions, it almost excludes low herbs and shrubs. When a tree seedling grows through it, the vine climbs upon it and bends down the slender stem, which, under the mass of honeysuckle, soon dies. Such lowland stands frequently have practically no tree reproduction beneath them. It is a matter of ecological interest as to how the natural development of these stands will progress. An economic aspect must be considered by the forester who is interested in regeneration of trees or planting these areas after cutting; for, unless the land is cultivated, the honeysuckle cannot be eliminated without considerable trouble. Kudzu (*Pueraria lobata*), also introduced in the south for erosion control and forage, has aggressive characteristics similar to Japanese honeysuckle and is rapidly losing favor.

An even more extreme change in community structure resulting from introduction is illustrated by species of *Eucalyptus*, in California. Unless controlled, they reproduce and may spread to form forest communities where grasses once were dominant.

Parasites. A parasite, dependent upon its host for its existence, thereby becomes a factor in the environment of a community. When conditions are favorable for the host, a certain amount of parasitism

Fig. 94. *Dead chestnut, killed by blight, in a forest stand of which they once were important members. Cherokee National Forest, Tenn.*—U. S. Forest Service.

can be tolerated with little apparent effect. Parasitic fungi and bacteria are almost constantly present but cause no serious disturbance of a community unless conditions become unusually favorable for their increase. Then they may cause death of enough hosts to produce a change in dominance or to destroy the community. Such occurrences are usually local and may be followed by gradual recovery of the original community. However, when a parasite is introduced from afar, it may be so effective in its new environment that disaster results.[253] Chestnut blight[9] has practically eliminated chestnut in the eastern United States, and oak is now dominant where oak-chestnut occurred before. Dutch elm disease[84] is gradually spreading from New England, where it first appeared, although its spread has been somewhat retarded by the drastic procedures used to check it.

Fig. 95. *A stand of scrubby oak infested with mistletoe* (Phoradendron flavescens).—U. S. Forest Service.

Parasitic seed plants are not usually of much ecological significance, but they are always of interest because of their peculiarities and relatively local distribution. A considerable range of degree of parasitism is possible.[106] The common dodder (*Cuscuta*) is representative of those parasites (*holoparasites*) completely dependent upon their hosts; but the mistletoes and others are termed partial parasites, because they are green and can manufacture food. Some species are attached to their hosts at a single point of contact, often by roots. A number of Scrophulariaceae are of this type. Others twine or sprawl over the host plant and are connected to it at intervals by absorbing

Fig. 96. *A striking witches'-broom on a young red pine in Michigan.*—U. S. Forest Service.

structures called haustoria, whose conducting systems may be in intimate contact with xylem and phloem of the host. Still others may be contained within the host and show only their reproductive structures externally. Effects upon the host are obviously physiological, and reduction of growth and vitality are usually apparent. Abnormal growth is also common in the presence of a parasite. It is often manifested as bushy masses, called "witches'-brooms," or is occasionally found in twisted, flattened, or distorted branches. Parasitic seed plants have little effect upon community structure in comparison with the drastic changes that may result from infestation with pathogenic fungi or bacteria.

Epiphytes. These include a wide variety of plants, all of which depend upon larger plants for physical support only. Algae, fungi, mosses, liverworts and lichens may be found growing on bark or, in some instances, even on leaves. Often their occurrence seems correlated only with the general humidity of the atmosphere in particular habitats, but they are frequently associated with certain communities and not with others, and, within a community, they may be distributed systematically.[111] Some may grow only on the bark of certain trees and, even more specifically, only in patterns related to drainage of water down that bark.[303] Others may be found only at the base, middle, or top of a tree trunk, and this may be correlated with moisture content of the bark.[30] The occurrence of the moss *Tortula pagorum*[8] illustrates how specific a habitat may be required by some epiphytes. This moss has rarely been found except in close proximity to man's habitations and then almost exclusively on the trunks of elm trees. The epiphytic lichens associated with evergreen forests of boreal and alpine regions are distinctive and characteristic.

In and near the tropics, higher and less variable humidity permits a greater variety of epiphytes to survive, and vascular species increase. In temperate regions, drought-resistant species, such as polypody ferns, are found occasionally, but farther south, first on swamp trees only and then almost anywhere, epiphytic vascular plants become the rule. Orchids, bromeliads, and ferns are especially abundant. Structures that catch or conserve water are characteristic of many of these species. Stratification at different levels in the forest, as controlled by light, air movement, and water supply, is common, and succession of epiphytic communities may be observed as organic "soil" is accumulated.[277] Occasionally their weight may increase sufficiently to break down the branches supporting them. Massive growths such as are produced by the well-known Spanish "moss" (*Tillandsia*) of the southeastern United States must reduce the normal tree foliage and its functioning (see Fig. 8). In general, however, the epiphytes and their "hosts" seem surprisingly well adapted to their relationship.

Symbiosis. The most generally accepted concept of symbiosis includes only the relationship of intimately associated, dissimilar organisms that live together to their mutual advantage. By appending descriptive adjectives, the concept has been expanded by some to include almost any relationship between organisms whether actually in contact or merely in competition with each other (e.g., cattle grazing in a meadow would illustrate antagonistic nutritive disjunctive symbiosis[256]). But the conservative interpretation recognizes only a few plant symbionts as significant in community life. The intimate association of unicellular green or blue-green algae with a fungus mycelium, termed a lichen, is an example of plant symbionts that is familiar to all who have any botanical interest. Lichens, however, can hardly be considered of general importance in community relationships. Although they often play a part in the establishment of communities on bare rock,[284] they probably influence mature, stable communities very little. Fungi and bacteria living symbiotically on plant roots are less noticeable but of far more importance.

Mycorhizas. When a young root and the mycelium of a fungus grow together, the fungus may form a feltlike layer around the root and by digesting the middle lamellae, penetrate the spaces between cells (ectotrophic mycorhiza), or the fungus may occur within the cortical cells of the root (endotrophic mycorhiza). The whole body, root and fungus, a united morphologic organ, constitutes a mycorhiza. Such root-fungus relationships are far more common than was once supposed. Actually, there are no woody plants (except parasites) which are definitely known to be without mycorhizas in all environments.[179] They occur on most forest trees and shrubs, and many herbaceous plants may have them.[318] They form during periods favorable to root growth and are practically restricted to the young roots in the surface strata of the soil.

Whether mycorhizas represent a mutualistic relationship or merely parasitism on the part of the fungus has been strongly argued by numerous investigators. The conflicting evidence makes interesting, if somewhat confusing, reading. However, the evidence that mycorhiza must be present for the successful growth of many species, particularly forest species,[31] is sufficient to suggest that the mycorhizal condition is desirable under most situations even though the reasons, apparently involving mineral nutrition, are not too obvious.

Pot cultures of certain tree seedlings in poor soil have been unsatisfactory until inoculated with mycorhizal fungi. On a larger scale, unsuccessful forest nurseries on prairie soil or long deforested agri-

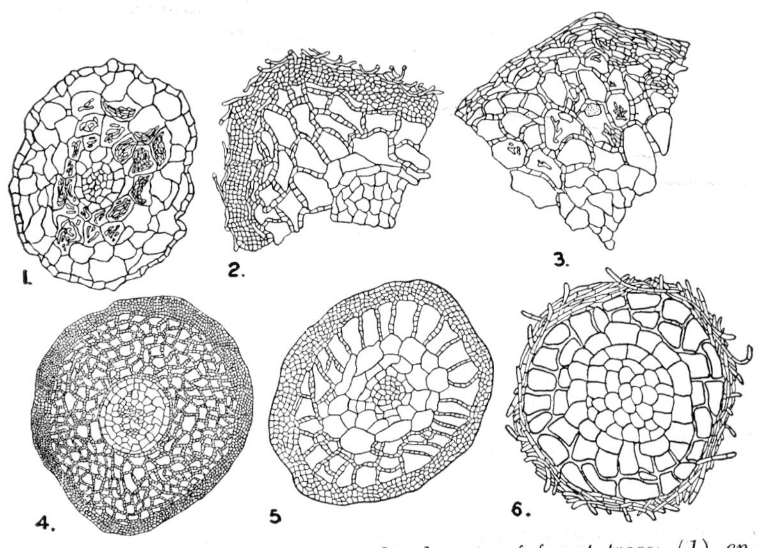

Fig. 97. *Transverse sections of mycorhizal roots of forest trees: (1) endotrophic, (3) ecto-endotrophic, others all ectotrophic. (1 and 4)* Pseudotsuga menziesii, (*2 and* 3) Pinus contorta, (5) Populus tremuloides, (6) Picea rubens.—After McDougall and Jacobs.[257]

cultural soil have been saved by bringing in small amounts of forest soil, which started the formation of mycorhiza. Tree seedlings transplanted without mycorhiza to treeless areas have been saved from gradual death by the application of small amounts of soil containing mycorhizal fungi.

Several members of the heath family (azalea, rhododendron, blueberry) are dependent upon the presence of mycorhizas that cannot tolerate alkaline conditions. Disappearance of the mycorhizas leads to death of the plants; and, consequently, the soil must be acid for successful propagation of these species.

Many orchid seeds germinate normally only in the presence of mycorhizal fungi and were difficult to propagate until it was found that proper nutrient media could compensate for the absence of the fungus. Such evidence indicates that, regardless of what the fungus may take from the root, the vascular plant is benefited by the presence of the mycorhiza or may actually be dependent upon it. Probably the benefit is derived through some nutritional improvement provided by activities of the fungus.

Nodules. Certain saprophytic bacteria, living free in many soils, enter the root hairs of most legumes when available, and induce a proliferation of cortical cells sufficient to appear as a small nodule on the root. Although the plant provides food for the bacteria and produces the nodule in which the bacteria multiply, the relationship is truly symbiotic. These nitrogen-fixing bacteria are able to take free nitrogen from the air, unavailable to most plants, and to combine it with other elements to form compounds that can be used by the plant during its lifetime. After death of the plant, the accumulated nitrogenous compounds are released in the soil and are used by other plants growing there. Legumes and nitrogen-fixing bacteria are, therefore, important factors in maintaining soil fertility in natural or cultivated soils. Plant communities becoming established on poor sites, such as eroded slopes, invariably include a number of legumes, which are, of course, particularly adapted to colonizing sterile or nitrate-depleted soils and contributing to their improvement. Agricultural practice includes legumes in most crop rotations, and worn-out lands are rebuilt by cropping with legumes of some sort.

Nitrogen fixation has been reported for a variety of non-leguminous vascular plants, and a number are believed to assimilate atmospheric nitrogen symbiotically (perhaps mycorhiza) or by some other means.[370] Several non-legumes regularly have nodules that usually are modified or arrested lateral roots. Regardless of origin, present evidence indicates that those containing bacteria do fix nitrogen (*Ceanothus americana, Elaeagnus* spp., *Casuarina* spp., *Podocarpus* spp., *Cycas* spp.). *Alnus glutinosa* and *Myrica gale* fix more nitrogen per unit weight of nodule than do several legumes.[146]

Nodules containing bacteria are also formed on leaves of a number of tropical plants, mostly in the family Rubiaceae. These bacteria are associated with the plant tissues in all stages of development from seed to maturity, but nodules form only on leaves. Although these bacteria have been credited with nitrogen-fixing ability, it is certain that the plants are not dependent upon them for their nitrates. Certain products of their presence are necessary, however; for without the bacteria, seedlings do not mature.[192] The relationship is, therefore, truly symbiotic (Fig. 98).

Fig. 98. *Two seedlings of* Psychotria punctata, *about three and one-half months old. The plant on the right is normal both as to growth and the presence of bacterial nodules dotting every leaf. The one on the left, grown bacteria-free, has reached its maximum development.*—From Humm.[192]

Other Soil Flora. In addition to the symbiotic fungi and bacteria, great numbers of bacteria, fungi, and algae occur free in the soil. Their importance to natural plant communities cannot be evaluated accurately, but their significance is indicated by their general functions of making nitrogen available by fixing it, or releasing it with other nutrients through their activities in decomposing organic matter.

The fixation of nitrogen as nitrates by free soil organisms is known to be accomplished by a number of bacteria under both aerobic and anaerobic conditions and even in practically sterile soils. Some are inhibited by acidity or chemical constituents of the soil, and temperature ranges may affect their activity, but, in general, some are present almost everywhere. Certain algae are also known to be capable of nitrogen fixation. It has been demonstrated for *Nostoc* as a symbiont in lichens and even in association with a liverwort.[34]

All nitrates appearing in the soil from sources other than fixation are the products of organic decomposition, particularly of proteins.

The breakdown involves a series of chemical changes accomplished by a succession of bacteria and fungi. The first of these causes the proteins to break down into the less complex proteoses, peptones, and amino acids. This digestive process allows the bacteria and fungi to use a part of the nitrogen for themselves, and, in so doing, they release ammonia as a waste. Few plants can use ammonia directly, and many are injured by its accumulation in the soil. Ammonification is followed by nitrification, in which a group of nitrite bacteria convert the ammonia to nitrites by partial oxidation. Subsequently, the activities of nitrate bacteria cause further oxidation and the formation of nitrates. Now, finally, the nitrogen is usable by higher plants. Digestion of proteins, ammonification, and nitrification must all take place before organic nitrogen can be used by plants, and the succession of bacteria must be present if the processes are to occur.

The activities resulting in available nitrates produce partial breakdown of organic materials, which are further decomposed by other bacteria and fungi acting upon the remaining non-protein plant materials. The partially decomposed plant remains, or humus, may be broken down completely in a single season if relatively high temperatures and sufficient moisture occur most of the year and permit more or less continuous functioning of the organisms. If the organisms can operate for only a few summer months, the deposition of litter usually exceeds the rate of decomposition, and humus tends to accumulate.

Animals as Factors

Pollination. Insects are by far the most important animals involved in pollination, and bees, wasps, moths, and butterflies are particularly concerned. A few birds, especially hummingbirds, contribute to pollen transport, and even some small crawling animals may be effective at times. Most animal-pollinated flowers have certain characteristics in common, such as conspicuousness in size and color and the production of an odor as well as nectar. It has been shown that all of these characters serve more or less to attract insects. In general, the flowers are more elaborate than those of wind-pollinated plants.

Devices that insure insect pollination are common and often of intricate design. Adaptations may occur in both insect and flower limiting pollination of a particular species to a single type of insect. Some adaptations are so extreme as to produce complete dependence of plant and insect upon each other.

Dissemination. Plant parts, called disseminules, give rise to new individuals in new places. Their food content is an attraction to various animals, which, consequently, often act as agents of dissemination. Many seeds that are eaten are indigestible and retain their viability after they are dropped at considerable distances from their sources. Locally, the flora of fence-rows usually includes numerous

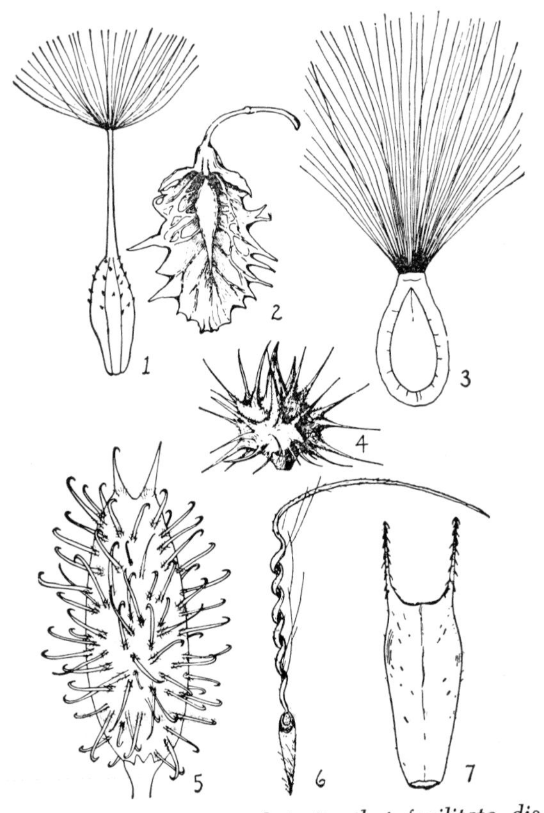

Fig. 99. *Structural modifications of seeds and fruits that facilitate dissemination by wind or animals.* (1) *The parachute fruit of common dandelion* (Taraxacum); (2) *winged fruit of dock* (Rumex pulcher); (3) *the silky-haired seed of milkweed* (Asclepias mexicana); *spiny, hooked, and awned fruits of* (4) *sandbur* (Cenchrus pauciflorus), (5) *cocklebur* (Xanthium canadense), (6) *red-stem filaree* (Erodium cicutarium), (7) *beggar's tick* (Bidens frondosa).—By permission, from *Weed Control* by Robbins, Crafts, and Raynor, copyrighted 1942, McGraw-Hill Book Company.

examples. Others, not immediately eaten, are carried by birds, rodents and even ants to places of storage or concealment, where they may germinate. Of course, great numbers of seeds are eaten or destroyed by animals, but dissemination is effective nevertheless.

Vegetative structures may be effective in the same way. Aquatic animals, such as muskrat, tear up rhizomes and bulbs, some of which float free and establish new communities elsewhere. In this connection, it is worth mentioning the importance of water as an agent of dissemination, especially of floating propagules, even though some do not retain their viability for long when saturated.

Finally, the hooks, spines, and other devices characteristic of many seeds and fruits insure their attachment to almost any animal contacting them and thus make possible their transport for some distance. Animals with long, soft hair are the most effective agents. The clothing of man is likewise well adapted to such transport, as anyone knows who spends time in the field during late summer and fall. Some of these devices are simple hooks, effective because of sharpness or strength; others are elaborate structures with several features insuring their transport. The fruits of awn and needle grasses are illustrative, since they have sharp-pointed, retrorsely-barbed fruits, which easily penetrate cloth, fur, or wool, and an awn which twists with changes of moisture and thus pushes the fruit forward to a secure anchorage. These may cause severe damage to grazing animals by penetrating skin, lips, or even internal organs.

Soil Animals. The microfauna of the soil, concentrated in the upper strata, consists of great numbers of protozoa, nematodes, and rotifers. In addition, there are various macroscopic worms and insects.[414] In general, the numbers of animals vary in response to the same factors affecting the microflora, and the greatest numbers are always found in soil with high organic content. All contribute to organic decomposition and use a part of the products for food.[145] Several protozoa probably consume bacteria, and some nematodes are parasitic on the roots of plants, causing much trouble in cultivated soils where they are present.

Of the macroscopic fauna, earthworms are most active. Their constant burrowing facilitates aeration and drainage and their use of fresh or partially decomposed organic matter as food contributes to decomposition. Since mineral matter is also ingested in feeding, the earthworm moves quantities of soil about, and this tends to mix mineral and organic materials. In cultivated soils this has no great significance, but for natural soils the advantages are obvious. Earthworms are found in the best soils and best sites but rarely in poor soils. It would appear, then, that they serve to make good soils better but that poor soils derive little from them.

Insects. The numerous insects present in all plant communities contribute substantially to the return of organic material to the soil by feeding upon and tunneling in it. A very high proportion spend part of their lives in the soil, where their activities facilitate organic decomposition and distribution.

Others, dependent upon living vegetation, are usually not of major consequence. However, when circumstances favor large increases in numbers, they may be extremely destructive. Actually, in commercial forestry, insects are ranked close to fires as destructive forest agents. Just what their effects were in original forests can only be surmised. Recently, Engelmann spruce beetle has caused great damage in Colorado, spruce bud worm has infested almost three million acres of fir forest in Oregon and Washington, Douglas fir beetle is presently a problem in Oregon, and outbreaks of the southern pine beetle in Texas and Mississippi resulted in heavy losses of pine. These are examples of extremes. To a certain extent insects are causing some damage to plants of every community, either directly by feeding, or indirectly by creating access for and by transporting lesser organisms of decay or disease.

Larger Animals. The principal effect of larger animals upon plants results from burrowing, trampling, grazing or other feeding habits. Carnivorous animals affect communities only indirectly by keeping down the population of herbivores and thus maintaining a balance in food relationships. In spite of this, the feeding by herbivores may sometimes be excessive enough to cause serious disturbance or even destruction of community structure.

Under natural conditions, grazing was undoubtedly greatest when buffalo ranged throughout our grasslands. Locally, as around water holes, their feeding and trampling sometimes destroyed most of the vegetation but otherwise probably did little damage, since they were constantly on the move and distributed themselves where grazing was best. Moderate grazing by cattle does not change the essential nature of a grassland community. A succession of dry years in the time of the buffalo could have resulted in local conditions similar to those in overgrazed range or pasture areas today.

Deer and moose similarly have little effect on the plant communities where they browse, unless there is an overpopulation. Then, especially as a result of winter browsing, the complete destruction of young woody plants sometimes occurs.

Prairie dogs may consume all the forage for some distance about their villages. The total consumption of food by such relatively small animals is sufficient to reduce considerably the value of a range for

Fig. 100. *Distinct browse line on stand of ironwood resulting from deer feeding on low branches. Note the uninterrupted view under stand, and absence of shrubs and tree seedlings. Such damage commonly results when deer population is high, and especially when winter supply of food is inadequate.*—U. S. Forest Service.

larger herbivores. The same may be said for jack rabbits, but their feeding is less localized.

The feeding of cottontail rabbits ordinarily affects natural vegetation but little. However, if a peak in their fluctuating population comes at the time of a bad winter with much snow, they can do serious damage to seedlings and even to larger trees from which they eat the bark. A large rabbit population on grazing land substantially reduces available forage, since it is estimated that five rabbits eat as much as one sheep. In Scotland, heavy rabbit grazing has been shown to produce marked changes in grassland composition and under extreme conditions to reduce them to moss dominance.[144] Because of selective feeding, snowshoe rabbits may change the course of forest succession.[93]

Rodents that eat bark by preference may cause considerable damage, especially if their feeding is selective as to species. Porcupines are in this category; and beavers are even more destructive, because their activities are concentrated around their dams. Here they cut down and strip the bark from the trees they most prefer nearest their ponds and then gradually extend their operations to surrounding slopes. Their dams, too, affect conditions locally, for they maintain ponds that

Fig. 101. *Injuries to seedlings and saplings resulting from feeding by rodents and larger animals may strongly influence the development of stands and the nature of future vegetation. (1) Young ponderosa pine girdled by porcupine. (2) Scotch pine browsed by deer the year after planting. All needles and buds eaten. (3) A pine seedling eaten back by rabbits in three successive winters. Such seedlings can never make normal trees.*—U. S. Forest Service.

sometimes flood large areas, modify drainage, and even affect the water table. This may sometimes be desirable, sometimes not.

Man. The effects of man upon vegetation are fundamentally similar to those of lower animals. The greater the concentration of population, the greater the modification of natural communities by use and destruction. Whereas man was once essentially a dependent in community structure, he is now more and more becoming the dominant organism everywhere. By cultivation, he has eliminated natural vegetation from vast areas. Logging, even without subsequent cultivation, has changed the forests; and stands equaling the original virgin forests will probably never occupy most logged areas again. Cities, highways, airfields, and similar products of man's living mean serious disturbance of natural vegetation. Drainage and irrigation projects, canals, road fills, and dams result in soil moisture changes that promote the development of quite different communities. Many similar disturbances can be noted as a result of animal activities but always on a more localized scale and, consequently, with less permanent effects.

Fire is not peculiar to man's activities and, undoubtedly, occurred here and there in North America before the white man came. However,

Fig. 102. *Center of a burned swamp in Maryland that once supported mature cypress-gum forest. Intense fire destroyed the forest and burned deep into the peaty soil, which had accumulated through the years. Rebuilding soil in the depressions, now filled with water, will require centuries and numerous generations of plants.*—Photo by G. F. Beaven.

the conditions provided by lumbering operations, and the constant use of fire, often with too little concern for its effects, have made it an important factor associated with man's presence. Local small fires occur almost everywhere occasionally, and the destruction of vegetation followed by gradual replacement is characteristic. Under the right conditions, fire may be so common as to become a major factor controlling the vegetation of a region. This is true of much of the coastal plain of the southeastern United States.[157] Prolonged dry periods and little attempt to control fire in these flatlands result in many areas burning almost every year. Only fire-resistant species predominate, and only a limited degree of vegetational development is possible before fire occurs again and sets back that development. As a result, grassy savannahs with longleaf or slash pine are characteristic instead of the potentially possible hardwood forests. California chaparral, Mediterranean garigue and macchia, south Australian mulga, and sclerophyllous scrub vegetation elsewhere, all fire-resistant and fire-induced, increase when burned, with proportionate decrease of forest. Illustra-

Fig. 103. *What fire can do to a mountain forest. Such fires are usually followed by erosion, and it may require centuries for the re-establishment of forest vegetation. Coconino National Forest, Ariz.*—U. S. Forest Service.

tions of fire-maintained vegetation types may be found in many parts of the world.[366]

The immediate economic loss from an intense forest fire is paralleled by other less obvious losses. Such fires in the temperate zones may destroy practically all the humus accumulated through the years and necessitate the slow rebuilding of the soil before forest can occupy the area again. Leaching and erosion, which follow such fires, may delay revegetation for years. Thus the productivity of the soil may be indefinitely impaired.

It is of interest that light, controlled burning has been found beneficial for certain purposes. On some grazing land, certain undesirable species may be kept down or eliminated to the advantage of more palatable plants. More vigorous growth of certain forage types is sometimes obtained after light burning in the proper season, probably because of the nutrients released and made available. Prescribed burning of some pine forests reduces the hazard of destructive fires by changing subordinate vegetation at the same time that succession to

Fig. 104. *A subalpine flat denuded by intense fire that killed all trees and burned off organic material down to mineral soil. Fire occurred many years before picture was taken. Obviously soil sufficient to support forest cannot be rebuilt until the remotely distant future.*—U. S. Forest Service.

hardwoods is almost stopped and reseeding by the desirable pine is favored. Some disease organisms may also be eliminated or controlled.[238] It would appear that under some circumstances fire is being used as a beneficial tool.[177]

Man, like lower animals, transports seeds and fruits, but to far greater distances and with resulting changes in vegetation of a more drastic nature. It is hard to believe that 60 percent or more of our weeds are not native but introduced species that have come from all parts of the world.[270] Some were brought in as ornamentals and almost immediately escaped and spread from gardens. Others came in accidentally with seeds of desirable plants. Many introductions have been useful and extremely valuable. Most of our cultivated plants have been improved by crossing with strains of foriegn varieties at some time, or they were themselves originally introduced. In recent years, such introductions are not made haphazardly.

Unfortunate experiences with unconsidered or accidental introduc-

Fig. 105. *An introduced weed, tumble mustard* (Sisymbrium altissimum), *dominant over the entire extent of a sagebrush burn, one year after the fire. Washoe County, Nev.*—Photo by W. D. Billings.

tions can be listed for all parts of the world. The water hyacinth, introduced from South America, spread throughout the lowland waterways of our southern states, where it choked canals, impeded drainage and navigation, and destroyed wildlife. A similar problem, now controlled, resulted with the introduction of Elodea in the low countries of Europe. The Eurasian water chestnut (*Trapa natans*), grown in ponds on the Mall in Washington, D. C. some 50 years ago, appeared in the Potomac River about 1923 and spread over 10,000 acres before it was effectively controlled. Only constant vigilance prevents its becoming a pest here again and in parts of Massachusetts and New York as well. Only recently it was found in Maryland waters with the immediate initiation of a program of elimination. An unfortunate, accidental introduction was that of *Halogeton glomeratus,* an Asian semi-desert annual so well adapted to the desert shrub lands of the Intermountain Region of the West that it has spread over 1.5 million acres of range land since it was first reported in 1934. Not only does it reduce the native available forage but it is poisonous to stock. Aggressive study of control measures has not yet produced adequate results. Its complete extermination is improbable.[402]

Fig. 106. *Massed water hyacinth covering the water in Louisiana swampland. The dusting by airplane is part of an eradication program.*— Courtesy of Department of Wildlife and Fisheries, Louisiana.

Animals may cause similar difficulties. Everyone is familiar with the now ubiquitous introduced English sparrow and the starling in the United States. The muskrat has become a pest in central Europe; and rabbits, introduced into Australia, increased in enormous numbers in only a few years. In 1950, myxomatosis, a virus transferred by biting insects and fatal to rabbits, was introduced with great initial success in reducing the population, but more recently there is evidence of increasing numbers that survive infection.

Natural communities are made up of groups of species adapted to living together within a particular complex of environmental factors. The requirements of the organisms are in balance with, and an expression of, the potential productivity of the environment. If the ecosystem is disrupted by the elimination of a species, for any reason, others of the community may increase in size or numbers to take its place, or this may provide the opportunity for an incidental species to become a part of the community. The perfection of the system usually is sufficient to prohibit the success of an introduced species, normally because it does not reproduce and, therefore, dies out. But, occasionally, an introduced species has the necessary characteristics to compete successfully and to reproduce regularly. Then adjustments must be made within the community, and a new balance among its members must be established. Such a species might even become a dominant, and then the adjustments would result in an entirely new community. The prickly pear (*Opuntia inermis*), introduced in Australia, became a dominant and made useless more than 30 million acres in Queensland alone. In turn, more recently, a deliberately introduced cactus moth, causing death of *Opuntia*, has virtually eliminated it.

When man has tampered with the balance among the species of a community by eliminations or introductions, he has not always considered the possible effects upon the community as a whole. If large carnivores are destroyed, herbivores increase, and, if their reproductive capacity is great, they may soon become so abundant that their grazing destroys the community or changes it radically. If a predator is introduced whose prey is some native species that is a pest, the predators may eliminate the pest and then become pests themselves.[10]

Only a few examples are necessary to illustrate these points. The Indian mongoose was introduced into Haiti, Jamaica, and other West Indian islands to rid them of rats and snakes. This the mongoose did most effectively, but its numbers increased, and, with its natural prey disappearing, it turned to robbing birds' nests of eggs and young. Now it may even be a problem in the raising of poultry. The gypsy moth was accidentally introduced into Massachusetts when it escaped from cultures being reared to test its silk-producing ability. It is now a serious pest of fruit and shade trees in most of the eastern United States, although much money and effort have been expended to control it. On the other hand, introductions of about sixty foreign predators or parasites of the gypsy moth have resulted in the establishment of a dozen or more that are aiding in its partial control. The destruction of coyotes in some western states has resulted in such marked increase of rabbits that their winter feeding on tree seedlings modifies vegetational development (see Fig. 101).

On game reserves where predators have been eliminated and no hunting is permitted, the population of herbivores, such as deer, usually increases rapidly. When the number of deer exceeds the natural carrying capacity of the region, a shortage of food results during unfavorable seasons. Then, especially in winter, many animals die unless they are fed by man. As a result of supplementary feeding, the population is still larger the next season, and the problem is not solved. Controlled hunting is now permitted on several such reserves where the population capacity has been determined. The effects on the vegetation of such overcrowding are very conspicuous. All young woody plants protruding above snow are eaten off, and the lower limbs of young trees, even conifers, are "pruned" to the height the animals can reach, standing on their hind legs. Obviously, community structure and development in such areas is completely out of balance.

Disturbance of natural communities should not be undertaken without a reasonable appreciation of the end results. Management or manipulation of the balance among species of a community may frequently be possible but should offer the best prospects of success when the ecology of the individuals and the community is well understood.

Man's unconcern for natural resources built up through the years has led to economic losses and a reduction of those resources, some of which can never be replaced. Soil erosion, often quite unnecessary if cropping is properly handled, had reached a shameful point before we began to do anything about it. Only recently have we attempted to correct overworking of poor soils, mismanagement of others, overgrazing, and other destructive practices. Contour plowing, strip cropping, terracing, and similar procedures check runoff, hold water, and permit the rebuilding of rundown soils. On wild lands and some submarginal cultivated lands, the re-establishment of natural vegetation is being encouraged where it should never have been removed. Application of ecological principles in such reclamation has generally paid good dividends.

Not only has man disturbed or destroyed natural vegetation, but he has also modified the environment, sometimes to his advantage. By irrigation or drainage, the soil moisture has been so modified that great acreages have been brought under his control and use. Enormous dams hold water in artificial lakes. When this water is properly supplied to the surrounding soils, it transforms almost worthless desert to highly productive agricultural land. Elsewhere drainage systems put into lowlands have changed swampy, untillable soil to some of the best

Fig. 107. *Drained swampland in the Everglades of Florida. Many acres of these muck soils are producing winter truck crops in quantities, now that problems of drainage, tillage, and fertilizing have been worked out.*—U. S. Soil Conservation Service.

truck and farming acreages. Not all drainage projects have been profitable, however, especially those of muck lands. Some drained mucks have caught fire and burned disastrously. Not all are equally productive, and cost of maintaining drainage may be out of proportion to the crop yields. Many such projects have been abandoned—to the joy of sportsmen and conservationists, who objected to the extensive destruction of homes and feeding grounds of all kinds of wildlife associated with these swamps.

General References

R. M. ANDERSON. Effect of the Introduction of Exotic Animal Forms.

J. M. COULTER, C. R. BARNES and H. C. COWLES. *A Textbook of Botany.* (Vol. III: Ecology, pp. 1-471.)

H. C. HANSON. Fire in Land Use and Management.

W. A. MCCUBBIN. Preventing Plant Disease Introduction.

S. A. WAKSMAN. *Soil Microbiology.*

Part Four: COMMUNITY DYNAMICS

Chapter 10 Plant Succession

Historical Background

When a cultivated field is permitted to lie fallow, it produces a crop of annual weeds the first year, numerous perennials the second year, and a community of perennials thereafter. In forest areas, the perennial herbs are soon superseded by woody plants, which become dominant. After any disturbance of natural vegetation—such as cultivation, lumbering, or fire—a similar sequence of communities occurs with several changes in the dominant vegetation through the years.

Such relatively rapid vegetational changes are familiar to most people today and must have been observed hundreds of years ago. It was not until the seventeenth century, however, that any systematic study of such changes was made, and those studies dealt primarily with the development of peat bogs. Bog studies were continued in the eighteenth century, and, in addition, some attempt was made to apply the principles to burned and disturbed upland areas. It was then that the term, *succession,* was first applied to the vegetational changes involved. During the nineteenth century, succession was considered rather frequently but invariably as incidental to other problems. Several writers hinted that succession occurs in all habitats, but apparently they did not recognize its universality. Kerner (1863), however, included a discussion of "genetic relationships" in his description of the vegetation of the Danube Basin, which showed a clear understanding of the principles of succession and stabilization. He received little credit for these ideas until recently when his work was translated.[92] It was not until 1885 that Hult made the first vegetational study of a region, in Finland, using succession as a basic consideration.[78]

Between 1890 and 1905, the modern concepts of succession were clarified through the efforts of several writers. Two, whose influence has been as great as any, were Americans. In the first comprehensive application of successional principles in the United States, Dr. Henry C. Cowles (1899) described the development of vegetation on the sand dunes of Lake Michigan. Later (1901) he described the vegetation of Chicago and vicinity, as it is related to physiography, in so logical a fashion that a pattern for studies of community dynamics was established. His papers also served to stimulate similar investigations by others. Beginning at about the same time, the publications of Dr. F. E. Clements, then working in Nebraska, included much that served to shape our present concepts of succession. The culmination of his ideas appeared in his exhaustive treatment of the entire subject of plant succession,[78] which remains a basic source of reference today.

The Concept

No plant community can be completely stable. Since its components vary in age and potential longevity there are always some weak or overmature individuals dropping out, and as they die others of the same stratum may expand to occupy their space; younger, suppressed individuals, released from competition, may rapidly replace them, or the openings in cover and soil may allow establishment of new individuals, or even species, which were previously excluded. Thus, a community is characterized by constant change[98] rather than stability. Sometimes these changes are radical and abrupt, as when a disease strikes one of the important species, or when woody plants invade and gain dominance over herbs. Again, the changes may be so slow as to be scarcely discernible over a period of years, as is true of lichen and moss mats on rocks.

These changes are not haphazard. There are elements of chance, particularly related to the vagaries of local climate from year to year or the availability of a seed supply, which may affect rates and even, temporarily, the nature of changes in particular habitats. Nevertheless, within a climatic area, the trend to be expected in a community in a given habitat is predictable. Not only do similar habitats support similar communities but they have a sequence of dominants that tend to succeed each other in the same order. Contrasting habitats do not support the same sequence of communities. As a result, a region with several distinct types of habitats will have an equal number of possible successional trends.

Causes

A detailed consideration of the relationships of organisms to their environment should suggest that major changes in the composition of a community can only follow changes in the environment. The specific, immediate cause of a particular change of species may not always be obvious because of the complex interrelationship of controlling factors and because the competitive characteristics of the species are incompletely known. Two general types of habitat change may result in modification of community structure or composition. (1) Development of the community causes parallel developmental changes of the environment, and (2) physiographic changes can likewise modify the environment materially.

Developmental changes of the environment result from reactions upon the habitat by the organisms living there. To illustrate: Under any vegetation, light and air movement are reduced and temperature and humidity are modified. Also, there will be accumulation of litter which affects run-off, soil temperature, and formation of humus; this, in turn, contributes to soil development, modifies water relations, available nutrients, *p*H, and aeration, and affects soil organisms. To a different degree, directly or indirectly, every organism in a community may have some reaction upon the habitat. By these reactions, the habitat becomes modified, and usually it becomes less favorable to the organisms responsible for the changes; while, at the same time, it is now more favorable for some species that could exist there previously only with difficulty. Under the changed conditions, new species are able to compete successfully with the established species and often even to replace them.

The habitat may also be modified by forces quite apart from the effects of organisms. A flood plain or swamp may become better drained as a stream cuts more deeply into its channel. Silting-in of a lake or pond raises the level of mineral soil. Chemical changes in the soil may result from leaching or accumulation of salts. Such modifications of the habitat also produce vegetational changes.

It is of more than passing interest that Clements restricted his concept of succession only to changes resulting from biological reaction, while Cowles included topographic (physiographic) factors as causal agents. Actually, a clear distinction can be made between vegetational development, as emphasized in the first instance, and a succession of communities;[393] but it is not always necessary, desirable, or even practical,[98] even though the implications may have a bearing on the classification of the resulting vegetation.[275] These two types of causes of succession, biotic and physiographic, are commonly in operation at the same time, and their effects cannot always be readily separated. Since they both result in vegetational change, it seems unnecessary to distinguish between their effects in a general consideration of plant succession.

It is probable that both reaction of vegetation and small-scale physiographic-edaphic changes can, and usually do, contribute to potional succession in every environment. Knowing, then, that each species has its own, usually limited, ecological amplitude, we have a broad explanation of why succession takes place. Yet, when particular stages of succession are studied intensively, as is increasingly being done, interesting exceptions may be found. In some arctic, alpine, and desert habitats, succession, as ordinarily conceived, does not occur. Rather, the numbers of individuals of the pioneer species merely increase as the habitat matures. With such restricted floras of ecologically related species, succession is barely apparent. For contrast, consider the marked and rapid successional changes often occurring on abandoned agricultural land under temperate conditions. Immediately after cultivation ceases, a sequence of communities with dominants sometimes changing from year to year is common. Surely, one might assume that here reaction is most effective and that each of these distinct communities prepares the way for the next as it inhibits its own continuation. Actually, it has been shown that neither reaction nor edaphic change need be involved.[306] In some instances the clear-cut sequence of dominants results primarily because phases of their life cycles are completed at different times.[203] Seeds of annual and perennial herbs may be abundant and widely distributed, but when the annuals germinate at once, they are the first dominants. The perennials may require after-ripening before germination, and may also go through a rosette stage to delay them for a year. Other perennials may have few seeds or low rates of dissemination, further to delay their appearance. Although, obviously, the species of later stages are competitively successful they do not necessarily require the early stages to prepare or modify the environment for them.

A further complication in the understanding of succession and its causes is the community of several dominants whose apparently equivalent components are sometimes segregated in pure stands on what appear to be similar sites. These communities with a single dominant occupy the same position in a successional trend as does the complex community. To explain such situations it may be necessary to study the complete ecological life history of each of the species. Perhaps the answer may be as simple as the difference in periodicity of good seed years that coincide with availability of the site. However, more than likely, the conditions of the site for a very short period during seed

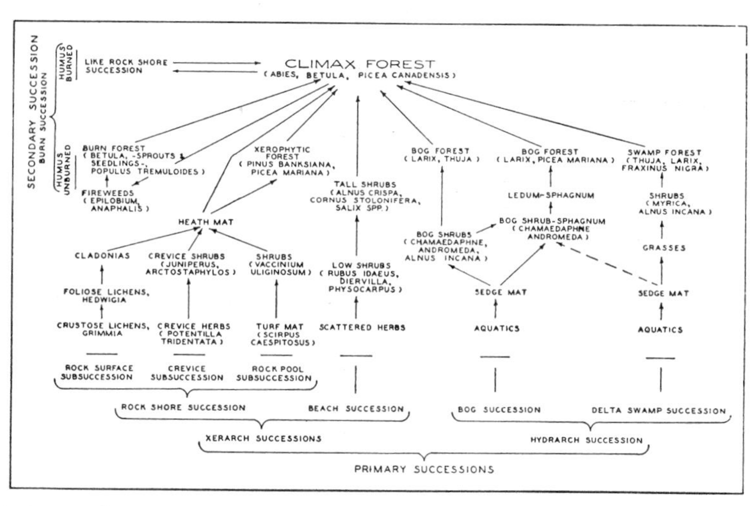

Fig. 108. *A diagram of the trends of succession for the principal habitats on Isle Royale, Lake Superior. This is one of the early complete condensations of a successional story for an entire region. On this pattern, similar diagrams have been worked out for many sections of the country. Note that the system shows at a glance the kinds of habitats in which succession originates, the interrelationship of trends, and the major dominants in each of the stages of succession. Study of the diagram should help to clarify concepts of succession and climax. It must be remembered that not all trends progress with equal speed.*—After Cooper.[95]

germination and establishment may be selective. Then, an explanation is possible only if the moisture, light, and temperature requirements[254] and tolerances[255] of seeds and seedlings are known.

Although succession is a universal phenomenon, its causes are not always the same. Consideration of specific situations indicates that generalized explanations do not always suffice and that they should be used with discretion. Actually, specific knowledge of successional causes is limited and sometimes difficult to obtain. The subject offers areas of profitable investigation that will not be exhausted for some time to come.

Kinds of Succession

Primary succession is initiated on a bare area where no vegetation has grown before. It may be observed on glacial moraine exposed by recession of the ice, a new island, an area of extreme erosion, newly deposited volcanic ash or rock, or any similar habitat newly exposed to colonization. Such habitats are apt to be unsuitable to the growth of most plants, and, consequently, the pioneers that do establish themselves must have adaptations permitting survival under extreme conditions. Moisture relationships usually control their ability to invade the new area. If the habitat is extremely dry, it is described as xeric; if wet, hydric; and if intermediate, mesic. The successional trends are similarly referred to as being xerarch, hydrarch, or mesarch succession. These terms are relative. Mesic habitats in two differing climatic regimes would not necessarily have the same moisture conditions. In each instance they would be intermediate between the habitat extremes represented under the given climate.

Whatever the condition of the initial habitat, reaction of vegetation tends to make it favorable to more plants by reduction of extremes, which is always reflected in improved moisture conditions. Thus xeric habitats become moister and hydric ones become drier as succession progresses. Because of the diversity of habitats upon which succession may begin, there are an almost equal number of possible pioneer communities. Within a climatic area, however, the variety of communities decreases as succession progresses, because the trend is toward mesophytism from both hydric and xeric habitats. Thus unrelated habitats may eventually support similar vegetation and may even undergo identical late stages of succession.

Secondary succession results when a normal succession is disrupted by fire, cultivation, lumbering, wind throw, or any similar disturbance that destroys the principal species of an established community. To what extent the development of vegetation on the secondary area resembles primary succession is determined by the degree of disturbance. Although the first communities that develop may not be typical of primary succession, the later stages again are similar. When disturbance is extreme, as after severe fire, many of the effects of previous

Fig. 109. *An illustration of relatively rapid secondary succession. The fire that destroyed this Oregon forest (above) did not appreciably affect the soil organic matter and was not followed by erosion. As a result, Douglas fir soon became established and, when fourteen years old, formed a closed stand 10 to 15 feet tall (below).*—U. S. Forest Service.

vegetation upon the habitat are eliminated. All organic matter may be destroyed, and then soil structure and soil water relationships are modified, the released minerals may quickly leach away, and erosion may be greatly accelerated. Subsequent vegetational development as a result may be relatively very slow. After wind-throw or lumbering, many of the products of community reaction remain, and succession is rapid. If seedlings and young trees are not destroyed, progress of succession tends to exceed that of the original trend.

Most of the settled parts of North America have little evidence of primary succession today, and even unsettled areas have largely been disturbed by grazing or lumbering. The least disturbed examples of vegetation are apt to be on the poorest or most extreme sites in an area, and these never were representative of the general conditions. Thus primary succession must often be interpreted in terms of small and often poor examples of what once occurred. Studies of secondary succession may, however, have the greatest practical value, because we are involved with secondary successions in any problems of applied ecology; yet their interpretation may be partially dependent upon an understanding of primary successions.

Representative Successions. Because water and bare rock represent the extremes in types of habitats upon which primary succession is initiated, the growth form of early stages of each is remarkably similar everywhere, and even genera and some species are often duplicated regardless of the region. It is, therefore, possible to present a general description of such successions, which can be applied almost anywhere and which will illustrate what has just been discussed.

Hydrarch succession begins in open water wherever vegetation can become established. It progresses in response to any environmental change that reduces the water (depth or saturation) and improves aeration in the soil of the habitat. The direction of change is, therefore, from an aquatic toward a terrestrial habitat. Initiated in a lake, pond, or stream margin, where water movement is not too great, the pioneer vascular plants are submerged aquatics with thin, dissected, or linear leaves. They may grow in any depth of water to which there is adequate light penetration, but they are almost invariably limited in a shoreward direction by a zone of floating-leaved species. These latter (water lilies, etc.) exclude submerged species by shading but cannot advance into the zone of submerged forms unless the bottom is built up or the water level falls. In still shallower water, emergent species predominate. These have their roots and rhizomes in the mud and extend upward, sometimes through water, into the air (rushes, reeds, cattails, sedges). The close growth in this zone serves to hold sediment, and the bulk results in substantial accumulation of partially decomposed organic matter called peat. Especially in glacial lakes, emergent vegetation may form a floating mat that slowly advances over the lake surface at the same time that it adds to the bulk of accumulating organic material which is gradually building up underneath it. Sphagnum moss invariably makes a substantial contribution to this accumu-

Fig. 110. *Hydrarch succession as illustrated by girdles of vegetation around a shallow lake in northern Minnesota. In what remains of open water are submerged and floating-leaved aquatics, the pioneer angiosperms. A marginal, floating sedge mat is gradually filling the lake with peat and advancing over the water. On the mat are a few bog shrubs, behind which is a girdle of tamarack forming a closed stand. The oldest part of the bog is marked by the spires of black spruce, which succeed the tamarack. On the upland, behind the spruce, is a mixed white pine-hardwood forest. Eventually, the entire depression will be a peat-filled bog supporting a forest of black spruce.*—Photo by Verona M. Conway.

lation. Peat-filled bogs of this type may be many feet deep and will have required hundreds of years to fill. Regardless of the circumstances, the shallower margins fill first and eventually are built up sufficiently to allow establishment of a zone of shrubs. Much later, the soil will be firm enough and sufficiently raised above the water table to support lowland or bog trees. The usual effect of such forest is to lower the water table, with consequent improved soil aeration, and thus, commonly, a sequence of other trees comes in whose requirements are successively more and more like those of surrounding uplands.

This entire sequence can sometimes be seen as a more or less continuous series of zones girdling a lake that is gradually filling in. Borings of the soil under any zone will show the partially decomposed remains, in vertical sequence, of each of the previous stages of succession that contributed to the development of that zone.

Xerarch succession on rock follows a definite pattern, whose progress

Fig. 111. *Hydrarch succession illustrated by swamp vegetation. The zone of cattails occupies the partially flooded, muddy margins. When soil builds up or drainage improves, bog shrubs (buttonbush, alder, willow) appear as in the middle background. On wet, but drained soil a swamp forest of mixed hardwoods develops as in background.*—Photo by H. L. Blomquist.

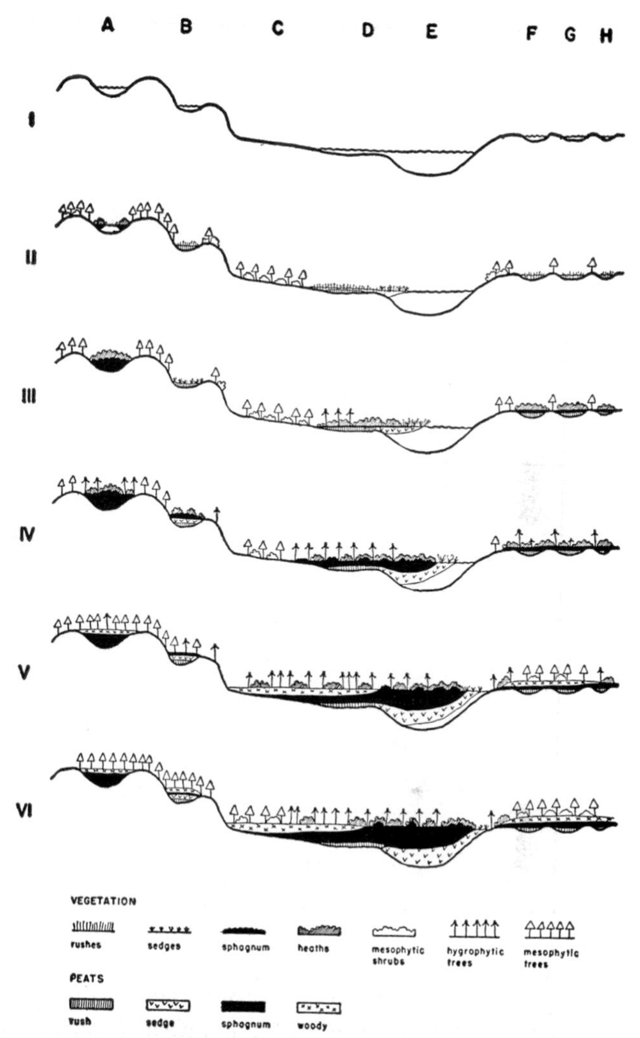

Fig. 112. *A diagram of the probable evolution of bogs on the Laurentian Shield, reading from top to bottom (I to VI) and showing the processes that operate in the case of closed drainage (A,B,F,G,H) and in situations where there is more seepage or water movement and wind action (C,D,E). Not shown is the accumulation of washed-in material in the main depression (E) and elsewhere, which would be filling the lake in its deeper parts before and after the mat advanced from the periphery and would begin as early as stage II.*—From Dansereau and Segadas-Vianna.[118]

Fig. 113 (a). *Xerarch succession as illustrated by vegetational development on granitic rock in the Piedmont of the southeastern states. Early stage (upper) of mat formation initiated by the pioneer moss* (Grimmia laevigata) *upon which a lichen* (Cladonia leporina) *is well established. As mat thickens (lower), herbs come in, with eventual* Andropogon *spp. dominance.*[284]

is controlled by the rate at which soil forms and accumulates. Pioneers on rock surfaces are either lichens or mosses capable of growing during the brief periods when water is available to them and lying more or less dormant through the usually longer periods of drought. The pioneer lichens are crustose and foliose types, which contribute little to

Fig. 113 (b). *Shrub stage of rock succession, mostly* Rhus copallina *here. Note fringe of* Andropogon, *smaller herbs, and finally mosses at periphery (upper). Tree stage (lower) on an old mat, forming an island on bare rock. Oak-hickory forest in background is growing on shallow soil overlying rocks.*[284]

succession since they are not mat-forming.[284] However, they may cause corrosion of the rock surface and thus provide some anchorage for other species. Pioneer mosses, on the other hand, are in tufts or clumps, which catch dust and mineral matter from wind and water. This material, combined with the remains of mosses, forms a gradually thickening mat with a periphery of young plants that spreads over bare rock (and the pioneer lichens) and with a central area that may become thick enough to support fruticose lichens (*Cladonia* especially), larger mosses such as *Polytrichum*, or often species of *Selaginella*. Such bushy plants catch and hold still more mineral material, and their death adds much organic soil to the mat.

When soil has built up sufficiently to provide the necessary anchorage and water-retaining ability, seed plants appear on the mats. A number of hardy, annual herbs, often weeds of field and garden, appear first and are followed by biennials and perennials, of which grasses are most abundant. Later a shrub stage becomes dominant, which commonly includes some species of sumac (*Rhus*) and various ericaceous shrubs. By this time, the mats may be several inches or even a foot thick, and then trees make their appearance.

Just as a series of girdles of vegetation usually surrounds a lake and indicates the sequence of succession from open water to solid ground, so the progress of succession on rock may be seen as a series of girdles of vegetation from the periphery to the center of an old mat. Pioneers are at the outer margin of the mat, and each successive stage of dominance is nearer the center, where on the thickest soil trees may be present.

The early stages of primary succession in both hydric and xeric habitats are apt to be extremely slow, but later stages speed up considerably as reaction of the vegetation becomes more effective. The final stages, as when trees become dominant, are again very slow. Changes of currents or drainage in the lake and wind-throw or fire on the rock may disrupt either of the trends and result in secondary succession. The result of succession in both habitats, however, is that they tend, by gradual change, to become relatively mesic and to support a community that is mesophytic for the climate of the region.

Rate of Succession

If succession is to be recognized as universal and occurring in all habitats, it becomes necessary to ignore time to some extent. A mesic habitat in a given climate will obviously produce a forest much more quickly than a xeric one, especially if the initial habitat is bare rock. Yet the potential ultimate communities of the two sites may be very similar, for all successions in a climatic area progress toward relative mesophytism. Two habitats of apparently similar characteristics might support the same successional sequence, but progress of the successions might be at different rates because of the type of soil and the difference in its response to reaction. Or, if seed sources were not equally available to both sites, one might develop more rapidly than another. This could result from an oversupply of seed, producing overstocking of certain species and consequent delay in development of the next stage because of competition; on the other hand, poor seed sources or a series of poor seed years might materially delay the initiation of a community that otherwise could have started. This should make it clear that the rate of succession is extremely variable. Pioneer stages of primary succession are commonly very slow because they can progress only with soil development. An extreme example is that of succession on bare rock, which must wait not only upon soil development but also upon the disintegration of the rock for soil formation. In contrast, the early stages of secondary succession, especially on fields abandoned after cultivation, are remarkably rapid, for often the dominants change every year for several years.

Stabilization and Climax

The discussion of succession up to this point has been generalized. Probably most ecologists would accept what has been said even though a different emphasis might be put on some points. Now, however, it becomes impractical to continue on this basis because of the wide discrepancies in interpretations held by different schools of thought. Early basic ideas are still in wide current use; but, as the science of vegetation has developed, intensive studies have provided new information that has led to modification of these ideas and new interpretations whose acceptance is on the increase. An introduction to at least two general philosophies is therefore imperative for an understanding of present day work and publications on vegetation.

One of these must, of necessity, be that of Clements whose ideas of community dynamics were presented so effectively[78, 79] that they influenced thinking everywhere and provided a foundation for development of an American or Anglo-American school of thought regarding community relationships and classification. In the second consideration, some opposing interpretations will be summarized.

All interpretations of succession recognize that the successive communities, and particularly the dominants, are more and more exclusive. Eventually, succession terminates in communities whose complex of species is so adjusted to each other, in the environment that has

Fig. 114. *Herb stages in secondary succession on abandoned upland fields in the Piedmont of the southeast. (1) Horseweed* (Erigeron) *dominance on a field abandoned one year. (2)* Aster *dominance indicating two years of abandonment. (3) Broom sedge* (Andropogon) *dominance in a field abandoned five years, and young pine well established.*

Fig. 115. *Forest stages of old-field succession (continuing Fig. 114.) (1) Fully stocked 15-year loblolly pine, which has eliminated all old-field herbs and under which hardwood seedlings may be found, (2) 26-year pine, under which saplings of gum, red maple, and dogwood are noticeable, (3) 50-year pine stand in which hardwoods, including oak and hickory, have formed an understory, (4) oak-hickory climax forest, of the type that could develop on an old field after 200 years or more.*[280]—Photo (1) by C. F. Korstian.

developed, that they are capable of reproducing within the community and of excluding new species, especially potential dominants, from becoming established in it. Both as to floristics and structure, the community is now stabilized. All community processes continue, possibily at a greater rate than in earlier phases of succession, but they result in no major modifications to the organization. Presumably, the progression has been from fractional utilization of available environmental resources in early stages of succession to a maximum utilization on a sustained basis.[117] The energy system (the ecosystem) represented by environmental potential and the functioning of interacting organisms is now at the peak of productivity, which continues in a condition of dynamic equilibrium, or, a steady state. Barring disaster or a change of climate, the community may continue indefinitely, because individuals that are lost for any reason are replaced by their own progeny. Regardless of the qualifying terminology characteristic of various systems of ecological classification, such a community is recognized as *climax.*

Monoclimax

The Clementsian interpretation postulates that a climatic region has but one potential climax; the most mesophytic community that the climate can support. It will be found on sites with average or intermediate environmental conditions, particularly regarding moisture relations, and the vegetation and environment will have developed together under the control and within the limits of the general climate.

Given sufficient time, with accompanying stability of climate and land surfaces, succession will have proceeded to such terminal, relatively mesophytic communities over much of the area. The stands will not be identical, yet they will have such a high degree of similarity that they are obviously related. The similarity is first apparent in the physiognomy or growth form and, on closer inspection, the recurrence of the same genera, and often species, as dominants. This vegetation type, an abstract community, is called the climax of the region. In this sense the term refers both to the ultimate or terminal nature of the vegetation and to the correlation with climate. The individual stands upon which the climax is based are, of course, climax stands. They are the local examples of the regional climax to which they belong. In describing the vegetation, the expressions *regional climax* and *formation* may be used synonymously and interchangeably. Reference to climatic climax, a redundancy, is sometimes made for emphasis.

The concept of a regional or climatically controlled climax necessarily includes recognition of the convergence of successional trends toward a similar end. In its simplest statement, it implies that any habitat in a region, given enough time, could ultimately support a community representative of the formation. From this statement it might be inferred that a region of fairly uniform climate would eventually have a continuous and equally uniform vegetational cover throughout. Actually, this is never true. In the same sense the geologist recognizes that baseleveling could produce featureless topography, but never does; only a peneplain ("almost a plain") results. Locally, there are always edaphic or physiographic situations whose complex of environmental factors differ to such a marked degree from those of the general climate that they cannot support the regional vegetation type and probably never will. To be sure, succession occurs on these sites, but the terminal communities may be very different from the regional climax although they have the same evidences of permanence.

Monoclimax theory does not ignore these extreme situations but rather emphasizes that they are to be expected. Furthermore, variations of the regional climax are also looked upon as normal, thereby necessitating a system of categories for their classification.

Variations of the Formation and Their Classification. One of the earliest ecological classifications divided all vegetation into desert, grassland, and forest and thereby unwittingly applied the regional climax or formation idea. Major formations acceptable today are as easily recognized: desert, semi-desert, with shrubs predominating, and forest, distinguishable as coniferous, deciduous, or broad-leaved evergreen. Each of these vegetation types predominates as climax over wide areas which, even in different parts of the world, have obviously similar climates. The formation is thus a product and an indicator of climate whose inter-acting factors must be broadly equivalent throughout the range of the similar vegetation. Although the individual factors may differ, their biological effectiveness must be similar. This fact may be illustrated by the wide band of similar prairie vegetation extending down the center of the continent where several factors, and especially temperature, differ greatly from the northern to the southern limit of the range.

Although the growth form of the formation is uniform and binding species are included among the dominants throughout, there are floristic variations that are normally to be expected. To illustrate, the wide-ranging deciduous forest of eastern North America has transitions to coniferous forest on the north, grassland on the west, and, to some extent, tropical vegetation to the south. Such transitions are not abrupt but there are very noticeable gradients of vegetational change between such extremely different formations. Again, because of the great area involved, climatic effectiveness will vary some; and, accordingly, species will assort themselves in different combinations of relative importance, because they are not all identical in their responses or adaptations to environment. Therefore, within the range of a formation, recognizably different climax communities are to be expected.

The usual great extent of a formation may well include areas of differing geological history; and, thus, time becomes an additional consideration in vegetation variation. The deciduous forest, already used as an illustration, is an excellent example of this point. Most of its northern extent occupies glacial soils and topography, and it has occupied the area only in relatively recent geological time. Unglaciated areas to the south supported deciduous forest throughout the period of glaciation and still do today. Thus there are differences in age of vegetation, topography, and soils in the two areas; and all contribute to variation in the deciduous formation.[64]

In the monoclimax system, such variation is accepted as normal to the extent that a formation is recognized as having at least two major subdivisions, called *associations*. These are distinct, but two or more dominant species will have the growth form of the formation, and some dominant genera and one or more species will be present in all the associations. An association is, therefore, one of at least two formational units of climax status, floristically and usually geographically distinct, but physiognomically similar and ecologically related.

Just as formations are made up of associations, so may further subdivisions be recognizable. A geographical variant with two or more, but less than the total number, of associational dominants is termed a *faciation*.[82] If there is but a single dominant, the unit is called a *consociation*. Faciations may be further subdivided into *lociations*, local units that differ from each other in the relative abundance and the grouping of dominants.

This system of classification, as presented in outline, uses the association, a climax community, as the basic unit of vegetation and is developed by dividing and subdividing this major unit into lesser categories. Because climax communities are the basic units, the classification incorporates succession into its every interpretation of vegetation. Unfortunately, the term association is not always used in this sense, although other systems of classifying vegetation invariably include it. European ecologists especially use several systems of classification in which associations are basic units, but as units that can be grouped into successively higher categories. Their associations are the simplest recognizable units and, therefore, the lowest ranking. This concept is, of course, in absolute contrast to the Clementsian association, which is a community of the highest rank, inclusive of, and divisible into, numerous lesser categories. The two divergent uses of the term have become so well established and so widely used that there is little hope of clarification in the near future. Attempts to standardize the association concept have been made at recent International Botanical Congresses but with little success. The two lines of thought are so well established that rulings and recommendations simply are not followed. The diverse points of view and their applications have been summarized by Conard[91] together with a helpful bibliography for those who would look further into the subject.

Special Climax Terminology. It is probably appropriate to repeat that succession does not proceed directly to the climatically controlled climax in every habitat. It may be halted or slowed down temporarily in almost any stage of its progress by a variety of circumstances. Diseases, fire, insects, or man may produce conditions that prevent completion of succession and hold it indefinitely at some stage preceding the climax. Additionally, as already pointed out, edaphic or physiographic conditions, usually relatively local, may be such that, although succession may proceed to a stable and permanent community, it will not be typical of the regional or climatic climax.

In the monoclimax system, these special situations are not ignored. Rather, they are assigned to separate categories[82] which indicate that development and stabilization do take place but as a product of local environmental features that are more effective than the regional climate. This leads to a further consideration of terminology, but only the concepts basic to an understanding of monoclimax theory need be presented. Although Clements has been criticized for introducing an excessively elaborate ecological vocabulary into descriptions of vegetation, he cannot be blamed entirely for the complex of terms currently being applied to climax concepts.[64, 429] The situation has become so involved that something might be gained by dropping the term climax entirely, as has been suggested.[140] This being improbable, only terms and concepts will be considered that are rather widely used or are necessary to a point of view. Those who are interested will find no dearth of literature on the subject.[429] The concepts that follow are basic to the expression of monoclimax theory, but they are not necessarily always restricted to it.

Subclimax. When, in any succession, a stage immediately preceding the climax is long-persisting for any reason, it can be called *subclimax.* It may be the result simply of extremely slow development to climax, or of any disturbance, such as fire, that holds succession almost indefinitely in its subfinal stage. In the eastern United States, most pine forests are subclimax to hardwood climax because of the relatively slow elimination of pine in the progression toward hardwood dominance. In the coastal plain, subclimax pine forests are maintained indefinitely by the constantly recurring fires to which the pines are resistant and that keep down hardwoods.

Disclimax. When disturbance is such that true climax becomes modified or largely replaced by new species, the result is an apparent climax, called *disclimax.* The disturbance is usually produced by man or his animals and the introduction of species that, under the existing conditions, become the dominants over wide areas. The prickly pear cactus thus has formed a disclimax over wide areas in Australia. A grass, *Bromus tectorum,* forms a disclimax in much of the Great Basin, where, because it burns readily, it facilitates fires, which reduce dominance of desert shrubs and increase the area of grass. The short grasses of the Great Plains, long considered as climax, are now rather generally interpreted as disclimax resulting from grazing and drought, which have practically eliminated the midgrass climax. The ravages of chestnut blight illustrate how disclimax may result from disease. Oak-chestnut climax of the Appalachians and southern New England is today an oak disclimax.

Postclimax and Preclimax. A climatic area is normally bordered, on the one hand, by one that is drier and warmer and, on the other, by one that is moister and cooler. The contiguous climatic areas, as a result, each has its own climax, distinct in species and, often, in growth form. On a large scale, this is apparent in latitudinal zonation from the tropics to the arctic. Often it is noticeable in the climaxes along a line from oceanic or maritime climate to the interior of a continent. It is most conspicuous on mountains where there is an altitudinal zonation of climates and climaxes. Each of the climatic areas in such a sequence has a bordering climate with a more favorable water balance, usually on the north, toward the coast, or at higher altitudes; while the climate to the south, toward the interior, or at lower altitudes, usually has less favorable water relationships.

For any particular climax the contiguous climax produced by a more favorable climate, usually cooler and/or moister, is termed *postclimax*, and the one produced by less favorable conditions, usually drier and/or hotter, is termed *preclimax*. To illustrate on a broad basis, deciduous forest climax has grassland as preclimax and northern conifer forest as postclimax. At the same time, deciduous forest holds a postclimax relationship to grassland that has desert as preclimax. The use of the concept is not restricted to formations as illustrated above; since it is just as applicable to associations, even within the same formation. For example, within the deciduous forest formation oak-hickory is preclimax and hemlock-hardwood is postclimax to the beech-maple association. Likewise, oak-hickory is preclimax and beech-maple (or hemlock-hardwood) is postclimax to the oak-chestnut association.

Should the present phase of relatively stable climates be interrupted, the climate of any given area would undoubtedly tend to become more like that of one of its contiguous areas, and a migration or shift of climax would result. Such a shift occurred during the glacial period when the northern coniferous forest moved southward, and the northern extent of the deciduous forest was proportionately constricted. When the climate ameliorated, the ice receded, and again, the ranges of the climaxes were readjusted. When such shifts occur, remnants of the previous dominants are left behind in locally favorable habitats, where they may maintain themselves indefinitely as relicts of a previous climax. These relicts are either preclimax or postclimax depending upon their relationship to contiguous climaxes and the direction of the climatic shift. The habitats in which they survive must have edaphic or physiographic characteristics that differ so markedly from the average for the region that conditions for growth are similar to those of a contiguous climatic area. Deep valleys or canyons with steep bluffs and contrasting exposures, poorly drained flood plains, bogs, ridges of rock or gravel, areas of deep sand, or other peculiar soil conditions are specific examples.

Where there have been shifts of climax, it is apparent that preclimax and postclimax communities should occupy such habitats. Not all preclimax and postclimax communities, however, need be relicts. Within the general range of a climax, there are bound to be local habitats such as those mentioned above that will continue indefinitely to be somewhat more favorable or less favorable, wetter or drier, than the conditions controlled by climate in the region as a whole. As a result, when vegetational development proceeds to a condition of stability on such a site, it will have characteristics of the contiguous more, or less, favorable climate. Such localized stable communities are likewise postclimax or preclimax for the region. In, or approaching, transition zones, such areas are particularly noticeable, and here the concept is especially useful for interpreting relationships of climaxes, both present and past.

Since communities such as these exist to some extent in every climatic area, they must be recognized. Use of preclimax and postclimax is, therefore, essential to the application of monoclimax theory.

Recognition of Formations and Associations. It has been mentioned that the major climax regions (formations) are fairly obvious because each has its distinctive physiognomy. This makes for clear demarcation so that their number and limits are not a matter of controversy. But, additional criteria, both static and developmental, must be met to corroborate the apparent unity based on physiognomy.[82] For convenient reference these are briefly summarized below.

Static Criteria

1. Life form must be uniform throughout.
2. All associations must include one or more of the same or closely related species as dominants or subdominants.

Developmental Criteria

3. Late stages of succession must be essentially identical for a climax; and distinct from those of another climax.
4. Postclimax should show relationships to contiguous climax or subclimax.
5. Historical records as to composition and structure must conform to the modern picture.
   a. Recent historical—old records and land surveys.

b. Historical development reconstructed from pollen statistics.
c. Geological record, physical history, and fossils.

Associations are not always as obvious as the formation to which they belong. There are transitions from one association to another that may be confusing. There may be no more than small, often disturbed samples of climax vegetation available for study; because man's activities have eliminated all virgin stands, even though they once were widespread. Sometimes even these poor examples may be lacking, and then climax must be judged on the basis of successional knowledge. There may, therefore, be different interpretations, and errors are possible. To illustrate: It was generally believed for years that short grasses constitute the climax of the plains. But, added evidence has led many ecologists to conclude that mid-grasses are climax and short grasses are disclimax, maintained by modern grazing practices under conditions of periodic drought.

A climax association must, of course, conform to the criteria that delimit the formation of which it is a member. To check these criteria, it becomes necessary to know the successional trends of the vicinity in detail, to know the composition and structure of the postulated climax and subfinal stages of succession, and to distinguish preclimax and postclimax communities and habitats. Thus it becomes necessary to know something of related associations as well as the one involved. Finally, the history of the region, both recent and geological, is desirable for proper interpretation of observations.

The local climax examples of an association must be capable of maintaining themselves indefinitely under existing climatic conditions. They must occur throughout the area under average conditions, or the evidence from succession must indicate their potential presence. All successional trends must lead to such communities except those of the isolated edaphic and physiographic situations that end in postclimax or preclimax.

Polyclimax

In a polyclimax interpretation of vegetation the word climax is used simply in its ordinary connotation—the culmination of development. In this sense, it can be applied to every relatively stable, self-perpetuating community that terminates succession, and it does not imply a connection with climate. Adherents of polyclimax do not necessarily rule out successional trends and climax developing primarily under the control of climate, since their systems of classification may include climatic climax[394] as a category. However, those terminal communities that develop in habitats whose usually local features have a controlling influence are given equal status and are distinguished by an adjective referring to the special circumstances of their development.

Thus, edaphic climaxes are the result of soil conditions, usually differing from those of the area as a whole, and with physical or chemical characteristics that prohibit the possibility of soil development in the direction of a regional profile. Physiographic climaxes are a product of local environment associated with peculiarities of topography—bluffs, ravines, bogs, etc. Pyric climaxes are maintained by recurrent fires. Salt spray climax may occur in maritime situations where only the species tolerant of salt spray can survive. These are illustrative of the various climaxes that have been distinguished.

The polyclimax interpretation of vegetational development is not new. Early studies of vegetation in America were so strongly influenced by the philosophies of Clements and Cowles that polyclimax ideas received little attention. Nevertheless, even Cowles considered both climate and physiography as determining climaxes although emphasizing the convergence of successional trends to terminal communities. An early dissenter among the American minority not accepting monoclimax was Nichols.[274, 275] His ideas for recognizing and classifying both physiographic and climatic climaxes do not seem extreme today, but they were substantial contributions to polyclimax thinking. Others objected even more strongly.[133] Gleason[159, 160] maintained continuous opposition of another sort by arguing for the individualistic nature of plant associations and the consequent impracticality of monoclimax. The British, although concerned with community dynamics, under the leadership of Tansley,[393] nevertheless, maintained a general skepticism of monoclimax while strongly crediting climate with general control of vegetation. The Russians had similar attitudes. Continental Europeans, however, were frequently little concerned with succession. Their systems of classification, developed on a descriptive basis, led naturally to polyclimax interpretations when, and if, climax was considered.

Through the years, American support of monoclimax has dwindled as the opposition has become more positive. It is of interest, however, that not long ago an impersonal summary of the differences between the two interpretations was concluded with the statement that "many criticisms of the monoclimax hypothesis resulted from incomplete understanding of Clements' point of view" and that the conflict is largely one of terminology.[64] Considering that there is agreement on the characteristics of climax communities and that the same communities are being dealt with in either instance, this evaluation does not seem to be inappropriate.

Nevertheless, polyclimax theory is gaining favor. A recent comprehensive survey of climax concepts[429] serves to emphasize the trend toward polyclimax interpretations; but it also emphasizes that there is much diversity in the ideas developed in recent years. The detailed consideration of evidence for and criticism of monoclimax indicates that two general assumptions are the focus of most objections. First, convergence to identity is held to be theoretically questionable and contrary to the bulk of evidence. Second, it does not follow because different self-maintaining populations occur on different sites in an area that these must be parts or subdivisions of the climax association. These are, fundamentally, the same bases upon which objections to monoclimax have always centered. With elaboration and time they have become more effective in themselves but certain other lines of thought have speeded the trend from monoclimax.

The discussion thus far has assumed that the association is a recognizable and definite entity, classifiable by a monoclimax or polyclimax system. Because of his belief in the individualistic nature of associations of species, Gleason[160] with little support in America but some elsewhere, contended that such classifications are impractical because every stand differs from every other stand. More recently these ideas have gained support from floristic geographers.[67, 163, 260] Their general thesis is that the genetic character of each species limits its range of environmental tolerance (or ecological amplitude) and that every environment has its own biotic potential. Thus, every example of vegetation results from the coincidence of biotic potential of the environment with the genetic potential of whatever species are available. Regionally, this leads to an infinite number of differing species-environment complexes of which none are strictly identical and some may differ widely. With a different approach, the situation has been described as a "mosaic of plant communities whose distribution is determined by a corresponding mosaic of habitats."[395]

Such generalizations, however true, provide no basis for describing the vegetation as a whole or for grouping any of its multiple related parts. The expression, *prevailing climax*, has been used to refer to the growth form or population occupying the majority of sites in an area[428] with no implication of monoclimax and permitting characterization of average environment and vegetation. A similar point of view was used in clarifying the problem of life-zones in mountains.[122] Related climax communities of wide distribution were considered to be the only sound indicators of climate; and the area of their occurrence was, therefore, termed a *vegetation zone.* Climax communities of local occurrence only and existing "as islands in the matrix of zonal vegetation" were termed *interzonal.* These are associated with habitats that inhibit zonal vegetation and are not restricted to one zone.

This general approach was effectively used in the study and classification of the complex of forest vegetation in northern Idaho.[124] Because the concepts and methods represent a conservative, practical application of modern polyclimax ideas, they will be outlined briefly.

It should be emphasized that the classification deals with climax vegetation of considerable diversity. It could not, therefore, be applied without a thorough knowledge of the general ecology of the region, its geological and recent history, the variability and extent of habitats, and particularly the successional relations in different habitats. Even with this broad background the final classification was not made until supporting quantitative data were obtained by sampling, particularly for constancy, a series of representative stands.

The union (or synusia), distinguished by floristic composition and not by physiognomy, was taken as the smallest structural unit and it was named after its dominant species. The association was used as the basic unit of classification. As used here, it embraces all unions superimposed on the same area, and each distinctive combination of vascular plant unions constitutes a separate association. Each is designated by a binomial. The *Thuja-Tsuga/Oplopanax* association signifies a dominant union typified by *Thuja* plus *Tsuga,* combined with a subordinate union typified by *Oplopanax.* This characterizes the association without reference to other less important unions that may be present.

This concept of the association embodies the characteristics of all the actual stands of a *habitat type* (variable as to factors but equivalent in effects on vegetation). Within the range of the study, differences in species composition of associations were attributable to such things as historical events and chance distribution rather than environments. However, it was recognized that widely separated, dissimilar stands might be shown to have connections through an intermediate series exhibiting continuous variation correlated with a gradual climatic gradient. It was concluded that such a unit might be indivisible and require associational recognition in spite of its diversity. This point is important to later discussion.

A closely related group of associations occupied, or potentially occupied, a *zone* that was characterized by its major dominants. Thus, the *Picea engelmannii-Abies lasiocarpa* zone includes three associations: *Picea-Abies/Menziesia, Picea-Abies/Xerophyllum,* and *Picea-Abies/Pachistima.* Each is characteristic of a habitat type.

In the above discussion, mention was made that associational rank might have to be given to climax vegetation with an indivisible gradient of variation between its extreme examples. Just such a situation was demonstrated for the physiognomically similar shadscale vegetation of the Nevada desert.[29] By plotting the ranges of the principal species it was shown that the extent of the vegetation type included only a small area of overlap of all species and extensive areas in which one or more species would be absent. Quantitative analysis of representative stands showed a variation in floristic composition throughout the range. Shadscale desert was therefore termed a vegetation zone in which a variable climax community of species with wide tolerances occupies zonal soils; while interzonal vegetation, consisting of species with narrower tolerances, occurs on interzonal soils as edaphic islands imbedded in the matrix. The zone is thus recognized as a climax entity in spite of its variation.

These ideas for community study recognize the individualistic nature of stands but accept the fact that practical considerations favor grouping the related ones into categories. Also, they suggest that variations in stands may appear as a mosaic or as a gradient of changing vegetational characteristics across the extent of a zone.

Recognition of such variation in climax vegetation has led to attempts to analyze and interpret it. This is further evidence of the changing attitudes of American ecologists toward vegetation studies.

In Wisconsin, detailed quantitative data were obtained for many stands of forest[114] and grassland.[112] By assigning weighted values (importance values) to each of the dominants, a summation index was computed for each stand. For both types of vegetation the indices formed a graded series or smooth curve of variability. Limited evidence showed a correlation with environmental factors. Each series was termed a vegetational *continuum,* in which the summation indices indicate the degree of difference among the stands and their positions in the continuum.

Such an analysis of communities in a successional series should give a similar gradation of values from pioneers to climax. Undoubtedly thinking along these lines has led to the objections that have been raised to the usual distinction made between climax and seral communities in vegetation study.

Yet another approach emphasizing continuous variability in the composition of vegetation considers it as a population within which there is a changing pattern associated with environmental gradients. By observing climax vegetation occurring along recognizable moisture gradients, a changing proportion and importance of species was demonstrated in coniferous forest of the Great Smoky Mountains.[430]

The accumulating evidence suggests that the combinations of dominants, both as to species and their relative importance, which may occur in a vegetation area, are potentially as numerous as the possibilities for minor variations of environment in the area. This leads to a questioning of the propriety of grouping these communities into units of any system of classification, the argument being that such discrete units do not exist and that it is improper to set up a system with categories whose limits are arbitrarily set and whose communities (never identical) must be selected to fit. Although such feelings have been held for some time, as yet they have inhibited but little the use of systems of classification that lump related communities. New quantitative evidence may result in more careful application of such systems, but it is not likely that their usage will decline abruptly. Present widely-used systems of classifying vegetation developed in spite of opposition and are so well established in the literature and habits of thinking that they will continue to be used. Regardless of personal opinions, every ecologist must be reasonably familiar with both the old and the new to appreciate what has been done and the work currently in progress.

General Procedure in Local Study. The desirability of familiarity with the area as a whole has been emphasized. Observation and note-taking should proceed at the same time that literature is searched to learn the historical aspects of the area, both geological and modern, and the relationships of its flora to that of surrounding areas. With continued observation, certain ideas will develop as to probable and possible successional relationships and the relative position of different habitats. Such methods alone have produced some excellent interpretations of vegetational dynamics. The general conclusions may be as good as those obtained by more detailed studies. However, there are reasons why supporting data are most desirable.

It is often possible for honest observation to be wrong, and only quantitative and qualitative data will demonstrate the discrepancies. Again, such data may bring to light pertinent information that could not be realized by observation alone. When questions of "why," "when," or "how" come up, they can be most satisfactorily answered with absolute data.

These things were soon realized by some early students of community dynamics, and quadrat methods were introduced as a part of their procedure. Early methods of sampling, however, were rarely adequate and it is unfortunate that they were not improved as rapidly as they should have been. Perhaps students of community dynamics were too much concerned with an overall picture rather than detail. As a result, much desirable information was not obtained and now may not be available because vegetation has been destroyed.

Phytosociological Methods in Studies of Succession. Because, from the first, European ecologists were interested in classification of vegetation units, they developed various sampling procedures that gave data for evaluating the contribution of species to community composition and structure and for the recognition and delimiting of communities. Objectives and uses of such procedures were outlined in our discussion of analysis and description of plant communities. Although developed for descriptive purposes such phytosociological methods may be applied with equal effectiveness in studies of succession[26, 280] or interpretation of climax.[124] In recent years studies of community dynamics have almost invariably incorporated phytosociological concepts and methods to great advantage. At the same time, systems of sampling have been improved both for greater statistical significance and for the time they require. The observational approach followed by description is a thing of the past. Supporting quantitative data are now a standard requirement. This is all to the good for, in addition to putting on record the sociological characteristics of the various communities, the data serve for substantiating observations, and as proof of conclusions.

Finally, it should be emphasized that sampling techniques and

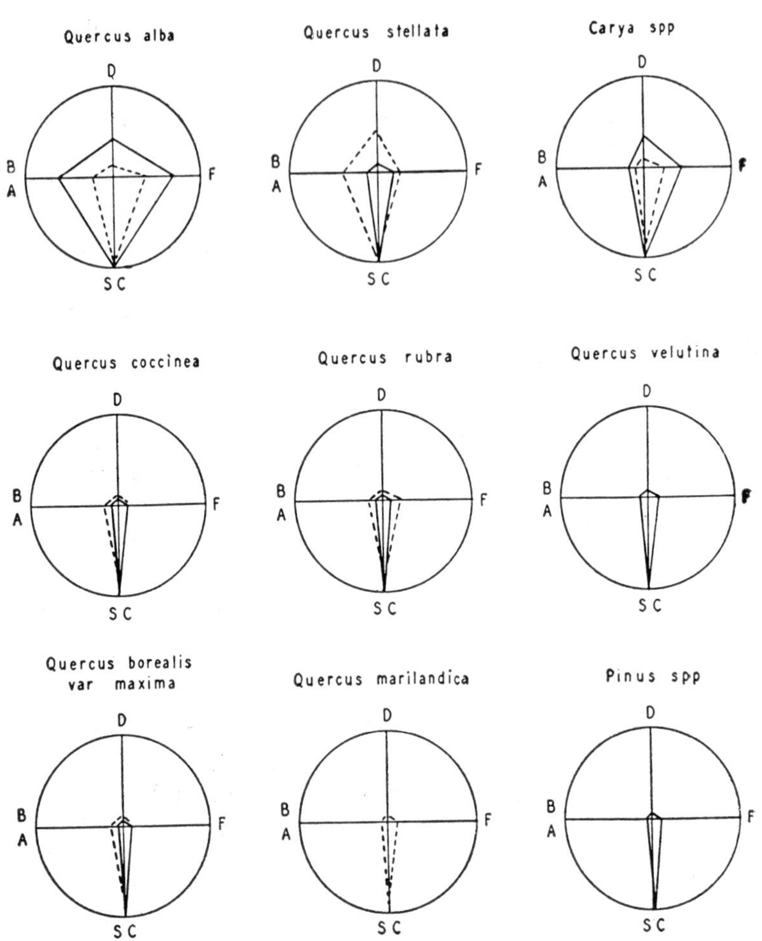

Fig. 116. *A phytographic comparison of the overstory species found in the two oak-hickory climax variants of the North Carolina Piedmont.*[280] *Values for the white oak type are indicated by solid lines, for the post oak type by broken lines. D—percent of total tree density, F—frequency percent, SC—percent of four size classes (overstory, understory, transgressives, seedlings) in which the species was found. BA—percent of total basal area. Zero is the center, 100 percent the periphery of the circle. Only quantitative data can give information such as illustrated by these phytographs.*

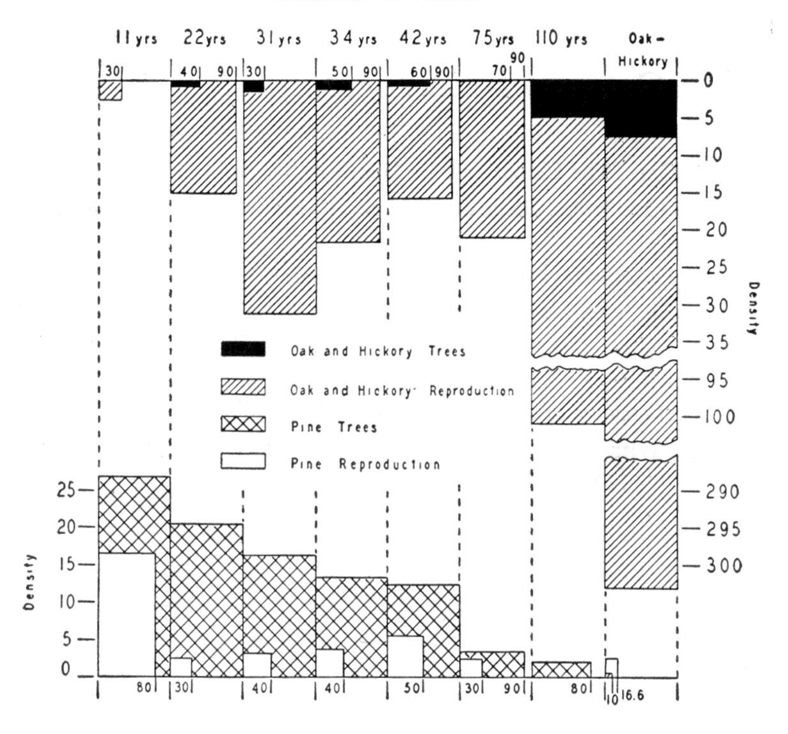

Fig. 117. *Relationships of trees and reproduction of pine and oak-hickory in old-field succession in North Carolina as shown by their density and frequency in successive ages of pine dominance leading to oak-hickory climax. Frequency is indicated by width of columns, density by height. Such phytosociological representations clarify relationships that might otherwise go unrecognized.*

phytosociological analyses are not a substitute for good judgment on the part of the investigator. All vegetation cannot be sampled; and the results, therefore, are meaningful and correct only for the stands he has selected for detailed study.

General References

S. A. CAIN. The Climax and Its Complexities.

F. E. CLEMENTS. *Plant Succession: An Analysis of the Development of Vegetation.*

F. E. CLEMENTS. Nature and Structure of the Climax.

W. S. COOPER. The Fundamentals of Vegetational Change.

J. PHILLIPS. Succession, Development, the Climax, and the Complex Organism: An Analysis of Concepts.

A. G. TANSLEY. The Use and Abuse of Vegetational Concepts and Terms.

R. H. WHITTAKER. A Consideration of Climax Theory: the Climax as a Population and Pattern.

Chapter 11 The Distribution of Climax Communities: Present Distribution of Climaxes

In the early nineteenth century, Humboldt drew attention to the importance of climate in determining the distribution and range of species, and Grisebach showed the possibilities of using communities, instead of species, as units of study. These were the beginnings of modern descriptive plant geography, which deals with the extent and distribution of vegetation types, particularly climaxes, and the reasons they occur where they do. The complex nature of climate necessitated from the first separate consideration of its components, and this led to oversimplified explanations of plant distribution based upon single factors. Even Warming,[419] to whom we are indebted for shaping the foundations of much of our modern ecological philosophy, was confident that communities and their responses are primarily controlled by water. Among the early geographers, Schimper[345] deserves special mention because he emphasized what is now generally recognized, namely, that a complex of interacting factors determines vegetation. There is still no simple means of expressing the effectiveness of the complex.

Merriam's[264] attempt to correlate all vegetational distribution with temperature is illustrative of the search for a single factor whose quantitative value would express climatic conditions. He showed that zones with similar summer temperature characteristics frequently have similar vegetation, but, unfortunately, he assumed that because there was a correlation there must also be a cause and effect relationship. His generalizations are, therefore, not acceptable, and too many exceptions remain unexplained.

A persistent search was made by Livingston, Shreve and associates[242] for a single quantitative value of physiological significance, which, when plotted to indicate isoclimatic lines, would closely match the distributions of major vegetation types. Summer evaporation rates, temperature coefficients, and temperature indices based upon physiological responses all were tried. The most successfully applicable value they found[239] was one that combined a physiological temperature index, precipitation, and evaporation. Actually this was a refinement of the precipitation:evaporation ratio proposed earlier,[405] but it is scarcely more useful. These and other studies serve to emphasize the complexity of plant-environmental relationships and the impracticality of expressing them as a function of a single variable. This becomes even more obvious when influences such as length of day, winter temperatures, and the season of precipitation are considered.

Climax Regions of North America

The vegetation maps available for North America[367, 374, 422] serve to emphasize by their similarities that the major vegetation types are fairly obvious; but their differences in detail indicate disagreement on the interpretations of climax relationships, especially within formations. An understanding of the bases for different interpretations can best be obtained by study of the many papers dealing with local investigations of vegetation. There are, however, several of a more comprehensive nature,[178, 367, 368] which give more detail than can be presented here.

The concept of climax formations and associations was discussed earlier (Chap. 10). This is a system for classifying vegetation locally and especially regionally, with growth form as its apparent major basis. The floral components become important only when the lesser categories or units of vegetation are considered. In addition to physiognomy, it is based upon community dynamics in respect to present day successional and climatic relationships, and it also permits interpretation of the past history of vegetation associated with changes of climate as well as any suggestion of such changes apparently now in progress.

The system thus has many advantages, especially when dealing

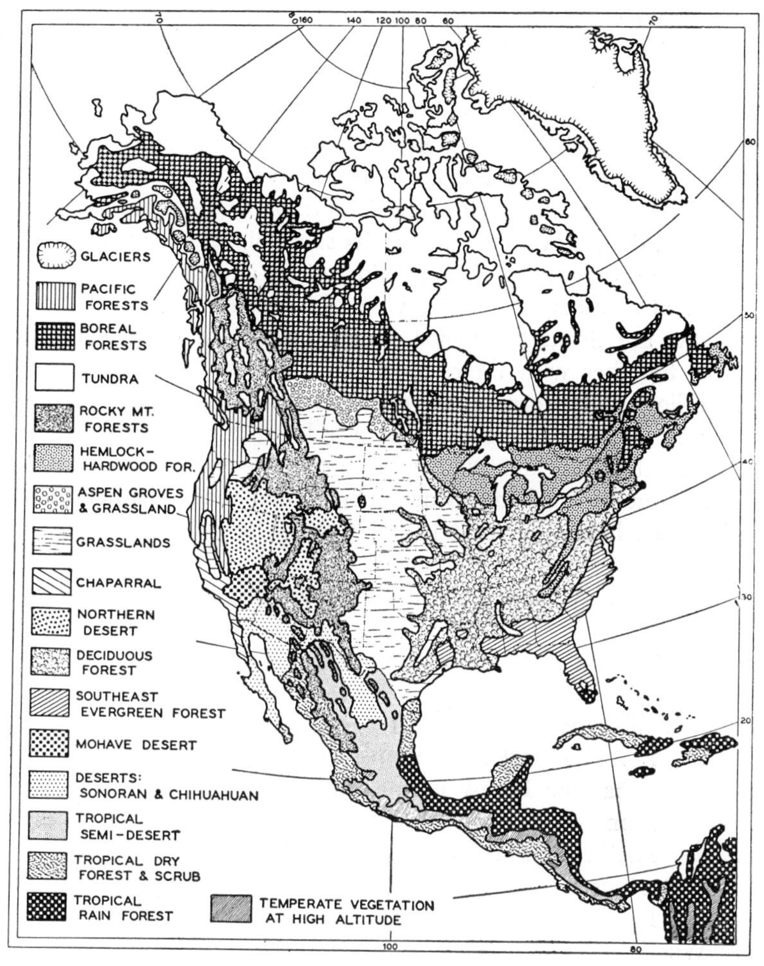

Fig. 118. *General ranges of the principal vegetation types of North America.*—Adapted from a privately printed map by Transeau, 1948

with regional vegetation types. It assumes, in basic form, an acceptance of monoclimax theory, and accordingly, postclimax and preclimax communities, which are not recognized by polyclimax adherents. Much of the literature on North American vegetation makes use of the system, as such or with modifications. For this reason it will be followed in this presentation intended for general use. For consistency, the discussion that follows will treat formations, associations, and some lesser subdivisions on a traditional basis throughout. However, the terminology sometimes used for special situations will be so minimized that reinterpretation or reclassification in terms of some other system should not be difficult.

The discussion will emphasize dominants, primarily in terms of climax. Outstanding successional trends will be mentioned, particularly when they result in conspicuous or widespread subclimaxes. However, details of local and special successional phenomena will not be introduced because of space limitations and the general objective of presenting a comprehensive view of North American vegetation.

Below are listed the major climax formations of North America. These, together with their associations, are discussed in the section that follows. The formations restricted to the mountains of the west occur in altitudinal zones, whose relationships should be clearly understood. Consequently, discussion of these zonal formations is centered about each of the principal mountain ranges rather than considering each zone separately, throughout its extent.

Climax Formations of North America

- Tundra
  - Tundra Formation
- Forest
  - Coniferous
    - Boreal Forest Formation
    - Subalpine Forest Formation
    - Montane Forest Formation
    - Pacific Coastal Forest Formation
  - Deciduous
    - Deciduous Forest Formation
- Woodland and Scrub
  - Broad-Sclerophyll Formation
  - Southwestern Woodland Formation
  - Cold Desert Formation
  - Warm Desert Formation
- Grassland
  - Grassland Formation
- Tropical Formations

Tundra Formation. Tundra lies between the northern limit of trees and the area of perpetual ice and snow in the far north, or above timber line in high mountains. In North America, it forms a broad band completely across the continent, and it also occupies the narrow low coastal area around most of the periphery of Greenland. It occurs on mountains as far south as Mexico if their elevation is sufficient to produce a timber line. Thus it is limited in its northern or upward extent by ice and bounded on the southern or lower margin by boreal or subalpine conifer forest.

Vegetation is low, dwarfed, and often matlike, and includes a high proportion of grasses and sedges. Even the woody plants, including willows and birches, are usually prostrate. The herbs are mostly perennial and of a rosette type, producing relatively large flowers, often with conspicuous colors. Mosses and lichens may grow anywhere and in favorable habitats form a thick carpet with the low herbs. The number of species is small compared with floras of temperate climates; and, even within the tundra, the number decreases northward. Most of the genera and many of the species are to be found throughout the Northern Hemisphere wherever tundra occurs.

The uniformity of the flora is undoubtedly related to the peculiarities of environment. The growing season is short and its temperatures are relatively low. Permafrost in the soil and its depth of thawing in summer is of great importance as is the extreme instability of soil, related to frost action.[378] Light is continuous throughout the growing season in the arctic, and is intense and high in ultraviolet rays in alpine habitats. Precipitation is largely in the form of snow and varies greatly. Drying summer winds, which are characteristic, produce high rates of evaporation and transpiration. As a result, water is often a critical factor, especially inland away from moist coasts. Local marked differences in vegetation are commonly related to minor variations in topography and the differences they produce in drainage and retention of snow. The poor, haphazard drainage associated with new topography is apparent everywhere.

Arctic Tundra. Although the flora of the tundra is fairly well known, its communities and their successional relationships have not been sufficiently studied.[317] In contrast with temperate vegetation, many species may occur in any type of habitat, and several that appear to be climax may also be pioneers in the newest of habitats. Even climax is not agreed upon, possibly because observations have been made in widely separated areas. Variations do exist, as has been shown for eastern Canada.[309] It is also possible that the climax criteria used for temperate climates are not always applicable in the arctic. Interpreted in terms of Greenland vegetation, Cassiope heath appears to be climax, and a Sedge-Dryas dominated community, of equal extent but on drier sites, is preclimax.[282] Two subclimaxes are frequent. Any habitat with sufficient moisture, whether it be pond margin, seepage area, or boggy ground, eventually is covered with a thick moss mat supporting several herbs, of which cotton grass (*Eriophorum* spp.) is most conspicuous. Xerarch succession on rock exposures eventually results in a lichen-moss mat, which may continue almost indefinitely.

Important climax dominants are *Cassiope tetragona,* one or more species of *Vaccinium, Arctostaphylos alpina, Empetrum nigrum, Andromeda polifolia, Ledum palustre, Rhododendron lapponicum,* and species of *Betula* and *Salix.* These and other species occur in varying combinations and degrees of importance.

Practically all habitats support some of the many species of *Carex,* of which the commonest include *C. capillaris, C. nardina,* and *C. rupestris.* The preclimax sedge community invariably includes *Elyna bellardi* in abundance. Some grow in mats, some are in clumps, but all are dwarfed. The same can be said for the grasses, which, although relatively abundant and widespread, are restricted to a few genera, of which *Festuca* and *Poa* are especially well represented. Many of the conspicuous herbs previously mentioned are included in the numerous species of one of the following genera: *Saxifraga, Potentilla, Ranunculus, Draba, Cerastium, Silene, Lychnis, Stellaria,* and *Pedicularis.* Conspicuous and widespread species typical of tundra are *Oxyria digyna, Papaver* spp., *Dryas octopetala,* and *Epilobium latifolium.*

Alpine Tundra. Mountains high enough to have a timber line support tundra, whose upward extent is limited by the snow line. In the east, as a consequence, tundra is found only on a few high peaks in New England. Farther south, the Appalachians are not of sufficient height to support tundra. That on Mt. Washington is representative of the type and is essentially similar to the not far distant arctic vegetation.

Alpine tundra in the western mountains mostly lies far to the south of the arctic and, consequently, is found at high altitudes only. In the Canadian mountains, it is found as low as 6,000 feet, but southward its lowest elevation steadily increases some 360 feet per degree of latitude to 30° north latitude and then, inexplicably, declines very gradually to the equator.[125] In the central Rocky Mountains, tundra is well developed between 11,000 and 14,000 feet. In the Sierra Nevada, where the snow line is lower, tundra lies mostly between 10,500 and 13,000 feet. In general, it is lower on coastal than interior ranges and on the wetter sides of mountains.

When climate changed and terminated the glacial period, vegetation similar to modern tundra must have followed the ice as it receded northward. This left only these high peaks and ridges where tundra could survive as relicts. The relict vegetation obviously belongs to the Tundra Formation because of the growth form and the duplication of characteristc genera as well as many species. The greater importance of grasses and the presence of numerous endemics in the

Fig. 119. *Alpine tundra in the Colorado Rockies.*—U. S. Forest Service.

western mountains suggest that these Alpine tundras might be classed as associations of the Tundra Formation.

Boreal Forest Formation. This great forest, often called "taiga" in its northern extent, spans the continent in a broad band to the south of the tundra. Along the Atlantic coast it extends from Newfoundland on the north to the New England states on the south. Westward, the southern boundary touches the great lakes region, trends northwestward across Saskatchewan and along the Rocky Mountains, and then to the Pacific coast only at Cook Inlet in Alaska. The band is, therefore, narrowed abruptly in the far west, although it extends much farther to the north there than it does over much of the continent.

Climate is scarcely less severe than that of the tundra. The short growing season from June through August is cool, and winters are very cold. Precipitation is moderate, averaging perhaps 20 inches, but on the east coast it may be 40 inches, and in the interior of Alaska, only 10-12 inches. The precipitation:evaporation ratio is, however, favorable because of the low temperatures. The topography is almost entirely that produced by glaciation. Lakes are scattered everywhere, and many of them have filled to form extensive bogs or muskegs. The mineral soils are either thin and residual, overlying the rock masses exposed by glaciation or, along the southern boundary, deep moraine and outwash. All are immature and often poorly drained. Subsoils, in the bogs especially, may not be frost-free even in midsummer.

Climax. The climax forest of white spruce and balsam fir is best developed in and about the St. Lawrence River valley, where the

Fig. 120. *Interior of boreal white spruce-balsam fir forest as it appears in northern Michigan*—U. S. Forest Service.

trees reach maximum size and grow in close stands under a variety of conditions. Here, and over much of the range, *Picea glauca* and *Abies balsamea* form dense stands under whose canopy there are relatively few dependent or secondary species. Paper birch (*Betula papyrifera*) is a constant associate, although it is successional after fire or disturbance and often occurs as subclimax in pure stands. Characteristic tall shrubs are *Viburnum alnifolium* and *V. cassinoides.* Typical lesser plants on the shady forest floor are *Aster acuminatus, Dryopteris spinulosa* var. *americana, Oxalis montana, Clintonia borealis, Cornus canadensis, Maianthemum canadense, Aralia nudicaulis, Coptis groenlandica,* and *Gaultheria hispidula.*

With increasing distance from the St. Lawrence center, both westward and northward, the number of species declines; and southward in New England, *Picea rubens* replaces *P. glauca.* Balsam fir is completely absent along the northern boundary and in most of the western range of the type. Beyond the range of fir, the subclimax species, otherwise found in bogs or on burned areas, often appear with white spruce or replace it as climax. Pure dominance of black spruce (*Picea mariana*) is common north of Lake Superior and westward. Along the northern and far western transition[269] tamarack (*Larix laricina*) may take an essentially climax position as does the black spruce, especially on high rocky ground. Both are bog species farther south. To the west, paper birch and jack pine (*Pinus banksiana*) have climax characteristics, although both are definitely subclimax nearer the center.

Successions. Primary succession occurs mainly on bare rock or in lakes.[95] The former is initiated by xerophytic mosses and lichens, which, after mat formation, lead to a heath mat stage. In the western part of the range, this is followed by the xerophytic jack pine, or black spruce to form a subclimax; but eastward white spruce-balsam fir may come in directly. Jack pine also occupies extensive areas of sand plains and gravelly soils.

Bog succession is everywhere apparent in the many lakes that are filling up. The usual submerged and floating-leaved aquatics are commonly followed by sedges and grasses, which may form a floating mat upon which a bog-shrub stage develops. This may include *Chamaedaphne calyculata, Andromeda glaucophila, Alnus incana, Ledum groenlandicum,* and *Vaccinium* spp. Larch is the commonest tree to come in after shrubs, followed by black spruce or, in less acid bogs, sometimes *Thuja occidentalis.* Any of these species may maintain their dominance for long periods, but they can be superseded by climax.[118]

Secondary succession is often caused by fire. If the burn is so severe that all humus is consumed, leaving bare rock, primary succession may be repeated. If a dry peat bog burns, it usually fills with water again,

Fig. 121. *Typical stand of jack pine* (Pinus banksiana) *on sand or gravel soils in northern Michigan.*—U. S. Forest Service.

Fig. 122. *Aspen stand* (Populus tremuloides) *at forty-five years of age in northern Minnesota. Its successional nature is clearly shown by the well-developed understory of spruce and fir.*—U. S. Forest Service.

and succession is reinstated at the aquatic stage. More often a burn results in pure stands of paper birch, which eventually give way to climax. Wind throw and lumbering of climax stands may also result in birch or aspen dominance but sometimes are followed directly by climax species.

Transitions. The northern border is irregular, depending upon topography. Forest extends far into the tundra in sheltered valleys, and tundra appears on the high ridges well within the forest area. Timber line seems definitely to be advancing in Alaska,[169] but evidences are strongly suggestive of retreat in eastern Canada. The southern transition is to deciduous forest in the east and to grassland in the west. From New England to Minnesota, the transition is marked by pure stands of white pine (*Pinus strobus*), a subclimax of long duration. In the lake states red pine (*P. resinosa*) and jack pine may also occupy similar positions on less favorable sites. Scattered individuals of white pine especially tend to persist well into the climax. Through much of the eastern transition, spruce, fir, and hardwoods may grow in mixture or in alternating stands.[274] The transition to grassland in the Middle West is marked by aspen (*Populus tremuloides*)[268] in a band some 50 miles wide. In spite of fluctuations produced by fire, grazing, and drought, the trees persist and, in some instances, seem to have advanced into the grassland. In the west, along the Rockies, the subalpine *Abies lasiocarpa* is associated with *Picea glauca*, and northward in Alaska there is a merging with the northwestern coastal forest.

Appalachian Extension. On the higher mountains of the Appalachian system, the northern conifer forest extends as far south as the Great Smoky Mountains of North Carolina. The growth form and associated species are in every way similar to the main body of the formation; but, from New Brunswick southward into New England, red spruce (*Picea rubens*) tends increasingly to replace white spruce. Still farther south, Fraser fir (*Abies fraseri*) takes the place of balsam fir so that the dominants in the southern Appalachians are ecologically equivalent to those elsewhere in the formation but are taxonomically distinct. It seems reasonable to consider the Appalachian extension as a distinct association whose limits are marked by *Picea rubens*. A northern and southern faciation are suggested by the presence of *Abies balsamea* and *Betula papyrifera* in the north but the substitution for them in the south of *Abies fraseri* and *Betula alleghaniensis*.[289]

The compensating effect of latitude is apparent in the altitudinal limits of the association, which increase southward. In the northern range of red spruce, it may be found anywhere, as is true of fir. Southward, the approximate lower limit of spruce-fir forest on Mt. Katahdin is 500 feet; in the White Mountains, about 2,500 feet; in the Adirondack Mountains, 3,000 feet; in the Catskills, 3,500 feet; and in the

Fig. 123. *Interior of red spruce-Fraser fir forest in the southern Appalachians. Compare with Fig. 120.*—U. S. Forest Service.

Great Smoky Mountains, almost 5,000 feet. At the highest elevations in the southern range spruce drops out entirely to leave fir the dominant.

Deciduous Forest Formation.[47] This formation occupies all of the eastern United States except southern Florida. Its northern transition to conifer forest extends into Canada along a line from northern Minnesota to Maine. On the west, forest gives way to grassland as precipitation:evaporation ratios become less favorable. The irregular line of transition runs northward from eastern Texas, with 35 inches of precipitation, to central Minnesota where precipitation falls to 25 inches.

The great extent of the deciduous forest includes soils and topography of diverse nature and origin. The northern portion was glaciated. There are mountains in the east. The great valleys of the Mississippi and Ohio Rivers are included as are the Piedmont Plateau and coastal plain of the Atlantic and Gulf coasts. Any and all kinds of topography as well as soil types are, therefore, represented.

Climate is temperate with distinct summer and winter, and all parts are subject to frost, one of the few environmental factors that applies

Fig. 124. *Mixed hardwood forest in Indiana. Large trees are white oaks.* —U. S. Forest Service.

throughout. Precipitation varies from 60 inches in the southern mountains to less than 30 inches northwestward, but it is everywhere fairly well distributed throughout the year. The ratio to evaporation is most favorable in the north, the east, and in the mountains and becomes decreasingly favorable approaching the transition to prairie.

The southern Appalachians represent the oldest exposed land surface in the region. Here the deciduous forest is more complex than in any other part. Practically all of the species found elsewhere in the deciduous forest are represented with several extra species. Numerous endemics occur as associates. Most of the trees also attain their greatest size here. Away from the mountains, the number of species declines, and habitat requirements become of increasing importance. It is believed that a forest similar to the present one has existed here since Tertiary time. Such evidence is taken to mean that the southern Appalachians are a center of dispersal for much of the widespread deciduous forest. The distribution and nature of the several associations of the formation give additional supporting evidence. In general, with increasing distance from the center, the associations are made up of fewer species and yet all are bound together or interrelated by several species that range throughout.

Mixed Mesophytic Forest Association. Throughout the Appalachian and Cumberland plateaus, the numerous species of this climax grow in varying combination. *Fagus grandifolia, Aesculus octandra, Magnolia acuminata, Tilia* spp., *Liriodendron tulipifera, Acer saccharum, Quercus alba,* and *Tsuga canadensis* are the most abundant trees, but there are 20 or 25 other species, any of which may have climax status. The differing sensitivity of the species to minor variations in environment result in their occurrence in all kinds of combinations, which may be referred to as association-segregates.[45] The best indicators of the association are large trees of basswood (*Tilia heterophylla*) or buckeye (*Aesculus octandra*).

The association prevails in the Cumberland and southern Allegheny mountains and in the adjacent Cumberland and Allegheny plateaus.[46] Away from this center, there is a progressively increasing tendency toward restriction to the most favorable habitats. To the south, the association is seldom found except in the moist coves of the high Appalachians. To the west, southwest, and east it is found only in ravines and deep valleys. To the northwest, it is represented in southern Ohio by a mixed hardwood forest of far fewer species.

Beech-Maple Association. The northward extension of the mixed mesophytic forest shows an increasing importance of beech (*Fagus grandifolia*) and sugar maple (*Acer saccharum*). North of the boundary of Wisconsin glaciation, they are the climax species over an area west of the Alleghenies from New York to Ohio and up into Wisconsin.[44] Virgin forest in Michigan showed beech predominating

Fig. 125. *Sugar maples (160-200 years old) in beech-maple forest association, Pennsylvania.*—U. S. Forest Service.

Fig. 126. *Sugar maple-basswood forest, illustrating the climax for much of southern Wisconsin and Minnesota.*—U. S. Forest Service.

over maple, and associates included red maple (*A. rubrum*), elm (*Ulmus americana*), northern red oak (*Quercus rubra*) and black cherry (*Prunus serotina*).[61] The original forests of southwest Michigan,[210] northern Indiana[311] and Ohio,[362] as reconstructed from land survey records, were beech-maple on good sites and oak-hickory on coarse soils with poor moisture conditions. This conforms with present conditions and can be interpreted as climax and preclimax. Nearing

Fig. 127. *Seventy-year-old jack pine with a strong understory of balsam, indicating the trend that succession may take in the Lake States region.*—U. S. Forest Service.

the margin of the prairie that extended into northwestern Indiana, the proportion of beech and maple in the forest declined rapidly.

Maple-Basswood Association. The natural range of beech does not extend to the northwest limits of the deciduous formation. Beech is replaced in the climax by basswood (*Tilia americana*), beginning in Wisconsin and continuing into Minnesota.[139] Otherwise the community is changed very little. Oak and hickory occur on rugged topography or poor sites and predominate along the prairie margin.

Hemlock-Hardwoods Association. Between the northern coniferous forest and the deciduous forest lies a transitional association of which hemlock (*Tsuga canadensis*) is an important and constant member, together with beech and sugar maple, and, in lesser numbers, yellow birch (*Betula alleghaniensis*), white pine, basswood, elm, white ash (*Fraxinus americana*), red oak, and other species. Although hemlock is less frequent toward the west the association extends from northwestern Minnesota through the Lake States to Nova Scotia. It has been given various names by authorities with different points of view. This is the

Fig. 128. *Virgin white pine* (Pinus strobus) *forest in Connecticut, of the type that once occurred over wide areas in the northeast.*—U. S. Forest Service.

area throughout which occurred the magnificent pine forests of the recent past—now mostly decimated by fire and lumbering. Where pine was dominant, *Pinus strobus* tended to occur on sites with more favorable moisture conditions than the sand plains and ridges occupied by *P. resinosa*. By some[422] these pure stands of pine have been considered as climax, but many more ecologists agree that the pines are successional species occupying inferior sites for long periods as subclimax. That white pine especially carries over into the hardwood climax[276] is undoubtedly true. Its long life and relatively low numbers suggest that these trees in the climax should be regarded as relicts even though they can maintain their numbers by reproduction under openings appearing in the hardwood canopy.[247]

Postclimax forests of the northern conifers—tamarack, black spruce, white cedar (*Thuja occidentalis*)—occupy the many bogs throughout the area. The extensive areas denuded by lumbering and fire are today largely occupied by second-growth forests of aspen or pine.

Fig. 129. *Virgin hemlock* (Tsuga canadensis) *as it once occurred in the hemlock-hardwoods association of the northeast and in mountain coves southward.*—U. S. Forest Service.

Oak-Chestnut Association. As the mixed mesophytic forest becomes restricted to special habitats to the east and southeast of its center, the slopes and uplands are occupied by what was, until recently, oak-chestnut forest. The almost complete elimination of chestnut (*Castanea dentata*) by blight has affected practically all of the original forest that extended along the mountains from southern New England to Georgia. Decline of chestnut was gradual enough so that associated species maintained dominance and excluded invasion by others. Chestnut oak (*Quercus prinus*) and scarlet oak (*Q. coccinea*) are important species everywhere with tulip poplar, red and white oaks, and hickory as common associates. Evidence points to increases of northern red oak[204] and tulip poplar[273] among the dominants and of red maple as a secondary tree, but equilibrium has not yet been reached. Judging by the northern forests where the blight first struck, no major changes are to be expected soon. None of this association remains in its original state today, for the remnants un-

Fig. 130. *The oak-chestnut forest that once occupied the lower slopes of much of the Appalachian system.*—U. S. Forest Service.

touched by extensive lumbering operations have been modified by the ravages of chestnut blight.

Pitch pine (*Pinus rigida*) is the important successional species throughout the range, but southward shortleaf and Virginia pine (*P. echinata, P. virginiana*) increase and white pine is locally abundant.

In its southern extent, the association is restricted to the mountains, occupying most of the favorable slopes. Northward it is found on progressively lower sites, occurring as far east as Long Island.[90] Through the foothills of the mountains, it grades into the oak-hickory climax of the bordering Piedmont Plateau.

Oak-Hickory Association. In all directions from the deciduous forest center, except northward along the mountains, precipitation decreases and becomes less effective. This results in dominance by the drought-resistant oak-hickory association, which consequently occurs as a fringe around all the margin of the formation except toward the north. Oak-hickory climax ranges through much of the Piedmont Plateau and the Atlantic and Gulf states coastal plain in an arc that widens westward to eastern Texas. North from eastern Texas and Oklahoma it is commonly savannah-like where it grades into prairie, but it is more or less continuous to western Minnesota.

Fig. 131. *Savannah-like transition from deciduous forest to grassland. Bur oak predominates in these scrubby clumps of trees on the Anoka sand plain northwest of Minneapolis. Note blowout in sand dune in process of restabilization by* Hudsonia.—Photo by W. S. Cooper.

Northwest of the Appalachian center, in unglaciated parts of Ohio and Indiana, oak and hickory occur in combination with numerous other species, forming a mixed mesophytic forest climax, which suggests, by its similarity, that the mixed mesophytic association may still be expanding its range. Throughout the association, various combinations of oak-hickory may occur as preclimax. Illustrative is the post oak (*Q. stellata*)—blackjack oak (*Q. marilandica*) community which occurs widely on poor sites and dry exposures. In Texas and Oklahoma the species occur in open, savannah-like stands, known as the Cross Timbers,[138] which mark the transition to grassland. Postclimax communities of mixed forest may occur within the oak-hickory area on sites, such as old flood plains, where moisture may be exceptionally favorable.[280] Beech, sugar maple, willow oak (*Quercus phellos*), overcup oak (*Q. lyrata*), swamp chestnut oak (*Q. michauxii*), and shagbark hickories are indicator species.

The dominants of oak-hickory forest are not the same throughout its extensive range, but several species occur consistently. *Quercus alba, Q. rubra, Q. velutina, Q. stellata, Q. marilandica, Carya cordiformis, C. ovata, C. tomentosa,* and *C. laciniosa* are species that may be found in the climax anywhere. Typical subordinate species are sourwood (*Oxydendrum arboreum*), dogwood (*Cornus florida*), black gum (*Nyssa sylvatica*), and sweet gum (*Liquidambar styraciflua*). Other oaks and hickories with more restricted ranges may be in association and produce local variations. Shingle oak (*Q. imbricaria*), not so important in the east, should be added for the western forest from Arkansas and eastern Oklahoma[53] northward.[4] Bur oak (*Q. macrocarpa*) is the characteristic tree of the sometimes extensive savannah-like transition from forest to grassland, as well as along the rivers in the prairie, from Texas to Minnesota.

Because of the amount of abandoned land throughout the eastern and southern range of the association, old field succession is particularly noticeable, and subclimax pine stands are conspicuous (see Figs. 115 and 117). Virginia pine (*Pinus virginiana*) predominates in the northern Piedmont, but southward and westward shortleaf (*P. echinata*) and loblolly pine (*P. taeda*), usually in pure stands, precede the climax in secondary succession on uplands. Successional trees in lowlands are sweet gum, tulip poplar (*Liriodendron tulipifera*), sycamore (*Platanus occidentalis*), river birch (*Betula nigra*), red maple, elms (*Ulmus* spp.), ash (*Fraxinus* spp.) and hackberry (*Celtis* spp.).

Fire and Swamp Subclimaxes of the Coastal Plain. The coastal plain, once covered by the sea, extends from New Jersey down into Florida and along the Gulf to Texas as a low-lying, relatively level area, mostly overlayed with sandy soil. Drainage is poor, resulting in much swampy ground, but any raised area between streams is apt to be very dry for a part of each year. The height of the water table during the wet seasons and the amount of fire in dry seasons are fundamental factors in determining the nature of the vegetation.

From the pitch pine barrens of New Jersey through loblolly pine and longleaf and slash pine in the more southern states, fire maintains pine dominance, usually in open stands, called savannahs, with the highly combustible wire grass (*Aristida stricta*) a common ground cover. These stands owe their origin and maintenance to their resistance to fire.[75] If protected from fire, they would unquestionablv be replaced by oak-hickory dominated forest.[423] No extensive areas exist where fire has been excluded for more than a relatively few years. Thus, by some, fire is interpreted as a normal part of this environment

Fig. 132. *Typical longleaf pine savannah (Georgia) as maintained by almost annual burning. Note that the only apparent ground cover is wire grass* (Aristida), *which is an important factor in facilitating fire.* —U. S. Forest Service.

and pine as a pyric climax. The successional evidence is clear enough, however; and if fire is considered as being irregularly catastrophic, pine dominance in the coastal plain must be classed as a fire-maintained subclimax within the oak-hickory association.

A possible preclimax is the scrub oak-hickory forest found on sand dunes near the coast and inland. Turkey oak (*Quercus laevis*), sand post oak (*Q. stellata* var. *margaretta*), blue jack (*Q. incana*) and black jack oak (*Q. marilandica*) are dominants. Wire grass may be present, but often the sand is bare, glaring white in the sun, except for a few characteristic herbs. These include *Euphorbia ipecacuanhae,*

Fig. 133. *Slash pine savannah after protection from fire for only a few years. With continued protection, the pine will soon form a closed stand with shrubs and hardwoods forming an understory.*—U. S. Forest Service.

Fig. 134. *Scrubby, open oak forest (mostly* Q. laevis *and* Q. incana) *of the southeastern sandhills areas. The open stand and expanses of bare white sand are typical.*—Photo by H. L. Blomquist.

Jatropha stimulosa, Stipulicida setacea, Polygonella polygama and *Selaginella acanthonota.*[425]

Undrained, shallow depressions in savannahs form upland bogs or pocosins, sometimes acres in extent, in which evergreen shrubs predominate. *Ilex glabra, Myrica cerifera, Cyrilla racemiflora, Persea borbonia, Magnolia virginiana,* and *Gordonia lasianthus* are representative of the numerous tall shrubs or small trees. With them are usually a large number of ericaceous shrubs of smaller size. All are

Fig. 135. *A hammock near Sebring, Florida, as seen from a clearing. Conspicuous trees are* Quercus virginiana *and* Sabal palmetto.—Photo by W. S. Cooper.

commonly overgrown with lianas, of which *Smilax laurifolia* is most abundant. The presence of *Pinus serotina* in the bogs explains its name of pocosin or pond pine. Sphagnum is the usual ground cover.

It is at the margins of pocosins and in wet savannahs in North Carolina that the venus fly trap (*Dionaea muscipula*) is found, sometimes in great abundance but never continuously over an extensive area. With it several other insectivorous plants may occur. Species of *Sarracenia, Drosera,* and *Pinguicula* are common.

The hammocks of Florida, in contrast with pocosins, are mesic habitats raised somewhat above surrounding, usually wetter, areas. Over much of Florida their dominants suggest postclimax to oak-hickory, but toward the southern tip of the state, the species are more and more tropical.

Any shallow depression in the flatland of the lower coastal plain fills with water. Permanent standing water results in open marshes,[301] sometimes miles in extent, dominated by rushes and grasses. If flooding is not continuous, subclimax swamp forests develop. Bald cypress (*Taxodium distichum*), which dominates where water

Fig. 136. *Cypress swamp in the North Carolina coastal plain.*—U. S. Forest Service.

normally stands most of the year, occupies stream and lake margins or entire shallow lakes to the exclusion of other trees. Gum swamps are usually flooded only seasonally. *Nyssa biflora* and *Nyssa aquatica* are the important species,[174] with ash (*Fraxinus profunda, F. caroliniana*), bald cypress, and red maple as associates. The less the flooding, the greater the number of pocosin species that may be present.[21]

Still another forest of undrained areas is formed by *Chamaecyparis thyoides,* which occurs on peat bogs where it apparently becomes established only after fire occurs when the water table is high. Although the stands have subclimax characteristics, there is evidence that they may be succeeded by species characteristic of pocosins.[54] These valuable trees have been cut so systematically that they remain only as small sample stands or in relatively inaccessible places.[218]

Perhaps the most extensive bog and swamp forest still remaining in virgin condition are to be found in parts of the Dismal Swamp in Virginia[201] and in the Okefenokee Swamp[437] in Georgia.

Fig. 137. *Southern white cedar bog* (Chamaecyparis thyoides) *in New Jersey.*—U. S. Forest Service.

The plant communities of the banks and islands along the coast, as well as a narrow fringe of the coast itself, are distinctive enough to merit more discussion[283] than can be given them here. The effects of salt spray on vegetation were considered earlier (p. 143). Live oak (*Quercus virginiana*) is the most important tree of the forested areas, and the associated shrubs include *Myrica cerifera, Ilex vomitoria, Vaccinium arboreum,* and several others, mostly evergreens.[302, 424] Thus, this maritime climax forest is an evergreen variant of the oak-hickory association.

Fig. 138. *Maritime live oak forest* (Quercus virginiana) *on Smith's Island, N. C. Once a characteristic of the banks and islands of the south Atlantic and Gulf Coast, much of it has been destroyed because of neglect. Note the dunes at right, which were once forested.*—Photo by C. F. Korstian.

Rocky Mountain Forest Complex. Changes of environmental factors with altitude and the resulting zonation of vegetation on mountains have been discussed earlier (see Fig. 45 and related text). The great height of the Rocky Mountains provides conditions for a discontinuous alpine zone on the peaks, a subalpine zone, a montane zone, and a zone of woodland forest, which grades into the surrounding desert or grassland. These zones are recognizable by their distinctive vegetation over an area extending latitudinally from northern Alberta to the southern end of the Sierra Madre of northern Mexico and from the Black Hills of South Dakota on the east to the eastern foothills of the Sierra Nevada and the eastern slopes of the Cascades on the west.

Climaxes with so great an areal extent would be expected to vary somewhat in different parts of their ranges, especially as to associated species. The zones are not always continuous, nor are they always all present. Near the northern limits of the area, the lower zones run out and the upper zones are found at relatively low altitudes. Southward all zones are, of course, found at successively higher altitudes. Because the prevailing winds are from the west and carry with them oceanic climatic influences, the entire eastern slope of the Rocky Mountain system has different growing conditions from those of the west slope and, accordingly, differences in vegetation. Within the system, the individual ranges likewise have similar east-west slope differences. North and south exposures produce marked irregularities in zonation. Narrow valleys permit the dominants of one zone to extend downward into a lower zone, and high dry ridges allow upward, fingerlike projections of dominants into continuous higher zones. Cold air drainage locally causes marked disruption of the zonal pattern.

The factors operative in producing and controlling vegetation and its zonation in the Rockies have been studied in a number of localities just as there have been many local studies of the vegetation. A review and synthesis of all these investigations is available[120] with an extensive bibliography. What follows is largely an adaptation from this report.

Vegetation Zones. The zonal climaxes may be grouped as follows:

Alpine Zone
  Tundra Climax (discussed earlier—pp. 274-275)
Subalpine Zone
  Engelmann spruce—Alpine fir climax
Montane Zone
  Douglas fir climax
  Ponderosa pine climax
Foothills (Woodland) Zone
  Piñon-Juniper climax
  Oak-Mountain mahogany climax

Each of these types of vegetation extends as climax over an altitudinal range of about 2,000 feet where fully developed. The foothill zones narrow down and then disappear entirely northward, where the upper zones are found at progressively lower altitudes.

Near the upper and lower limits of a zone, the characteristic species are more and more restricted to special habitats. Upward, the climax species do best on ridges and south-facing slopes, which are warmer and drier than the general climate. Thus, in its upper transition area, each association shows its preclimax relationship to the climax of the next higher zone. At its lower limits, the association tends to be restricted to relatively moist, cool sites and extends into the next lower zone only in such habitats. It, therefore, holds a postclimax relationship to the climax below. Subalpine and alpine zones are colder and may be drier[232] than the zones below. If this is true, preclimax and postclimax relationships would be reversed above the montane zone. A mid-altitude zone of maximum precipitation does not occur everywhere. On the Wasatch Plateau of Utah a linear increase of 4.94 inches per 1,000 feet of elevation is reported between sagebrush and alpine zones,[243] and similar observations have been made elsewhere.

Subalpine Spruce-Fir Climax. From timber line downward for about two thousand feet, the climax forest is made up largely of

Fig. 139. *Virgin Engelmann Spruce* (Picea engelmannii), *with some alpine fir* (Abies lasiocarpa) *of the subalpine zone in Colorado.*—U. S. Forest Service.

Engelmann spruce (*Picea engelmannii*) and alpine fir (*Abies lasiocarpa*), which grow in dense stands. The spruce is the larger, longer-lived and more abundant tree.[292] In Arizona, New Mexico, and southward, *Abies lasiocarpa* var. *arizonica* is as important as *A. lasiocarpa.*

In Montana and northern Idaho, mountain hemlock (*Tsuga mertensiana*) is often found in the zone; and still farther north, approaching the merging with northern conifer forest, *Picea glauca* and *A. lasiocarpa* may grow in association.

Subordinate species vary far more than do the dominants. On the relatively dry eastern slope of the central Rockies, ground cover is sparse and made up largely of dwarf Vacciniums, while the moister

Fig. 140. *Dense aspen stand* (Populus tremuloides) *that came in after fire in the subalpine zone in New Mexico. Spruce reproduction underneath.*—U. S. Forest Service.

west slope has an abundance of bryophytes and herbs. Northward, the bryophytes increase until they practically cover the ground, and the vascular plants, both herbs and shrubs, also increase.

The most conspicuous succession in the subalpine zone follows fire and may result in subclimax stands of lodgepole pine (*Pinus contorta* var. *murrayana*), aspen (*Populus tremuloides*), or Douglas fir (*Pseudotsuga menziesii*). Progression to climax is extremely slow.

Lodgepole pine is absent in the southern Rockies, but elsewhere aspen is favored over the pine on moist sites, and after light fires it has an advantage, probably because of its ability to regenerate from sprouts. Near timber line, burned areas are revegetated directly by climax.

The transition from subalpine forest to alpine tundra is usually gradual with a thinning out of trees, which near their upper limits commonly have the dwarfed and distorted form known as *Krummholz*. Characteristic of timber line are several trees that cannot survive in the tundra above and cannot compete with climax species below,

Fig. 141. *Bristlecone pine* (Pinus aristata) Krummholz *at timber line of the subalpine zone in Colorado.*—U. S. Forest Service.

where they are only found on dry and windswept ridges. Bristle cone pine (*Pinus aristata*) occupies this position in the southern Rockies, limber pine (*P. flexilis*) in the central Rockies, whitebark pine (*P. albicaulis*), and alpine larch (*Larix lyallii*) in the northern Rockies, except in the far north where lodgepole pine occurs at timber line.

Douglas Fir Climax. Below the subalpine zone, Douglas fir (*Pseudotsuga menziesii*) is the climax dominant, growing in such dense stands that subordinate species are negligible. As in the subalpine zone, climax associates differ in the north and south. In the southern

Rockies, white fir (*Abies concolor*) and blue spruce (*Picea pungens*) are found in relatively small numbers and mostly on moist sites. In the north, grand fir (*Abies grandis*) is an associate west of the continental divide and principally on west slopes. East of the divide, *Picea glauca* of the northern conifer forest shares dominance with

Fig. 142. *Montane zone climax forest of Douglas fir* (Pseudotsuga menziesii) *and white fir* (Abies concolor) *in Colorado.*—U. S. Forest Service.

Douglas fir and extends southward through the montane zone as far as the Black Hills.

Dry, exposed ridges in both the montane and subalpine zones support open stands of pine, including several species characteristic of timber line. *Pinus flexilis* var. *reflexa* is important in the south. *P. aristata* occurs in northern Arizona and southern Utah and Colorado, while *P. flexilis* is more common northward to where *P. albicaulis* takes over in the northern Rockies.

Fig. 143. *Subclimax stand of lodgepole pine in Montana.*—U. S. Forest Service.

Fire in the Douglas fir climax results in the establishment of lodgepole pine or aspen stands, which bear the same relationships here as in the subalpine zone.

Ponderosa Pine Climax. Below the Douglas fir is a belt in which *Pinus ponderosa* or a close relative forms a relatively open climax forest that becomes savannah-like with decreasing altitude. The widely spaced trees form little shade so that the ground cover is made up of grasses, among which numerous species of *Festuca, Agropyron, Poa,* and *Muhlenbergia* are important. Between the zone of Douglas fir and the drier, lower altitudes with pure stands of ponderosa pine is a fairly broad transition where the two trees may share dominance.

Although the climax is termed ponderosa pine, the species is dominant only in the northern Rockies to the west of the continental divide. Elsewhere it is replaced by or in association with closely related varieties whose ecological characteristics are similar. The tree on the east slope in the north and throughout the zone southward is commonly recognized as a variety (*P. ponderosa* var. *scopulorum*). In the southern Rockies, the substitutes are *P. ponderosa* var. *arizonica*, *P. leiophylla* var. *chihuahana*, and *P. contorta* var. *latifolia*.

Fig. 144. *Climax forest of ponderosa pine* (Pinus ponderosa) *in typical open stand. Montane zone, Arizona.*—U. S. Forest Service.

The only exceptions to ponderosa pine dominance are found along streams and drainage lines where narrow-leaf cottonwood (*Populus angustifolia*), the commonest tree, forms postclimax stands with *P. acuminata* and *P. sargentii* in association. Aspen (*P. tremuloides*), in glades, and box elder (*Acer negundo*) may also occur frequently on these moist sites. Although fires are common, in dry summers, favored by the grasses of the forest floor, they are rarely severe enough to kill the fire-resistant older trees. That pine seedlings are destroyed is indicated by the even-aged groups of saplings, which can be related to favorable fire-free summers after good seed years. Severe fires in the upper part of the ponderosa pine zone may be followed by stands of lodgepole pine. Lumbering and over-grazing often result in the

Fig. 145. *Characteristic open stand of piñon-juniper, and the transition from sagebrush desert.*—U. S. Forest Service.

development of a dense scrub made up of species from the oak-mountain mahogany zone.

Piñon-Juniper Climax. This open forest of widely spaced, small trees (10 to 30 feet) forms the lowest coniferous zone in the Rockies and on many of the low ranges of the Great Basin represents the only zone present. It is, therefore, typical of the intermountain region as well as forming a distinct zone in the southern Rockies.[434] In some areas, especially in Arizona and New Mexico, under the present conditions of climate and grazing, the zone, and particularly juniper, is expanding at the expense of grassland. Although it is fairly constant in appearance and characteristics over its wide range and extensive acreage, several species with restricted ranges are involved, and there are marked regional differences in taxonomic and sociologic composition. The junipers include *Juniperus scopulorum, J. osteosperma, J. monosperma, J. occidentalis, J. deppeana,* and others, and the piñons, or nut pines, are sometimes treated as varieties of *Pinus cembroides* (*edulis, monophylla, quadrifolia*) or as distinct species.

The type extends from northern Mexico along the west slope of the Rockies to the Snake River in Idaho, beyond which it continues into southern Alberta with piñon and desert junipers replaced by *Juniperus occidentalis* or *J. scopulorum.* Along the east slope, it extends to Fort Collins, Colorado, where piñon drops out. Northward through Wyoming it is represented by *J. scopulorum,* often with sagebrush in association. Piñon-juniper is completely lacking in northern Sierran zonation, which goes directly from *Artemisia* and *Purshia* to *Pinus jeffreyi.* In the southern Sierra it is present on low slopes of the east side, usually not well-developed and often in mixture with other species. However, almost without exception, it occurs on every westernmost range and mountain of the Great Basin, often lying just across a valley from the base of the Sierra Nevada.

The openings between trees support a grass cover (*Bouteloua, Stipa, Agropyron, Poa*) and numerous other herbs, together with a few shrubs (*Ceanothus, Cercocarpus, Purshia, Cowania, Artemisia, Opuntia*) characteristic of the next lower zone. Over-grazing or fire may result in the temporary dominance of these shrubs.

Oak-Mountain Mahogany Climax. The transition from the conifer forest of the lower slopes to the treeless plains and plateaus may be marked by a zone of broad-leaved scrub. The zone is widest and best developed in the southern Rockies, narrows and becomes discontinuous in the central Rockies, and fades out entirely farther north. The components of the community vary, but oaks (*Quercus gambelii, Q. gunnisonii, Q. undulata, Q. fendleri, Q. emoryi,* and others) are the largest (up to 35 feet) and most conspicuous dominants in the south. North of the latitude of Denver, Colorado, the oaks are spottily represented, and mountain mahogany (*Cercocarpus montanus*) is dominant. Other important associates include *Rhus trilobata, Purshia tridentata, Fallugia paradoxa, Amelanchier* spp., and *Symphoricarpos* spp., any of which may assume local dominance. The vegetation does not form a continuous cover but occurs in dense clumps, or even as individual plants, separated by areas of grassland or desert vegetation.

The Black Hills. Although they are now isolated, the Black Hills are geologically and ecologically related to the Rockies. They deserve especial mention because of their mixture of eastern, western, and northern species. Because the highest elevation is only a little over 7,000 feet, the montane zone is chiefly represented. There is no Douglas fir present. Here, instead, *Picea glauca* var. *albertiana*, which

Fig. 146. *An example of the scrub oak-mountain mahogany zone in the foothills near Colorado Springs, Colo.* Quercus gambelii *predominates here with* Cercocarpus montanus *and* Rosa arkansana *as associates. Although the scrub is sometimes taller, its open, irregular distribution is typical.*—Photo by R. B. Livingston.

extends southward from Canada along the east slope of the Rockies as an associate of Douglas fir, is the only dominant on the high slopes at the southern limit of its range. Paper birch from the northern conifer forest is also present. Ponderosa pine dominates most of the lower slopes, which include most of the area, and lodgepole pine in small numbers is an additional representative from the Rockies. Species from the eastern deciduous forest are ash, hackberry, elm.

birch, and bur oak, of which only the last attains substantial size. The scrubby appearance of the community, as well as its distribution along the lower margin of the conifer forest, suggests the oak-mahogany zone of the Rockies.[120]

Sierra Nevada Forest Complex. The area here considered includes the southern portion of the Cascade Mountains and the Sierra Nevada, which together extend from Oregon southward along the eastern boundary of California as the innermost ranges of the coastal mountain system. The long west slope of the Sierra rises gradually to elevations of 14,000 feet and more, but the east slope drops abruptly to the floor of the Great Basin, which lies at about 4,000 feet. At the base of the west slope, there are only 10 to 15 inches of rainfall and a long, unbroken, dry summer season. Upward precipitation increases, temperatures decrease, the dry summer season shortens, and a larger proportion of precipitation falls as snow.

Because the general north-south axis of the range lies across the path of the prevailing westerly winds, climatic conditions for the region as a whole are influenced by them and east slopes are much drier than west slopes. Winter precipitation makes up 80-85 percent of the total, and at high elevations, most of the moisture falls as snow (35 to 70 feet in the subalpine zone). The greatest total precipitation has been reported to occur in the middle slopes,[232] between 5,000 and 7,000 feet, which support the luxuriant mixed coniferous forest of the montane zone. The subalpine zone coincides with the altitudes of greatest snowfall, where precipitation equals about 40 to 50 inches a year. Actually this may be the zone of highest precipitation, since snow gauges exposed to high winds give low values and spring snow surveys show a higher water equivalent in snow on the ground than reports of total precipitation.[214]

Subalpine Zone. This zone extends through an altitudinal range of little more than 1,000 feet, its limits, varying with latitude, being between 6,500 and 9,500 feet. The climate may be described as cool, winter-wet, summer-dry, with a short growing season.

Red fir (*Abies magnifica*) is the important climax species, growing in dense stands and making up 80-90 percent of the forest.[287] Of the associated species, none is an important component of the climax. Although western white pine (*Pinus monticola*) is constantly present in small numbers, it is only a minor constituent. Lodgepole pine

(*Pinus contorta*) is often present, especially at the margins of wet meadows, but its role is primarily successional. Mountain hemlock (*Tsuga mertensiana*) and white fir (*Abies concolor*) occur in an extremely irregular fashion. The former is always near the upper limits of the zone while the latter is scattered in stands at lower elevations.

Fig. 147. *Interior of the red fir* (Abies magnifica) *forest that occupies most of the subalpine zone of the Sierra Nevada.*

Of the shrubs, which are few, *Ribes viscosissimum* and *Symphoricarpos rotundifolius* are the most abundant and most constantly represented. The herb flora is also sparse. Constant species are *Chrysopsis breweri, Pedicularis semibarbata, Gayophytum ramosissimum, Pirola picta* and *Monardella odoratissima.* The yellow-green lichen (*Evernia vulpina*) is conspicuously present on the trees throughout the zone.

Although the altitudes in the Sierra are often greater than those of the Rockies, environmental conditions are severe and timber line is lower, varying from about 7,000 feet in the north to some 10,000 feet in southern California. The characteristic trees are *Pinus albicaulis, P. flexilis,* and *P. balfouriana.*[381] On exposed, bare, rocky slopes, *Juniperus occidentalis* is common at timber line and, especially on the west slope, at much lower altitudes.

Fig. 148. *Lodgepole pine at 8,800 feet in the subalpine zone, Carson Range of the Sierra Nevada.*—Photo by courtesy of the Agricultural Extension Service, Univ. of Nevada.

The upper margin of the red fir forest does not commonly extend to timber line but, instead, grades into a relatively narrow band of *Pinus contorta-Tsuga mertensiana* dominance. Although *P. contorta* is successional to *Abies magnifica* at lower altitudes, it, with hemlock, has climax characteristics near the upper limits of the zone. Thus, although the major part of the subalpine zone, from the lower margin upward, is occupied by the red fir forest, there is a distinct upper altitudinal band of hemlock-lodgepole pine dominance, which is replaced nearing timber line by white bark pine.

Montane Zone. The altitudinal range of this zone lies between about 2,000-6,000 feet in the Cascades, 4,000-7,000 feet in the central Sierra, and 5,000-8,000 feet or more in the south. Five or six principal species have climax characteristics and may appear in any com-

Fig. 149. *Virgin forest in the Sierran montane zone of California, in this instance made up of sugar pine* (Pinus lambertiana), *ponderosa pine* (P. ponderosa), *and white fir* (Abies concolor)—U. S. Forest Service.

bination at any altitude. However, the upper and lower parts of the zone tend to have consistent vegetational differences.[94, 217] White fir (*Abies concolor*) is usually the important dominant in the upper part of the zone, sometimes in pure stands, and decreases markedly at lower elevations. Lower down, incense cedar (*Libocedrus decurrens*), predominating on the most favorable sites, sugar pine (*P. lambertiana*), Jeffrey pine (*P. jeffreyi*), ponderosa pine, and Douglas fir are the species of importance. Douglas fir is more abundant in the north than in the south.[228] Sugar pine and Jeffrey pine are more conspicuous than ponderosa pine at the upper altitudes, a logical arrangement since the latter is the most drought-resistant of the major species. Dense chaparral communities of species of *Arctostaphylos, Ceanothus, Rhamnus,* etc., may appear after fire and sometimes last for years.

Fig. 150. *Giant Sequoia* (Sequoia gigantea) *of the Calaveras grove, Calif.* —U. S. Forest Service.

Included in the montane zone, on the western slope, are the forests of *Sequoia gigantea* at altitudes of 4,500-6,000 feet. Once widespread, they now occur only southward from the latitude of San Francisco in a disrupted zone. Their present best development is in the southern Sierra Nevada where they reproduce but do not spread. Sugar pine, ponderosa pine, and incense cedar are common associates.

Foothills (Woodland) Zone. As in the Rockies, the vegetation of the lower slopes and foothills is made up of coniferous and scrub

Fig. 151. *Characteristic open oak woodland of the Sierran foothills. Sequoia National Forest, Calif.*—U S. Forest Service.

associations, but they are not as sharply separated here. The zone ranges between about 1,500 and 4,000 feet. In the upper part, digger pine (*P. sabiniana*) and blue oak (*Q. douglasii*) are the dominants, forming typical open, or woodland, stands. The lower altitudes are characteristically covered with close-growing, evergreen scrub, or chaparral, in which *Ceanothus* spp. and *Arctostaphylos* spp. predominate. Common associates are several oaks (*Q. wislizenii, Q. chrysolepis,* and the scrub oak *Q. dumosa*), *Aesculus californica, Rhamnus californica,* and numerous other species are represented.

East Slope. Although the same zones are present on both the west and east slopes, many of the generalizations made above must be qualified for the east slope because of its less favorable conditions. The red fir forest occurs only in restricted areas on the east slope, as in the Carson Range east of Lake Tahoe and locally in the northern Sierra. The subalpine zone is represented, therefore, largely by the timber-line pines and patches of lodgepole pine. The montane and foothill zones extend to high altitudes, and the vegetation is poorly developed. *Pinus jeffreyi* is the important species of the montane zone, in which the open forest has little resemblance to that of the west slope. The woodland forest is practically absent. Although piñon-juniper occurs as a major zone on the next ranges across the valley, it is not found on the east slope of the Sierra except where an occasional high spur extends eastward. The scrub zone is sometimes made up of oak and mountain mahogany as in the Rockies, but is more often represented by species from the desert below (*Artemisia, Purshia, Chrysothamnus,* etc.), which, especially on areas of disturbance or fire, may be found high up in the montane zone as well as on the lower slopes.

Pacific Conifer Forest. This area parallels the coast from northern tree limits in Pacific coastal Alaska southward to central California. Coastal mountain ranges with varying altitudes are included throughout its length. The climate, tempered by the Pacific Ocean, is mild and without extremes. Although interior Alaskan winters are cold, subzero temperatures are uncommon along the coast. Southward, temperatures are progressively less severe until, in Oregon and California, frosts are rare. Precipitation is adequate to heavy (30 to 150 inches or more), and the humidity is always high, producing an extremely favorable P/E ratio. The southern part is winter-wet with no snow; here fog partially compensates for the summer drought. Northward, the summer dry season shortens until, in Alaska, there is none. Northward, too, there is an increase in the proportion of precipitation falling as snow. In the higher mountains, it may be entirely snow with falls as great as 50 to 65 feet a year.

The coastal forest is primarily montane in character, although ranging from sea level to altitudes of 5,000 feet. Only in the Cascades, in the United States and for a short distance into British Columbia does it include a subalpine forest. Here it is well developed, but the

Fig. 152. *Douglas fir* (Pseudotsuga) *and western arborvitae* (Thuja plicata) *in the coastal montane forest. Snoqualmie National Forest, Wash.*—U. S. Forest Service.

dominants are derived from the Rockies (*Abies lasiocarpa*), the Sierra (*Tsuga mertensiana*), as well as the coastal forest (*Abies amabilis, A. procera*). Northward, the zone becomes fragmentary or disappears entirely.

Fig. 153. *Pacific coastal forest in California showing redwood* (Sequoia sempervirens) *predominating and Douglas fir in association. Conspicuous subordinate species are* Lithocarpus densiflorus, Rhododendron californicum, Gaultheria shallon, Vaccinium *spp.*, Polystichum munitum.—U. S. Forest Service.

Species of the coastal forest are most fully represented in the general vicinity of Puget Sound, and the best development of the forest is indicated by the luxuriant vegetation of the Olympic Peninsula. Here the ranges of all the major species overlap and most of the trees attain their maximum size. The climax dominants are western hemlock (*Tsuga heterophylla*), western arborvitae (*Thuja plicata*), and grand fir (*Abies grandis*). Subordinate broad-leaved species and many herbaceous species are associated in abundance.[199b] Douglas fir, which reaches its greatest size here, is the most abundant and widespread species, but it occupies drier sites, is relatively intolerant of shade, and is the major dominant after fire. It is, therefore, subclimax in nature.[188]

To the north of the Puget Sound region, Sitka spruce (*Picea sitchensis*) is increasingly important as the forest becomes more closely associated with coastal conditions. Although it has subclimax characteristics near its southern limits, Sitka spruce becomes, with *Tsuga heterophylla* and *T. mertensiana*, an important climax dominant in the northern extension of the forest.[100] At its extreme limit in Alaska, the coastal and boreal forests meet and both *P. sitchensis* and *P. glauca* are found at timber line advancing into the tundra.[168]

Southward, the important species of the Puget Sound center extend down the low coastal mountains into Oregon with Port Orford cedar (*Chamaecyparis lawsoniana*) as an added climax species and Douglas fir of relatively greater importance.[299] Along the coast, however, Sitka spruce is replaced by redwood (*Sequoia sempervirens*), which, commonly in pure stands, closely follows the limits of the fog belt[96] to below San Francisco and fades out southward.

If the ranges of the principal species of the Puget Sound area are mapped, they appear in the form of a peninsula extending eastward across northern Washington and southern British Columbia and expanding north and south on the west slope of the Rockies.[120] The coastal dominants extending into this area are *Tsuga heterophylla*, *Thuja plicata*, and *Pseudotsuga menziesii*, which occupy a zonal position between the normal Douglas fir and spruce-fir zones of the Rockies. Although the importance of hemlock and arborvitae decreases eastward and Douglas fir increases, the zone remains distinctive largely because of the species peculiar to the forests developing after fire. The two principal successional trees are western larch (*Larix occidentalis*), which is endemic to the peninsula area, and western white pine (*Pinus monticola*), which grows more abundantly here than anywhere else. The presence of *Abies grandis* in association with these species is another indication of coastal affinity.

It has been pointed out[120] that this eastward overflow of coastal species marks an area in which steady winds blow inland from the coast, following a well-developed storm track, and thereby extend the coastal climate far inland. This theory is supported by the superior

Fig. 154. *Successional community of western white pine* (Pinus monticola) *and western larch* (Larix occidentalis) *in Idaho. Understory of* Thuja plicata *and* Tsuga heterophylla.—U. S. Forest Service.

development of the coastal species in the "peninsula" on westward slopes at intermediate altitudes and their occurrence in the Rockies only in the storm path and west of the continental divide.

Broad-Sclerophyll Formation. As the name indicates, major species in both associations of this formation have thick, hard, evergreen leaves, but all are angiosperms. One climax is dominated by trees and termed *broad-sclerophyll forest.* The other, called *chaparral* (in its Spanish origin applied to oak-dominated shrub communities), is a shrub climax. Both reach their best development on the coastal ranges of southern California, but their ranges extend from southern Oregon southward through the coast mountains, as well as through the Sierra Nevada foothills, into Lower California. Several of the species are found on the east slopes of the Sierra, and some appear in the desert woodland zone on the lower slopes of the Rocky Mountains.

The climate of the sclerophyll region is mild-temperate to subtropical with long, dry summers and heavy winter rainfall. Total precipitation is not less than 10 nor more than 30 inches, and, of this amount, no more than 20 percent falls in summer. In this area, desert vegetation occurs where precipitation is less than 10 inches, and, if it is over 30 inches, conifer forest is dominant.[97]

The two climaxes may be found in alternating patches in almost any part of their more or less coinciding ranges. However, chaparral occupies the greatest area and is climax in the south where it grades into desert, and sclerophyll forest is climax in the north and at the margin of montane conifer forest where its variations may be a part of the woodland zone. Where found together, the two communities bear no successional relationship to each other. The forest consistently appears on north slopes and the better sites, chaparral on south slopes and drier sites. The forest is postclimax in the south, and chaparral is preclimax in the north.

Sclerophyll Forest. The important evergreen forest trees are *Quercus agrifolia, Q. chrysolepis, Q. wislizenii, Lithocarpus densiflorus, Umbellularia californica, Arbutus menziesii, Castanopsis chrysophylla,* and *Myrica californica.* Several deciduous trees are almost as characteristic, as are a number of shrub and herb associates. The dominants may occur in various combinations related to altitude and exposure.

Chaparral. This community extends its dominance over a wide area and a diversity of habitats, and its composition is proportionately diverse. It includes at least 40 species of evergreen shrubs with varying degrees of dominance and importance, which may occur in many combinations but which invariably form low, dense thickets. The most important and constant species is chamiso (*Adenostoma fasciculatum*). The numerous species of manzanita (*Arctostaphylos*) are scarcely less characteristic. *A. tomentosa* is the widest ranging,

Fig. 155. *Broad sclerophyll forest* (Quercus agrifolia, Arbutus, *etc.*) *on north-facing slope (foreground and right). Chaparral on south-facing slope (left). Santa Lucia Mountains, Calif.*—Photo by W. S. Cooper.

Fig. 156. *Chaparral in the Santa Lucia Mountains, Calif. Smooth cover at top, mostly* Adenostoma. *Light-colored shrubs in shallow ravine at left,* Arctostaphylos glauca. *Grades into broad sclerophyll forest in deep ravine at right.*—Photo by W. S. Cooper.

Others with high constancy are *Heteromeles arbutifolia, Ceanothus cuneatus* (9 other spp.), *Quercus dumosa,* and *Cercocarpus betuloides.*

Fires. The long, dry summers and the nature of sclerophyllous vegetation make frequent fires the rule. A study in the Santa Monica mountains showed that chaparral stems were mostly about twenty-five years of age, and a stand without fire for fifty years was considered old. An ordinary fire causes chaparral to sprout profusely, and then, come back to normal within ten years.[16] Fire usually favors the spread of chaparral at the expense of sclerophyll forest. Too frequent fires, however, may cause the death of chaparral and its replacement by grassland. Undoubtedly, the original extent of sclerophyll dominance has been much reduced by fire, since, once they are destroyed, the return of the sclerophyll species is long delayed.

Desert Formations. The major area of the North American desert extends from southeastern Oregon and southern Idaho southward through the Great Basin, including most of Nevada and Utah except high elevations, continues southward into southern California and western Arizona, down most of the peninsula of Lower California and, on the mainland, through Sonora as far south as the Yaqui River. The highlands of eastern Arizona and western New Mexico interrupt the continuity of desert, but from south-central New Mexico, there is almost continuous desert through eastern Chihuahua and most of Coahuila in Mexico.[376]

In spite of the great extent of this area, there are certain environmental features characteristic throughout.[377] Precipitation is low and erratic; temperatures of air and soil are extremely high by day, drop abruptly at night and have great seasonal ranges; atmospheric humidity is usually low, winds are strong, and bright sunny days are the rule. These factors serve to explain why predominating plants are those that can survive desiccation without injury or that store water in their succulent tissues. This is not to imply that desert vegetation is uniformly similar throughout. Climatic differences, associated with latitude and altitude, are accompanied by differences in species and life forms. Locally, the physical differences in topography, exposure, and soils produce distinct vegetational variations just as in moister climates. Finally, there are numerous undrained depressions into which the water of winter rains flows and, upon evaporation, deposits the silts and clays it has transported as well as salts of various kinds. The resulting mud flats (playas) in themselves constitute a special habitat with associated species, but the nature and concentration of salts in the soil is even more effective in controlling the communities there.

Four desert areas are distinguishable on the basis of regional environments and, likewise, by the nature and importance of the major dominants:[376] namely, the Great Basin, Mojave, Sonoran, and Chihuahua deserts. In each of these areas there are stabilized communities that occur throughout with minor variations in the least extreme environments. Other communities, seemingly equally permanent, are associated with special habitats only. The vegetation is thus a complex mosaic of communities that may be of great or limited extent and almost all of which have climax characteristics. Because the effects of biological reaction on environment are seemingly negligible, primary succession is hardly apparent. However, secondary succession does occur, especially after the often extensive fires in shrub types. Without such marked disturbance, most communities remain indefinitely unchanged and dominant, each in its own special habitat. It is not surprising, therefore, that polyclimax interpretations have been used widely in their description.

The presentation that follows will emphasize the characteristics distinguishing the four deserts and the dominant vegetation of major differing habitats with little consideration of succession or climax. The discussion is adapted largely from Shreve's[376] excellent summary of desert vegetation, except for the distinction made here between Cold Desert and Warm Desert Formations.

Cold Desert Formation. This was termed the Great Basin Desert by Shreve because of the physiographic, climatologic, and vegetational unity that extends north from southern Nevada and southern Utah. The topography of the area is variable. The wide valley floors, lying at about 4,000 feet elevation, are interrupted by numerous ridges, often rising to more than 8,000 feet, and by depressions of the playa type. Nevertheless, climatic conditions are distinctive when compared with deserts further south, and they produce a characteristic vegetation.

The meager rainfall (4 to 12 inches) increases with altitude and toward the east side of the Basin. It is greatest in winter, and thus summers are very dry in the west and only slightly less so in the east. Average temperatures are much lower than further south and the frost-free season is very short. Killing frosts may occur in any month of the year; and in fall, winter, and spring, there are frosts almost every night, often with bitter cold; thus, the term Cold Desert. When the distinctive growth form of the dominants, mostly shrubby chenopods and composites, is taken in conjunction with the climate, it is apparent why the vegetation of the Great Basin should be considered as a distinct formation, separate from the scrub types of deserts to the south.

The two major communities are simple, with few dominants in each,

Fig. 157. *Sagebrush semi-desert* (Artemisia tridentata) *northwest of Reno, Nev.*—Photo by W. D. Billings.

and often extend uninterrupted for miles. The sagebrush association, dominated by *Artemisia tridentata* (common sagebrush), is climax in the northern portion of the Great Basin or at relatively high elevations. The shadscale association, with shadscale (*Atriplex confertifolia*) and bud sage (*Artemisia spinescens*) as its important species, ranges through the south and at low elevations. In its northern and eastern distribution, shadscale is found on heavy lowland soils containing some salt in the subsoil, but, to the south and west, it is climax on gray desert soils with a shallow carbonate layer and regardless of salts.[29] Sagebrush occurs most commonly on brown soil, either sandy or clayey, with the carbonate layer at a deeper level and with a minimum concentration of salts.[27]

The controlling effect of salts on community structure has been amply demonstrated for different parts of the area.[27, 148, 202] Zonal patterns around playa lakes are the same everywhere (see Fig. 92). Where flooding is periodic and salt content excessive, vegetation is absent or dominated by samphire (*Salicornia* spp.) or iodine bush (*Allenrolfea occidentalis*). With somewhat less salt, shadscale and

Fig. 158. *Typical dry desert expanse with shadscale* (Atriplex confertifolia) *dominance. Mineral County, Nev. Characteristic gravelly desert pavement shows here.*—Photo by W. D. Billings.

greasewood (*Sarcobatus vermiculatus*) or gray molly (*Kochia americana* and var. *vestita*) are dominant. Away from the playas on soils with a minimum of salts, sagebrush may be the major species.

Many other species occur, of course. They are mostly shrubs with the same growth form. There are numerous species of *Atriplex* and *Artemisia*. *Chrysothamnus puberulus, Grayia spinosa, Coleogyne ramosissima, Eurotia lanata, Purshia tridentata,* and others are variously associated with the major species or sometimes assume dominance under local special conditions. Several species of *Ephedra* are characteristic.

Warm Desert Formation. Mojave Desert.—This, the smallest of the desert units, lies almost entirely in California below and to the east of the southern end of the Sierra Nevada. Physiographic conditions are similar to the Great Basin but elevations are generally lower (1,000-4,000 feet). The irregular, meager precipitation of less than 5 inches comes mainly with winter cyclones or as occasional late summer

Fig. 159. *Creosote bush* (Larrea divaricata) *with* Franseria dumosa *in association as is typical of much of the Mojave Desert.*—Photo by W. D. Billings.

cloudbursts. Summers are very hot and dry. The area includes Death Valley which at its lowest point is 282 feet below sea level. Its infrequent maximum annual rainfall is 2 inches, and official records show one period of four years when temperatures were above 100° F. on 538 days.[294] Records of temperatures remaining above 114°, both day and night are not uncommon.

Conditions are not too different from those of the Great Basin although somewhat more extreme. This is borne out by the vegetation, which includes many of the same species, their distribution controlled here, too, by soil texture and salt concentration. Certain character species do stand out, however, and this justifies the vegetational distinction from the Great Basin. At the upper elevations (3,000-4,000 feet) and in the transition from sagebrush, with maximum precipita-

Fig. 160. *Joshua tree* (Yucca brevifolia), *characteristic of the northern Mojave desert, particularly in the transition from creosote bush dominance to shadscale of the sagebrush formation.*—Courtesy Univ. of Nevada Agricultural Extension Service.

tion, Joshua tree (*Yucca brevifolia*) is conspicuous. With decreasing altitude and precipitation, creosote bush (*Larrea divaricata*), with bur sage (*Franseria dumosa*) in association, becomes the major dominant. This community occupies 70 percent of the total area of the Mojave Desert.

Sonoran Desert.—The lowlands around the Gulf of California in Mexico and Lower California, which lie chiefly below 2,000 feet, constitute the Sonoran Desert. Much of the area is made up of dunes and sand plains. Precipitation is extremely uncertain, not exceeding 2 to 4 inches in the vicinity of the Gulf, although increasing some with altitude. Its effectiveness is counteracted by the extremely high temperatures.[375]

Fig. 161. *Sahuaro* (Carnegiea gigantea), *the giant of the columnar cacti that characterize the uplands of the Sonoran Desert.*—U. S. Forest Service.

The low plains are dominated by *Larrea-Franseria,* with various associates, as in the Mojave Desert. *Fouquieria splendens* is common. Because drainage here is not internal, margins of streambeds support a distinctive mixed community including species of *Prosopis, Cercidium, Olneya,* etc. In the higher elevations of Arizona and northern Sonora (1,000-3,000 feet), there is a great mixture of species and life forms. Although numerous species characteristic of the other deserts are present, *Cercidium microphyllum* is a dominant with numerous arborescent and columnar cacti, including *Carnegiea gigantea, Lemaireocereus schottii,* and many species of *Opuntia.* The variable topography of the peninsula of Lower California supports an equally variable flora including many species. Near the coast, there are more leaf succulents than in any of the other desert areas.

Chihuahua Desert.—Extending from southern New Mexico southeastward to western Texas and down into Mexico, much of this area is interrupted by high mountains and lies between 4,000 and 6,000 feet. Precipitation varies with altitude from 3 to 12 inches and falls largely in summer. Temperatures are somewhat lower than in the Sonoran Desert, and frosts are not uncommon.

Under these conditions, the communities are not as complex as those of the Sonoran Desert or as simple as those of the Great Basin,

Fig. 162. *Mesquite* (Prosopis juliflora), *a common ground-water indicator in the desert scrub formation.*—Courtesy Univ. of Nevada Agricultural Extension Service.

but there is much regional variation. Shrubs and semishrubs predominate with a great variety of inconspicuous stem succulents in association. Ocotillo (*Fouquieria splendens*), which is found throughout the area, creosote bush (*Larrea tridentata*), and mesquite (*Prosopis chilensis*) are the only three species common in the Sonoran Desert that are also important and widespread here. A number of species are conspicuous because of size or unusual form. *Yucca, Nolina,* and *Dasylirion* are large semisucculents. *Agave* and *Hechtia* are particularly abundant leaf succulents. Leafless, green-stemmed trees, columnar cacti, and *Dasylirion longissimum* with its six-foot, linear leaves, are examples of locally important species of striking appearance.

Grassland Formation. Grasses are climax dominants over all the vast area extending from southern Saskatchewan and Alberta to eastern Texas, and from Indiana and the western margin of deciduous forest westward to the woodland zone of the Rockies. Separated from this major area are the Palouse region of Washington and the grasslands of the great valley of California. The formation has the greatest extent of any in North America and consequently grows under a great diversity of conditions. This is possible because of the growth form of the species, their long period of dormancy, and the fact that their moisture requirements are critical primarily in spring and early summer.

The eastern transition to forest is marked by an annual precipitation of 30 to 40 inches from Texas to Indiana and 20 or 25 inches farther north. A high proportion of this precipitation falls as spring rain, but westward, as the total decreases to about 10 inches near the Rockies, the proportion falling in spring and summer also decreases. Temperatures are equally variable. In the north, the growing season is cool and short, and subzero temperatures occur for long periods in winter. In the southern part of the range, frosts may be almost unknown, and extremely high summer temperatures are characteristic.[36] Throughout the formation, late summer dry spells with high temperatures and drying winds are the rule; but, if there is sufficient moisture for the grasses during the spring growing period and summer maturation, such extremes affect them but little because of their long period of dormancy. The hot season with limited precipitation is probably of great importance in maintaining grassland climax against the advance of forest.

The increasingly severe moisture conditions from east to west are accompanied by changes in the dominant species whose combinations are distinguishable as associations of the formation. Three major regions are recognizable either by climate or vegetation, or both. Their limits, climate, and vegetation have been summarized and the important regional and local studies of grassland have been listed in a concise presentation.[71] This condensation of grassland information could well be used as the starting point for any consideration of the nature and distribution of grassland. The great number of classifications attempted for grassland communities and the disagreements as to major dominants and most important species implied by the terminology suggest the complexity of the formation. Probably, too, there is a suggestion of much more variation regionally than might at first be supposed. Of necessity, we are restricted here to a simple presentation. On this basis, the discussion will deal with only three major associations, which may be termed Tall Grass Prairie, Mixed Prairie, and Short Grass Plains. Some authorities recognize as many as seven associations,[79] and, even then, most of these can be divided into several faciations. Furthermore, a detailed discussion must recognize within

Fig. 163. *Tall grass prairie community in which* Andropogon scoparius, Bouteloua gracilis, *and* Sporobolus heterolepis *are the most important species.*—Photo by R. B. Livington.

each faciation the usually distinct upland, slope, and lowland variations.

Tall Grass Prairie. Sometimes called "true prairie,"[421] this association borders the deciduous forest, receives the most rainfall, has the greatest north-south diversity and the greatest number of major dominants of the formation. Bunch grasses are the conspicuous species, for many of them grow in excess of 6 feet tall, but sod-forming species are also dominants. Because of the generally favorable climatic and soil conditions, most of the area is cultivated and little of the original vegetation remains today.

The long list of major dominants includes tall grasses, such as *Stipa spartea, Andropogon furcatus,* and *Sorghastrum nutans;* medium grasses, such as *Andropogon scoparius* and *Bouteloua curtipendula;* and the short grasses, *Bouteloua gracilis* and *B. hirsuta.* The association of dominants with topography should be indicated at some point even though it is impossible to recognize it throughout our discussion. The following groupings are not uncommon for Tall Grass Prairie.

UPLANDS	SLOPES	LOWLANDS
Agropyron repens		
Bouteloua gracilis		
B. curtipendula		
Andropogon scoparius		
Poa pratensis	*Poa pratensis*	*Poa pratensis*
Sorghastrum nutans	*Sorghastrum nutans*	*Sorghastrum nutans*
	Koeleria cristata	
	Andropogon furcatus	*Andropogon furcatus*
	Stipa spartea	
		Agrostis alba
		Spartina pectinata
		Panicum virgatum

This distribution is not the same everywhere. In the north *Koeleria* and *Stipa* appear in the upland, and *Poa* and *Sorghastrum* do not appear at all. In the central region, *Panicum virgatum, Buchloë dactyloides,* and *Bouteloua hirsuta* are added to the uplands and *Sporobolus heterolepis* and *S. cryptandrus* to the slopes. The southern faciation, sometimes regarded as a separate association, is even more distinct, especially because of added species in the uplands, such as *Stipa leucotricha, Andropogon saccharoides, A. tener,* and *A. ternarius.*

Tall Grass Prairie once extended eastward as a "peninsula" into Illinois and Indiana and in isolated patches much farther east, but particularly in Ohio. Today these areas are almost entirely cultivated; but their distribution has been well-mapped,[406] and their vegetation is known from studies of relict stands. The typical species are those of Tall Grass Prairie, but their combinations, especially among the dominants, differ sufficiently for the community to be termed a separate association by some. The predominating tall grasses, as well as other basic similarities, make it reasonable to others to consider the prairie peninsula as a faciation of Tall Grass Prairie, to which it bears a postclimax relationship. The soils within the peninsula are prairie soils, although the climate is now that of forest climax. The community may, therefore, be regarded as preclimax to the forest, maintained by edaphic conditions.

A condensed discussion such as this must emphasize the dominants,

Fig. 164. *Mixed grass prairie in which* Bouteloua gracilis, Stipa comata, *and* Calamovilfa longifolia *are the principal species. Colorado.*— Photo by R. B. Livingston.

which are invariably grasses and sometimes sedges. Other species, commonly called forbs, are important in all grasslands. Seasonally, their abundant and conspicuous flowers may even give an impression of dominance. Composites and legumes are especially well represented, and the species, as varied as the dominants, are similarly characteristic of different parts of the range and particular habitats.

Mixed Grass Prairie. Although the mixed grasses occupy an area between that of the tall grasses and short grasses and the dominants are derived from both these communities, it is generally agreed that there is sufficient unity and distinctness to justify associational rank. Important dominants throughout the area are *Bouteloua gracilis, B.*

Fig. 165. *Mixed grass community in Arizona in which grama grasses predominate.*—U. S. Forest Service.

hirsuta, Andropogon scoparius, and, except in the north, *Buchloë dactyloides.* In the north, *Koeleria cristata, Stipa spartea,* and *S. comata* are added dominants, which suggest the recognition of a northern faciation. Other important species included among the dominants are *Andropogon furcatus, Sporobolus cryptandrus,* and several species of *Stipa.*

The western limit of the association may be taken as the line where tall grasses disappear and beyond which only short grasses are dominant. Since the tall grasses require available soil moisture to a depth of 24 or more inches during their active growing season, the limit of mixed grass prairie is a line beyond which precipitation is insufficient to provide moisture to this depth. The eastern limit is not as sharply defined but is also determined by soil moisture, since mixed prairie is marked by prairie grasses in bunch-grass habit sharing dominance with permanently established short grasses.[3] Thus the area forms a strip from Saskatchewan through the central Dakotas, Nebraska,

Fig. 166. *Short grass plains pastured to sheep in Wyoming.*—Photo by W. D. Billings.

Kansas, and western Oklahoma into Texas. The sand hills of Nebraska are an exception, for here soil conditions are such that postclimax tall grasses predominate. During protracted dry periods, the short grasses increase at the expense of the moisture-requiring tall grasses;[420] but during a series of years with favorable or normal moisture, recovery of original vegetation is assured.[421] Thus the boundaries of the association are not particularly static and are represented by a wide transition zone.

Short Grass Plains. Westward from the Mixed Grass Prairie to the woodland zone of the Rockies, the xeric short grasses are dominant.

On the basis of exclosure studies and other observations, the climax nature of short grasses has been questioned, and the community has been described as disclimax resulting from overgrazing an area that would otherwise support mixed prairie.[422] Regardless of terminology, the short grasses are, at present, dominant over the entire area.

The most important species are *Bouteloua gracilis* and *Buchloë dactyloides,* except north of the Dakotas, where the latter is absent. Several faciations are recognizable that result from combinations of the major dominants with *Stipa comata,* or *Agropyron smithii,* or *Aristida longiseta.*

Fig. 167. *A variation in Colorado short grass plains where local conditions, probably affecting moisture, permit growth of trees and Agropyron-Koeleria dominance under moderate grazing.*—U. S. Forest Service.

The desert plains area extending from western Texas across northern Mexico and southern New Mexico and Arizona supports short grasses, which, although including different species, are related to short grass dominance. Several species of *Bouteloua* and *Aristida* predominate.

Overgrazing has greatly increased the numbers of desert shrubs here, and these include *Larrea, Opuntia, Flourensia,* and several others of which widely spaced individuals occur everywhere.

Other Grassland Climax. There is evidence that the great valley of California was once dominated by grasses, which, because of fire and grazing, have been eliminated except for relict stands. The latter suggest that the dominants were bunch grasses, which produced grassland similar in appearance to mixed prairie. Throughout most of the area it appears that *Stipa pulchra* was the principal species, except near the coast. Today introduced annual grasses occupy most of the remaining grassland areas, especially species of *Avena, Bromus, Festuca,* and *Hordeum.*

The rolling hills of the Palouse region, as well as most of eastern Washington and Oregon and eastward into Idaho, supported prairie grasses before being cultivated for wheat production. Although numerous species characteristic of other grassland areas are present here, the major dominants are distinctive, including *Agropyron spicatum, Festuca idahoensis,* and *Elymus condensatus.* Possibly much of the sagebrush dominance in this region is only the result of grazing, and certainly the dominance of the annual, *Bromus tectorum,* results from fire and grazing as it does southward in the Great Basin.

The Palouse and California grasslands, in contrast with the major areas, are products of winter, rather than spring and summer, precipitation.

Aspect Dominance. Probably no other formation has such marked variations in appearance through the growing season. Since not all the grasses mature at once, there are times when simple observation might lead to incorrect conclusions as to their relative importance. Associated herbaceous species other than grasses, often called forbs, may be seasonally so conspicuous as to obscure the grasses and, temporarily at least, to give the appearance of dominance.

Tropical Formations. The truly tropical vegetation of North America, which occurs only in southern Mexico and Central America, probably includes as great a diversity of communities as is usually found in temperate climates. The major controlling factor in this diversity is moisture, as affected by topography, exposure, and seasonal distribution. Although numerous local studies of the vegetation of

American tropics have been made, it is only recently that a comprehensive classification of the plant communities has been attempted [19] and that detailed discussion is available[335] in the light of modern concepts.

A misconceived but popular idea of tropical vegetation is undoubtedly one that can best be placed in the category of rain-forest-in-its-jungle-form. But such tangled masses of vegetation are found only on areas of disturbance and "True rain forest always gives the impression of the vault of cathedral aisles." [19] It is made up of many species of tall broad-leaved, evergreen trees in several strata with the tallest sometimes rising 90 feet to the lowest branch. Undergrowth is sparse, lianas are few, and epiphytes are not abundant near the ground. Apparently, after disturbance of any kind, such forests are replaced by a tangled jungle of growth that is almost impenetrable. The rain forest is not widespread, because conditions for its development are by no means everywhere available. It occurs where temperatures are fairly constantly high, precipitation is plentiful (over 200 inches in some areas), and on good sites with proper drainage but with a continuous supply of available water.

It should be re-emphasized that not all tropical vegetation is rain forest, and to this should be added that not all broad-leaved evergreen forest is rain forest. The presence in the tropics of mountains of sufficient height to have permanent snow on their peaks insures altitudinal zonation similar to that of temperate regions. These mountains may interrupt moisture-bearing winds and so maintain desert conditions. Seasonal deciduous forests, pine forests, and even tundra are to be found on their slopes. The major variations in American tropical vegetation, including the West Indies, have been grouped into twenty-seven formations (physiognomic), each of which includes several associations (floristic), except Rain Forest, which stands alone.[20]

A. Optimum Formation
   Rain Forest
B. Seasonal Formations
   1. Evergreen Seasonal Forest
   2. Semi-Evergreen Seasonal Forest
   3. Deciduous Seasonal Forest
   4. Thorn Woodland
   5. Cactus Scrub
   6. Desert
C. Montane Formations
   1. Lower Montane Rain Forest
   2. Montane Rain or Cloud Forest
   3. Montane Thicket

4. Elfin Woodland or Mossy Forest
5. Paramo
6. Tundra

D. Dry Evergreen Formations
1. Dry Rain Forest
2. Dry Evergreen Forest
3. Dry Evergreen Woodland and Littoral Woodland
4. Dry Evergreen Thicket and Littoral Thicket
5. Evergreen Bushland and Littoral Hedge
6. Rock Pavement Vegetation

E. Seasonal-Swamp Formations
1. Seasonal-Swamp Forest
2. Seasonal-Swamp Woodland
3. Seasonal-Swamp Thicket
4. Savanna

F. Swamp Formations
1. Swamp Forest and Mangrove Forest
2. Swamp Woodland
3. Swamp Thicket
4. Herbaceous Swamp

The climate of southern Florida[127] and the Gulf coast down into Mexico permits the growth of numerous species with tropical characteristics and affinities. The palms, the many broad-leaved evergreens, the mangroves, the abundant epiphytes and lianas, and the sometimes jungle-like masses of vegetation are all suggestive of tropical conditions.

General References

E. LUCY BRAUN. *Deciduous Forests of Eastern North America.*

J. R. CARPENTER. The Grassland Biome.

F. E. CLEMENTS. *Plant Indicators: The Relation of Plant Communities to Processes and Practice.*

R. F. DAUBENMIRE. Vegetational Zonation in the Rocky Mountains.

J. W. HARSHBERGER. *Phytogeographic Survey of North America.*

B. E. LIVINGSTON and F. SHREVE. *The Distribution of Vegetation in the United States, As Related to Climatic Conditions.*

H. L. SHANTZ and R. ZON. The Physical Basis of Agriculture: Natural Vegetation, in *Atlas of American Agriculture.*

V. E. SHELFORD (ed.). *Naturalist's Guide to the Americas.*

F. SHREVE. A Map of the Vegetation of the United States.

J. E. WEAVER. *North American Prairie.*

J. E. WEAVER and F. E. CLEMENTS. *Plant Ecology.*

Chapter 12 The Distribution of Climax Communities: Changes of Climaxes with Time

The apparent distributional and floristic constancy of climax communities is relative and always has been. A degree of instability is characteristic for two general reasons, namely, climatic change, and the continuing nature of evolutionary processes.

In geological time there have been known changes of climate with parallel vegetational changes, sometimes so radical that whole floras were displaced from extensive areas. Even minor climatic modification must result in adjustments in the ranges of species. Some, with narrow tolerances, might be eliminated; others might survive with reduced vitality. Under these conditions, species from other populations could move in and take their places. The rates of migration would vary with the means of dispersal of the species, their competitive ability, and their adaptability. As the climate tended toward stability, the rate of migration into the area would slow down, but adjustments in the population could continue for a long time, even geologically speaking.

Genetic variants, typical of every species, are constantly being produced. Many do not survive, but those especially adapted to an environment may become established there. Whether they occupy the area of their origin or migrate elsewhere, an earlier, less well-adapted representative of the species will be supplanted. Such an evolution of species characteristics has probably always gone on with parallel evolutionary adjustments in vegetation. Geographic variants clearly adapted to local conditions are well known for many present day species.[77a] The processes of ecological evolution that produced them and favored their survival must be going on at all times to some degree in every species and every environment.

These evolutionary processes have certain important ecological implications. In a given area, when climate is relatively constant, all species must be evolving together if they survive and, therefore, all must be constantly changing. If successful variants of one species appear, there must be compensating variation in associated species if they are to maintain their relative status. Undoubtedly adjustments are continuous. With a slowly changing climate, only such ecological evolution could account for the continued maintenance of a type of vegetation in a given area, since variants adapted to the changes might otherwise soon predominate.

Furthermore, some migration and redistribution of populations probably results from these processes at all times. During periods of climatic stability the range of a vegetation type might well be slowly extended if species acquired a wider ecological amplitude through genetic variation. If climate changed rapidly, there might be an initial reduction of range, especially near the limits, but again, there would follow a redistribution of the species among themselves and a readjustment in transitions to other contiguous populations.

Within relatively recent geological time, glaciation in the Northern Hemisphere obviously must have produced such changes in climate that disruption of then existing lines of vegetational distribution were inevitable. Advance of the ice southward resulted in constriction of vegetational zones and retreat of species and growth forms as the climate changed. With the recession of the ice, there was again a northward advance of species and a readjustment of plant communities as the glaciated area was reoccupied by vegetation. Probably there were several minor advances and retreats of vegetation correlated with the shifting ice fronts and the similarly varying climate.

Within historical time, there have been major shifts of climate producing conditions that may have had serious effects on vegetation. There is evidence that early Norsemen who colonized Greenland were able to carry on a primitive sort of agriculture on lands along the southern coast. Between the twelfth and the fourteenth centuries the climate there deteriorated rapidly so that summers became shorter and colder, the soil remained frozen, and the colonists disappeared. Today, as for some time past, the receding glaciers in Greenland indicate an increasingly milder climate. Receding glaciers in Alaska have been similarly interpreted.[101] It has been shown that conifer forest has for some time been advancing into the tundra in Alaska.[168] Periodically, prairie vegetation is invaded for some distance by forest, and although drought often eliminates such advances, they may be permanent or, at least, appear so.

Climates have changed over long periods of time, and slow change continues today in certain areas. With shifts of climate, vegetational change is to be expected. Some modern changes are easily recognized, as indicated above. In highly populated areas the changes may be much less obvious, because natural vegetation has been disturbed by man.

Paleo-Ecology

This phase of ecology deals with the history of vegetation, especially the reconstruction of past climaxes and climates, their rise, decline, and migration[74] over long periods of geological time.[438] Its basic source materials are derived from paleontology and geology[200] and must be interpreted in terms of what is known of the ecology of modern organisms.

Tracing changes in modern climax vegetation is a complex procedure involving the use of every kind of evidence available. A reconstruction of the natural vegetation of Ohio[351] and its prehistoric development[354] illustrates how historical records and pollen statistics may be used to great advantage. Again a study of New England climate and vegetation[316] utilizes still other sources of evidence. Archaeology, zoology, botany, and geology all were drawn upon in a variety of ways before it was concluded that New England had had a warmer climate within recent years—probably no more than a thousand years—and that the trend has since been to the cooler and moister, with parallel vegetational changes.

Knowing that climates have changed, one may be equally certain that vegetation has varied accordingly. Major alterations in vegetation may likewise be assumed to indicate modification of climate. In some instances, however, such shifts have been interpreted as purely successional in nature, a point not to be ignored, since succession has gone on in the past as it does today. Change within historical time, if still in progress, may be observed, or may become apparent from detailed quantitative and qualitative studies of transition areas.[56] Another source of information is the historical literature, less reliable, unfortunately, because of the limited knowldege of the early writers. It is, nevertheless, a source from which much of value can be learned,[51, 316] particularly when the information is drawn from several sources and is correlated with other kinds of evidence.

The difficulties of reconstructing the vegetational picture during early historical time are as nothing compared with those involved in determining prehistoric climaxes.[67] Fossils, variously preserved, are

Fig. 168. *Interglacial forest relicts on beach below high tide, Glacier Bay, Alaska. These hemlock stumps, probably several thousand years old, represent forest that lived before the last major advance of ice, which buried them under glacial debris (above beach). Tide action has again exposed the stumps.*—Photo by W. S. Cooper.[100]

the chief source of our knowledge of ancient floras, many of which have disappeared completely. Considering that different species and even parts of the same plant are unequally preserved, it is surprising that we know as much of these old floras as we do.[73] Certainly we know that there have been extreme climatic changes on various parts of the earth and that with them have come modifications in vegetation, sometimes so extreme that entire floras have disappeared.

More recent vegetational history has been given greater attention because of its direct relationships to our modern flora and, possibly, because it offers greater probability of solution. Post-glacial climate and vegetation have been studied more intensively, therefore, than those of preglacial time. Plant remains, buried and preserved between layers of glacial drift, have yielded much information on the amount of time involved, the climate, and the vegetation. These deposits, often preserved in a natural state as wood, leaves, fruits, or seeds, have been

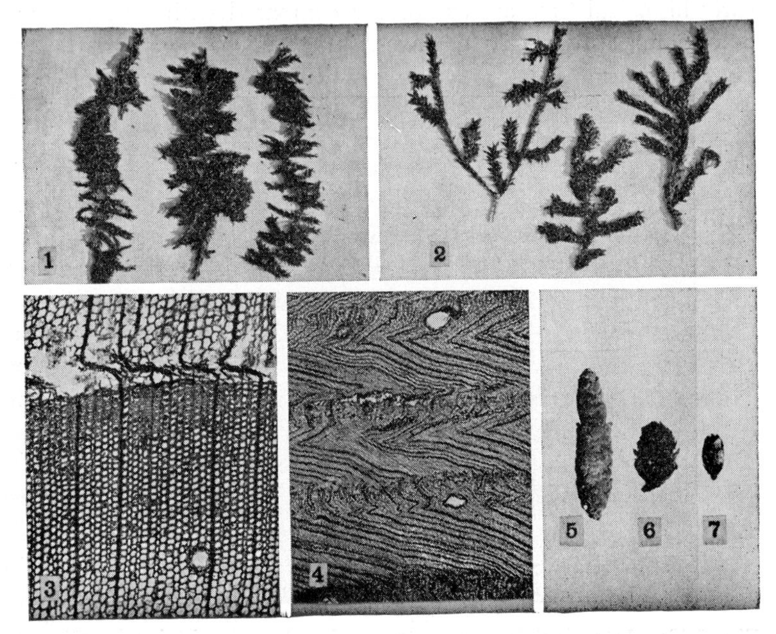

Fig. 169. *Well-preserved Pleistocene plant remains found in silt or peat layers buried under 10 to 12 feet of undisturbed moraine in Minneapolis, Minn.* (*1*) Calliergon giganteum, (*2*) Neocalliergon integrifolium, (*3*) Picea *sp., wood structure almost perfect,* (*4*) Picea *sp., wood structure distorted by pressure,* (*5*) *cone of* Picea glauca, (*6*) *cone of* Picea mariana, (*7*) *cone of* Larix laricina.—From Cooper and Foot.[102]

uncovered by erosion, excavation, and even in driving wells at considerable depth. Such findings have been fortuitous largely, since the deposits do not occur generally and, when stumbled upon, must be brought to the attention of those interested if they are to be of any scientific value. As a result, the information they have yielded is fragmentary and discontinuous both in time and space.

A more promising approach to the problem began with the study of the nature and composition of the strata of plant remains and other sediments that have accumulated in lakes and ponds as peat or related material.[115] These strata may give almost continuous records back to glacial time; and, since deposits are distributed over wide areas, their study makes possible the correlation of findings, particularly regarding climate, in one place with those in another. Obviously such studies can not be entirely satisfactory, since they indicate vegetation only within the bogs themselves, or at their immediate margins, and bog vegetation is not of the climax type.

Pollen Analysis

When, in 1916, von Post presented the results of his studies of pollen preserved in Swedish peat deposits, an entirely new approach to the reconstruction of prehistoric vegetation was begun.[143] Wind-borne pollen is deposited everywhere, and much of that which falls in a lake is preserved in its sediments because of the low rate of oxidation. Since the pollen of most dominant species, especially trees and grasses, is wind-borne, the pollen deposited at any one time should include that of the important species in the general vicinity and the numbers of grains of various species should be suggestive of their relative importance in the surrounding vegetation at the time. Because pollen grains of a species are constant in size and form, genera, and sometimes species, can be identified positively. Consequently, if samples of lake deposits are taken from the bottom upward to the present surface of the sediment, changes in pollen content of the successive strata should parallel the trend of changes in vegetation throughout the period of accumulation. Pollen counts cannot give precise evaluations of the genera and their abundance in the area; because some produce more pollen than others, weight, and therefore distance of transport, varies widely, and the strength and direction of winds is not known.

Sediments on lake bottoms as well as peat deposits have been studied. Samples must be taken with care to prevent contamination, and several types of augers have been devised for the purpose. With these, cores can be cut that, placed end to end, form a continuous column of material for the entire depth of the deposit. Borings are made in summer under most conditions; but, since it is desirable to have them from the deepest part of the depression, it is often advantageous to make them in winter from the frozen surface.

Identification and counting of the pollen grains must be done under a microscope. This necessitates treatment of the samples with one of the several methods recommended[65] to eliminate foreign material and to concentrate the grains. Identification is facilitated by reference to illustrations[432] and by comparison with a series of grains taken from modern plants. What constitutes an adequate sample in the count of grains is not agreed upon by all investigators, but fewer than 150-250 grains are rarely counted.

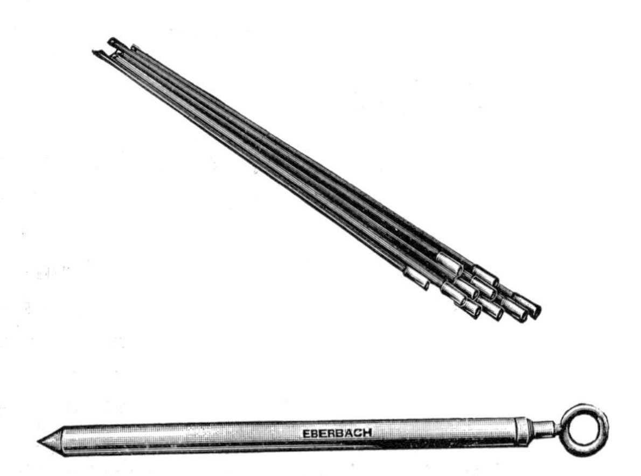

Fig. 170. *A type of sampler frequently used for pollen studies of peat and marl deposits. It consists of a jacketed plunger that completely closes the sharpened end of the jacket. After it is pushed down to sampling depth, using the four-foot extension rods, it is drawn upward a few inches. This partly withdraws and locks the plunger in the upper part of the jacket. Then, when forced downward, the jacket cuts a ten-inch sample core.*—Courtesy of Eberbach and Sons Company.

When the proportions for genera are known for each stratum, they are represented in a standardized form, known as a pollen diagram, in which pollen spectra—the relative importance of each genus in a stratum—are plotted on horizontal lines, one spectrum above another to show the progressive changes for genera, which are shown on vertical lines. A pollen diagram is no more than a means of visualizing the pollen spectrum of a section—a vertical series of samples from the bottom to the surface of a deposit. Changes in the spectra from the bottom upward are, of course, to be correlated with time.

The shortcomings and pitfalls of pollen analysis as a method of determining past vegetation and climate are appreciated by all who

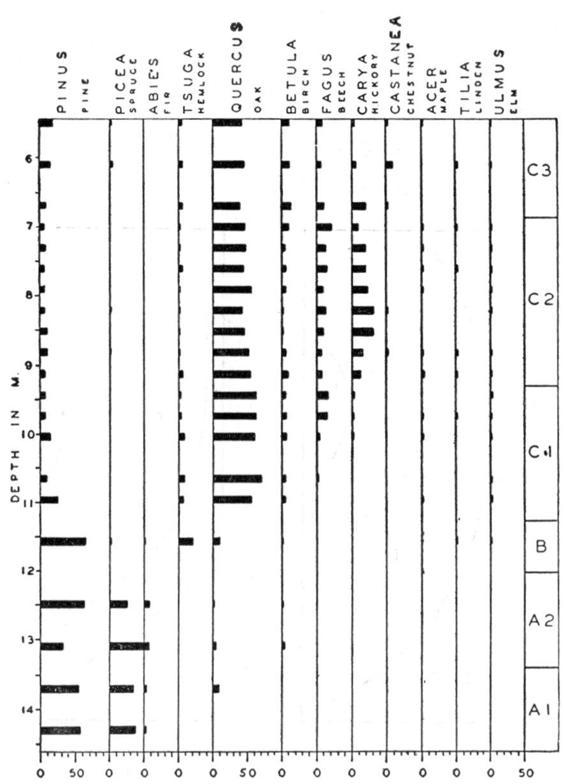

Fig. 171. *An example of a common form of pollen diagram, which also illustrates what we know of the vegetational history of the northeastern United States although derived from one place (Upper Linsley Pond, North Branford, Conn.) Zone A indicates a spruce-fir forest; the high values for pine are attributed to over-representation resulting from its light weight and its abundant production. In the northeast, a secondary maximum for spruce (A-2) is not uncommon in this period and is thought to represent a local readvance of retreating ice. Zone B, a warm dry period, shows a pine maximum and the beginning of warmth-loving, deciduous trees. Then followed deciduous dominance over a long period, in which hemlock-oak were first important (C-1), then oak-hickory (C-2), and, with cooler moister conditions (C-3), an increase of chestnut, followed by a reappearance of spruce in some localities.*—From Deevey.[130]

have used it.[65, 143] There are sources of error in methods, in records, which may be incomplete, and in identifications, which may not always be correct, and interpretations may be based upon inadequate data. Because of its simpler flora and greater amount of study, the pollen spectrum for Europe is better established and accepted than in North America. The majority of our studies have been made in gla-

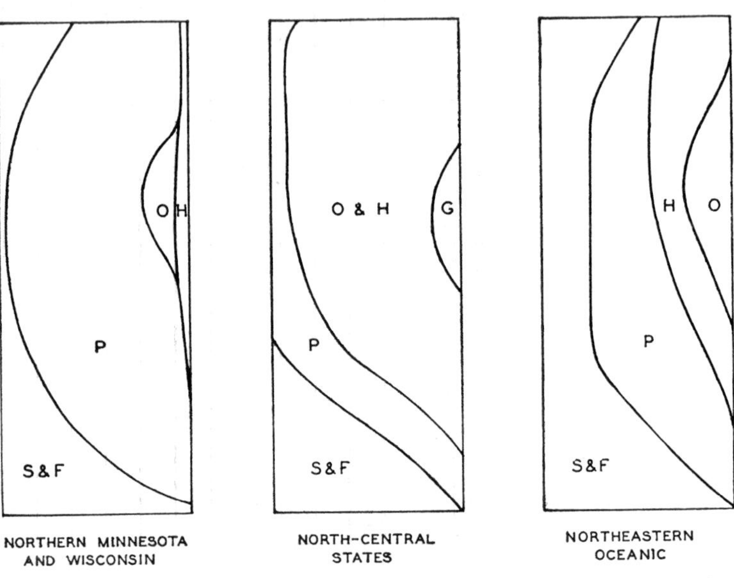

Fig. 172. *Schematic pollen profiles that show the general picture of what is known of vegetational history for the eastern United States. F—fir, G—grassland complex, H—hardwoods except oak, O—oak, P—pine, S—spruce. Depth shown vertically, percentages horizontally. Although there are differences relatable to continental and maritime climates, there is regional similarity in the indications of a middle warm, drier period, and the suggestion of subsequent cooler, moister conditions leading into the present, as well as the shift back toward early proportions of species in the upper portions of the diagrams. Succession may be a factor in these latter shifts.*—After Sears.[352]

ciated areas where bogs are common and concentrated in the northwest, the north-central states, and the northeast. Nevertheless, there is still need for more work within the glacial area to complete the picture.

It is somewhat surprising that investigators are in as close agreement as they are. Most generally accepted is a postglacial climatic series beginning with increasing warmth, followed by a period of maximum warmth and drought, followed by a period of decreasing warmth, the present.[357] This is applicable to both Europe and America. Some students would subdivide these three major periods, claiming that greater refinement is possible. Others contend that their data contain no evidence of a warm dry period in North America.

More studies are certainly necessary in North America before agreement can be reached as to all phases of the basic normal pollen spec-

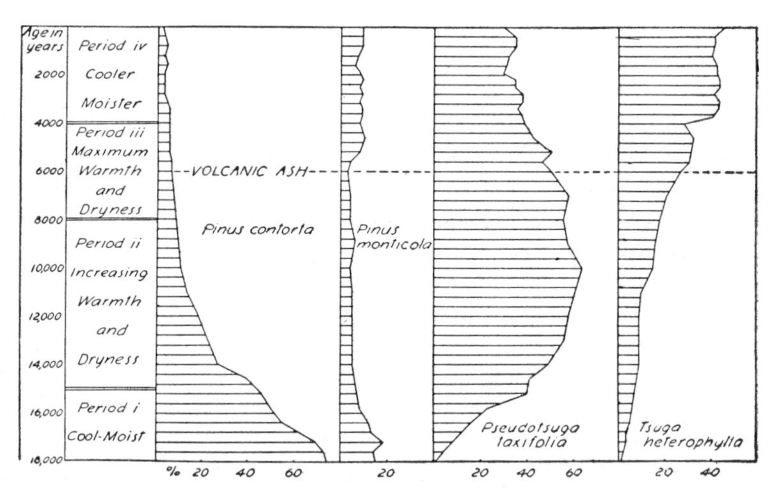

Fig. 173. *A composite of ten pollen profiles from the Puget Sound region, which is indicative of postglacial climate and vegetation in the northwest although not typical of all areas as to species. The volcanic ash level, present in all northwestern profiles, is considered to be of common age. Such composite profiles, because they eliminate the sharp fluctuations from level to level found in individual profiles, give a better picture of the trend of post-glacial vegetation.*—From Hansen.[175]

trum and its meaning in terms of climate and vegetation. The effects of glaciation on vegetation beyond its borders are still variously interpreted,[48] and greater knowledge of pollen profiles could be used effectively. The few studies of widely scattered deposits to the south of glaciated areas (particularly, New Jersey, the Carolinas, Texas) have contributed vegetational data that need further substantiation to support the possible interpretations. Likewise, there must be more efforts to correlate all sources of contributing evidence,[130, 316] a truly paleo-ecological approach: floristic, vegetational, zoological, geological, archaeological, as well as evidence from pollen analysis.

Radiocarbon Dating

The last statement in the previous section leads directly to consideration of the latest geochronometric technique that is of special biological interest because it can be used effectively for dating organic remains even of prehistoric age. The carbon assimilated by any living organism includes a known proportion of carbon 14, radioactive because of cosmic ray bombardment, and constant because it is replaced as it breaks down. After death, there is no replacement and therefore, knowing that the half life of carbon 14 is 5568 years, the radioactivity of the residual carbon in any organic matter is, within limits, indicative of its age. Original techniques permitted dating of material no more than about 12,000 years old but newer methods extend the range to some 30,000 years.

The necessary elaborate equipment and experience in its use are, as yet, available in only a few laboratories, some of which will make determinations for a fee. Nevertheless, much valuable and useful information has been obtained in the few years since the method was developed. Summaries of methodology[236, 438] and new lists of determinations[314] appear frequently. Some generalizations[147] from the latter illustrate their importance for reconstructing vegetational history. Wood buried in Carolina bogs was older than 20,000 years, but wood from bogs in the Lake States was only about half as old, a point of interest in relation to glaciation. With regard to bog-pollen sequences, carbon 14 determinations suggest that the time since maximum glaciation was less than has been supposed, that the pine period with a boreal type of climate was progressively more recent with increasing latitude (9,000 years in West Virginia, 6,000 years in Maine), and that the thermal maximum may have occurred no more than 6,000 years ago or even as recently as 3,000 years ago.

The method has many possible applications, especially when combined with other dating techniques. It is being used to advantage in archaeology and geology as well as in biology.

Dendrochronology

Another bioclimatic approach to past history was originated by an astronomer. Dr. A. E. Douglass, when he began studies of annual growth rings of trees in an attempt to correlate their differences with climatic variations, presumably related to solar activity. Cross-dating, or matching the growth patterns year by year, for modern trees in Arizona was first accomplished in 1904, but its significance was not fully appreciated until several years later.[134] Then a chronology was established from modern times back to A.D. 1400 by matching ring records of modern trees to the exterior ring records of earlier trees and so on with trees that grew still earlier. When these records were matched with rings in beams taken from ancient pueblos, the records became complete back to A.D. 1299, then to A.D. 700 and, more recently, successively to A.D. 643, A.D. 500, and finally to A.D. 11. Recent finds suggest that the chronology will be carried even further back.[181]

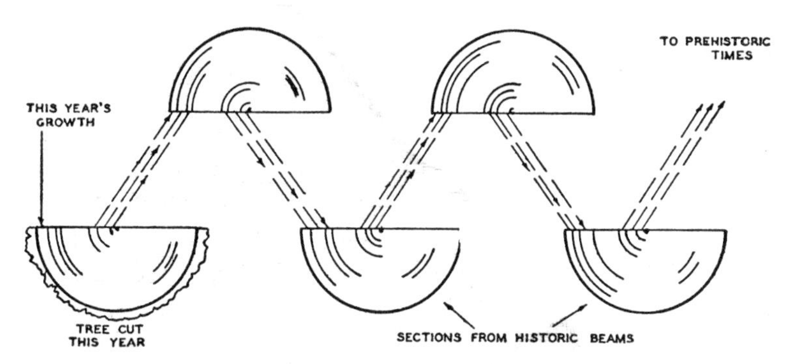

Fig. 174. *A diagram illustrating how the bridge method is used to extend knowledge of dated rings, an important part of the building of complete and continuous chronologies. The usual desirable overlap is fifty years.*—After Glock.[162]

Some of the record was completed and some of the cross-matching was made possible by fragments of wood from ancient pueblos and some even with charcoal, which was better preserved than wood. It should be noted that an even longer chronology has been worked out for the giant redwoods, which is complete for 3,000 years.

When the pueblo dendrochronology was completed, it was a major contribution to archaeology, since some thirty prehistoric ruins were immediately given absolute dates, and later many more were dated. This usefulness of the method in archaeology is apparent. At the same time, findings in archaeology have contributed to the establishment of dendrochronology as a means of studying past climate.

Recent ring studies in moist cool regions indicate that no better climate than the arid Southwest could have been selected for the initial investigations. In extremely dry regions, growth and size of rings are closely related to annual precipitation, and the correlation is not complicated by light or temperature effects. It is now known that in the north, or at high altitudes, tree growth is most responsive to temperature and that in temperate regions with adequate rainfall, both temperature and moisture factors are reflected in the rings.[131]

This does not mean that tree-ring studies are possible only in arid regions but rather that their interpretation is more difficult elsewhere. That cross-dating and correlation with climatic variation is possible in moist-temperate climates was demonstrated by Douglass' studies in Europe and several parts of the United States. In the Mississippi drainage area, the deviation from the normal annual precipitation has been shown to affect ring growth more than total precipitation, but the relationship is modified by temperature and wind as they influence evaporation.[181] Ring growth in New England has been shown, in a chronology from hemlock, to have close correlation with climate as indicated by exceptional and poor crop years.[251]

Since tree-ring analysis was originally begun with the hope that it would show solar-terrestrial relationships, it was natural that, with the establishment of long, dated chronologies, the data should be studied for cyclic characteristics. Permanent periods, or those of fixed length, showed no correlation; therefore, this idea was discarded for one of cycle complexes in which any obvious or significant recurrence of variation in data was considered to be cyclic. On this basis, definite relationships were demonstrable between sunspot activity in the past and terrestrial climate as recorded in certain long-time chronologies of tree rings. An eleven- (10 to 12) year cycle is especially pronounced throughout the old records and continues to be borne out, in a general way, for modern conditions. During periods of sunspot maximum, drought is characteristic, and sunspot minimum is associated with excessive precipitation. Motivation for making such correlations is the hope of applying them to climatic prediction if they are borne out by analysis of more long-time meteorological records and more tree records of great length.

Cyclic climatic phenomena and dendrochronology are both being used in the interpretation of glacier fluctuations in Alaska during the last several centuries. With no long-time records available, climatic conditions associated with advance and recession of the ice are being interpreted in terms of sunspot activity,[230] and dates are established from studies of growth layers of trees along the margins and below the termini of present day receding glaciers.[231] The exact year of maximum advance is determined from trees partially pushed over by ice but which continue growth from a tilted position. This is possible because, after tilting, growth in diameter is assymetric in non-conformity with the previous symmetric growth of erect trees. Landslides, windstorms, and river floods can also be dated in this way.

The Relict Method

As for several other phases of dynamic ecology, we are indebted to Dr. F. E. Clements for recognizing the potentialities of the relict

Fig. 175. *Community of ponderosa pine occurring as an isolated island in sagebrush desert wherever the special local soil conditions exist in Nevada. Often disjunct from nearest ponderosa pine forest of the Sierra by fifty miles or more.*—Photo by W. D. Billings.[28]

method, for demonstrating its usefulness, and for a clear and complete exposition of the entire subject. The brief discussion that follows can hardly avoid being a condensation of his ideas.[80]

"In the ecological sense, a *relict* is a community or fragment of one that has survived some important change, often to become in appearance an integral part of the existing vegetation." The concept may be applied to individuals or a species, but is more often used for communities. It may be used to describe delayed or lagging stages of succession, but it has far greater usefulness in connection with climax vegetation.

The usefulness of relicts lies in their indicator value of past conditions of habitat and vegetation as well as of the causes underlying

Fig. 176. *Effect of grazing on mixed prairie in central Colorado. Short grass, to left of fence, is typical over much of the region but, where cattle are excluded, to right, mixed prairie develops.*—Photo by R. B. Livingston.

changes that have occurred elsewhere in the area. A relict community having remained relatively unchanged because of peculiar local conditions is an actual sample of, or shows strong similarities to, previous vegetation. At the same time, the peculiarities of the relict habitat may be indicative of environmental conditions previously characteristic of the area as a whole and may, therefore, be suggestive of why vegetation changed generally there.

Relict communities occur where local edaphic, topographic, or biotic factors differ sufficiently to compensate for the effects of environmental conditions obtaining generally. Thus altitude, exposure, or soil may provide locally unusual moisture conditions. Ridges, streams, and lakes may constitute barriers to fire. Peculiarities of drainage may result in swamps, bogs, and low flood plains. Any such condition may be effective in maintaining relict communities, which, in terms of

Fig. 177. *Relict (postclimax) black spruce forest in a Minnesota bog.*—U. S. Forest Service.

climate, could not be anticipated. They are the relicts indicative of shifts of climax and climate over long periods.

A quite different kind of relict is one that is maintained by man, purposely or otherwise, after he has destroyed or modified the picture of climax generally. Overgrazing, cultivation, and lumbering have destroyed or modified climax over extensive areas to such a degree that its recognition and interpretation, even though its destruction was within historic time, are dependent upon remnants of the former vegetation. Such relicts may be found in fence rows, along railroad right-of-ways, in old cemeteries, and in any areas long undisturbed and may yield much information about the past. The deliberately protected areas, such as game and wildlife preserves, natural areas, Indian reservations, and national parks, offer still more possibilities because of their extent, frequently included virgin areas, and relative permanence.

Relicts related to climatic change are most abundant in the transitions from one climax to another but may likewise be found well within the general range of a climax, provided the local conditions are present that maintain the necessary compensating factors. Usually the local conditions are a result of topography, which, through its effects on precipitation, drainage, temperature, and air movements, permits the relict to survive. The resulting relict communities are the previously discussed postclimaxes and preclimaxes, whose presence greatly simplifies the interpretation of shifts of climate and climax in the past. The present condition of the relict, if free from disturbance, may furnish strong evidence of the present degree of climatic stability. Judgment of such evidence must, of necessity, be tempered by what is known of climatic cycles. In parts of the West, precipitation may be several times as great during a period of sunspot minimum as that during sunspot maximum. The condition of vegetation, particularly in relict communities, must be interpreted accordingly.

General References

S. A. Cain. Pollen Analysis as a Paleo-Ecological Research Method.

S. A. Cain. *Foundations of Plant Geography.*

F. E. Clements. The Relict Method in Dynamic Ecology.

A. P. Dachnowski. Peat Deposits and Their Evidence of Climatic Change.

G. Erdtman. *An Introduction to Pollen Analysis.*

W. S. Glock. *Principles and Methods of Tree-Ring Analysis.*

P. B. Sears. Climatic Interpretation of Postglacial Pollen Deposits in North America.

F. E. Zeuner. *Dating the Past, an Introduction to Geochronology.*

Part Five: PRACTICAL CONSIDERATIONS

Chapter 13 Applied Ecology

Man is rapidly becoming the earth's dominant organism. To an increasing extent, natural communities survive because he tolerates them, are modified to suit his purposes or fancy, or are destroyed, sometimes through his carelessness, but usually so that land may be used for agriculture, industry, or other activities. His dominance is of a different order from that characteristic of communities in nature, for, with his knowledge and technology, his activities are often so extreme and so rapid that their effects are like those of a series of catastrophic natural events. Thus he may not only destroy or modify natural communities, but he may also frequently modify the environment to a great extent. Suggestive of a different form of environmental modification are the recent experiments with rain-making by seeding of clouds. All this is necessary from our modern point of view and will continue, perhaps at an accelerated rate, as populations increase and the earth is more completely occupied and used.

Natural communities and their environments, particularly the soil, are natural resources. When they are destroyed or modified, they may reappear only after a long period of time or, with extreme disturbance, their reappearance may even be impossible. It becomes increasingly apparent that future generations may require these natural resources and likewise that man has been most wasteful of them, especially in modern times. A problem today, which will become greater in the future, is that of how to use such natural resources to the fullest extent without jeopardizing their continued availability for future needs. The problem is fundamentally ecological. Its solution depends upon the comprehension and application of ecological principles.

Since man is becoming the dominant organism and also is gifted with thought processes, his dominance should be such that he turns natural laws to his advantage or, at least, does not permit them to work against him. It is in this connection that applied ecology becomes useful. The characteristics and distribution of natural communities, the nature of the environment, and the interrelationships of organisms and environment are subject to natural laws, which the ecologist seeks to recognize and verify. The more completely man understands the pattern of these interrelated processes, the greater the probability that he will remain a permanent dominant, assuming that he restricts his activities to the limits of these laws. Only if biological laws are recognized in full can we hope to rebuild the natural resources we have destroyed, or even maintain those still available to us.

If we knew the ecology of all natural vegetation and that of all crop plants, strong recommendations for land use could be made in terms of its greatest contribution to society. Not only could agricultural, forestry, and grazing lands be positively recognized, but the details of management for maximum continuous production could be recommended with certainty. Quite obviously, ecological knowledge has not accumulated to this extent. The ecology of natural vegetation is still inadequately known, and the ecology of cultivated plants has not been sufficiently studied. If the ecologist is to contribute successfully to the direction of man's activities as a dominant, there is still much that must be learned. On the other hand, even though knowledge is incomplete, ecology has much to contribute that has not been fully utilized in applied fields. What is known should be applied when man destroys or modifies natural communities. Much progress has been made in the use of ecological principles in several fields, but their potential application is still great.

Forestry

The early history of lumbering in North America indicates, on the part of lumbermen, a complete disregard for forests as a natural resource and little concern for the future. Foresters have long been conscious of this improvident attitude, although until recently they were usually unable to change the lumberman's methods or point of view. Through the years forestry has become a respected profession as the necessity for scientific management has become apparent. An important part of a forester's training is forest ecology, or *silvics,* in which he learns the scientific background upon which silvicultural practices are based.

A generally accepted definition of silviculture states that it is that branch of forestry dealing with the establishment, development, care, and reproduction of stands of timber.[404] More often than not the silviculturist aims to control the establishment and development of forests so that they will be made up predominately of economically desirable species or so that merchantable timber will be produced in a minimum of time. Or, he may be interested in results not directly related to the production of lumber. Cultural operations may point to erosion control, watershed protection, dune stabilization, game encouragement, or recreational purposes,[385] or, if in the West, to a better balance between timber production, watershed control, and use of the forest for range purposes. If his methods are scientific, they will be based upon reasons derived from silvics. Consequently, the more completely forest ecology is understood, the more successful should be its application in silviculture.

Since the practice of silviculture almost invariably involves attempts to control forest communities and their development, a knowledge of successional trends and the climax of the region is all important. Knowledge of the principles of succession should make it obvious that the simplest form of management would be one that least modifies the natural development of vegetation. To maintain a successional community indefinitely requires considerable effort, if it can be done at all, but the nearer the desired forest type is to the climax, the easier it should be to maintain it. These may seem to be obvious generalizations, but they have not been, and are not, fully appreciated or applied.

In the past, artificial forest types have been attempted under a great variety of conditions. Species have been planted outside the limits of their natural ranges, even including several introduced from other continents. Often such trees are grown in pure stands or, if not, then in combination with native species to make quite unnatural communities. Even more common have been the attempts to grow species on sites to which they are not naturally adapted. The situation in New England is illustrative. Here the original forest has long been gone, and reforestation and silvicultural programs have been in progress for some time. Introductions include Scotch pine, European larch, and Norway spruce from Europe, and white spruce from the northern conifer forest. Red pine and white pine have been grown at the fringe of their range in pure stands on rich, heavy soils instead of the sandy soils on which they naturallv occur.

Fig. 178. *Eight-year-old plantations of pine on the same soil type (Duke Forest, Piedmont of N. C.) to compare growth of nothern species, (1) red pine, and (2) white pine, with that of native loblolly pine (3). The pictures speak for themselves.*—Photos by W. R. Boggess.

The production of artificial forest types in New England can, as a whole, be described as unsuccessful. An appeal for the application of ecological principles,[387] maintains that older trees are often of poor form, and growth is likely to decline sharply in later years. Very few artificial stands have been profitably brought to maturity. Furthermore, these types are especially susceptible to damage—from insects and other animals, from disease, and from the elements. Norway spruce is severely attacked by the white pine weevil; exotic larch plantations may be severely damaged by the porcupine and squirrel; red pine, south of its natural range, is particularly susceptible to *Tympanis* canker and to attacks by the European pine-shoot moth; crookedness of Scotch pine has been attributed to frost damage; weevils do more damage to white pine on heavy than on light soils. It is admitted that eventually, if sufficient knowledge is acquired, artificial types may be grown successfully. For the present, they cannot be recommended for New England because of previous lack of success, the risk involved, and the high cost of production. Probably similar generalizations can be made for most of the forest regions of North America but with less evidence, because there has not been as much experimenting elsewhere. Although forest species have been successfully introduced into new areas, as, for example, the eucalyptus into California, the results in New England are suggestive that such experimenting might be of dubious value and certainly would not yield the necessary information except at great cost over a long period of years.

If only natural forest communities are to be the objective, there are two general types to be considered. Silviculture has usually been given consideration only after the old forests were destroyed and, not uncommonly, after much of the land was used for agriculture and subsequently abandoned. Under these conditions, the abandoned land supports various early stages of secondary succession, and cutover land is in late successional or subclimax forest. The problem then becomes one of cultural practices designed (1) to maintain the temporary forests of successional nature or (2) to permit stands to develop to climax or near-climax conditions.

The relatively short-lived successional communities often include as dominants the most valuable trees (e.g., pine where hardwoods are climax) and, because of their rapid growth, the most desirable commercial species growing in the region. But, because of their successional position, when these species are removed, they are replaced by other species, representing later stages of succession, whose seedlings were there and released by the cutting. The problem of maintaining dominance of such temporary species has been given much study, but it is by no means solved. Without expensive cultural operations, usually including planting and periodic weeding, these temporary types cannot be maintained indefinitely. Even though the productiveness of a desired species in a stand may be extended by various types of cutting and treatment, its replacement is inevitable. Almost invariably, the succeeding stand tends to be nearer the climax than its predecessor and will include a higher proportion of economically less desirable species. Where successional species are fire resistant, there is the possibility of using controlled burning to hold back succession and maintain dominance of the temporary type. Under these conditions a temporary type could be cut selectively and provide a continuous yield. The merits of the method have been argued and are being tested for the longleaf pine forests of the coastal plain of the southeast.

The alternative would be to allow all forest land to develop toward the climax or at least to near-climax conditions. Once established, such forest would require a minimum of silvicultural attention. Continuous production would be assured, and with judicious selection of species for cutting, undoubtedly the proportions of desirable and undesirable species could be controlled. Additionally, permitting natural development of stands should result in a distribution of species in the habitats to which they are best adapted. Different conditions of soil, exposure, and moisture would support stands of different composition, but presumably these species would be making their best growth although a minimum of management would be involved. This is not to imply that silviculture is unnecessary. For example, artificial planting is frequently economically justifiable, since it assures uniform stocking and even-aged stands and may speed stand development by several years. If there are few seed sources of desirable species, succession may be so long delayed, by shrub stages, perhaps, that planting becomes an economic necessity.

Silviculture is usually desirable and often a necessity, but it should be emphasized that its practices, to be most effective, should be governed by ecological knowledge. The less cultural practices tend to modify the natural trends of succession, and the more nearly the desired forest is to the climax of the region, the less the effort and expense there will be in developing and maintaining it. Here is an economic reason for learning the nature of virgin forest wherever it still remains and for determining all that is possible of its variations with habitat. Similarly, successional trends must be known in detail for every major soil type and situation if cultural practices are to be adjusted accordingly. Secondary successions are of major importance these days, and they can be worked out for any region. Climax forest in virgin condition is rapidly disappearing, and usually only remnants remain for study. Their characteristics should be recorded at every opportunity. When possible, representative portions of these virgin forests should be saved intact for future study.

Range Management[391]

The objective of range management is to produce the highest possible forage yield while the condition of the range is maintained

Fig. 179. *To permit grazing to continue until range is entirely depleted and gullying has reached such extremes is obvious mismanagement, but it happens all too frequently. Note absence of gullies under protection of oak tree.*—U. S. Forest Service.

or actually improved. To this end, the methods of ecology have been used to such an extent that range management is largely applied ecology, and just as silvics is the basis of silviculture, so is range ecology the basis of range management.

Fig. 180. *Illustrations of blue grama-oak savannah range that tell their own story of good and poor range management.* (*Above*) *"One of the finest demonstrations of range and livestock management in the southwest."* (*Below*) *Range depleted by overuse and poor management. Note differences in condition of cattle, amount of forage, ground cover, and erosion.*—U. S. Forest Service.

Range ecology has, on the one hand, concerned itself with the purely ecological concepts of regional climaxes, with grazing value, and the patterns of succession for each. On the other hand, there has been the practical consideration of the quality and type of forage provided by each of the communities and of how they may be controlled or modified to advantage. Only suggestions of the nature of the research on these problems can be given here, but they should indicate the degree to which ecology is contributing to the solution of range problems.[295]

Fig. 181. *Two years before, this Idaho range supported* Wyethia *and sage. Seeding with timothy, smooth brome, and clover, and protection for one year produces this abundance of forage at the end of the first grazing season.*—U. S. Forest Service.

The seasonal variations of major species have been studied in terms of grazing value. Competitive relationships of grasses and forbs (associated herbs) have been investigated as well as their relative palatability. The effects of grazing on community structure have been given much attention, particularly with regard to criteria for the recognition of excessive use and the time and conditions necessary for recovery to normal. As a result, the carrying capacity of many forage types is well known, even for different seasons of the year. With regard to range condition and carrying capacity, the effects of rodents have been studied as well as the effects of predators upon the rodents. Effects of drought have been given much attention as well as the rates with which ranges recover from drought; and, in this connection, the water requirements of important individual species have been determined. In the consideration of water, the effects of different types of cover on runoff, flooding, erosion, and water supplies have been studied in detail. In attempts to rebuild depleted and eroded ranges, there have been studies of artificial seeding and planting to speed recovery. As in forestry, numerous foreign species have been tested with some successes (e.g., crested wheat grass) in an attempt to improve conditions.

Because grazing is a part of every question in range ecology, the exclosure method is an important technique in range research. Exclosures are especially useful for testing experimental conditions; but they, or equivalent isolated areas, are likewise necessary for determining the nature of climax and related successional communities. In experimental studies, exclosures, in combination with grazed areas around them, are one of the better means of determining the effects of conditions in progress on that range. If causes are to be investigated, they are tested separately, each with its controlled treatment, on individual plots within an exclosure. Such treatments may include clipping (for grazing), burning, trampling, seeding, etc. As indicated earlier, the installation of exclosures of sufficient size, which will keep out rodents and yet will not alter microclimate, presents numerous difficulties. Consequently comparative studies on ranges supporting different, but known, animal units are coming into use whenever possible, because they do not require exclosures.

When the results of such studies are evaluated and expressed in general terms, it becomes apparent that several principles have been established that appear to be universally applicable.[81] From an ecological point of view, these principles, determined by experiment, would seem to be self-evident, since they conform to ecological theory. It must be remembered, however, that these things were originally theory and now can be stated as principles supported by experimental evidence. The testing was necessary to establish them as tried bases for range management. In grasslands, no less than elsewhere, succession is operational; and all trends constantly proceed toward the climax unless they are modified by disturbance or are held in check by an unfavorable swing of climate, as during a series of dry years. Grassland is a climatic life form, which maintains itself in the absence of disturbance and which, if destroyed, reappears when the disturbance is removed. All evidence indicates that perennial grasses tend to become dominant and eliminate annual grasses, forbs, and shrubs in the absence of grazing, fire, or similar destructive agencies. The grasses of a particular climax are adapted to its climate and usually have an advantage in terms of competition over introduced ones.

From the above, it becomes apparent that, as in forestry, practices of management that least disturb the natural balance of grassland and its environment are most desirable. Those that take into consideration the trends of succession and local climax are likely to be most successful at the same time that they require a minimum of expended effort. Although a few exotic species have proved to be easier to propagate than native ones, the introduction of foreign species for range improvement or erosion control is likely to be unsatisfactory, unless those species are to be given extra care or special cultural conditions. In fact, there is evidence that seeding of native species should be done only with locally produced seeds, since the species may consist of geographic physiological races.

The establishment of general principles is being followed by more and more intensive studies of local variations in communities and environments. The productivity of most range lands has been reduced by man's domestic animals coupled with seasons of unfavorable climate, and to rebuild ranges to a higher level of productivity will require an understanding of the special conditions of local areas as well as the broad principles for the region. Our public lands in the West, most of which are grazed, have been divided for research and administration in a fashion that suggests a natural application of the above. Several grazing *regions* are designated, which correspond to the major differences in the grassland and scrub formations. These in turn are divided into several districts, which represent local variations in dominants and environment. Application of the general principles is possible for regional administration and management, but local application must be modified in terms of detailed local studies.

Agriculture

If a crop is planted and grown successfully, it follows that the methods applied, within the general region, to the particular field and for that season, were ecologically correct; since cultivated crops are as subject to ecological laws as are plants growing naturally. Study of the ecology of cultivated plants has progressed rapidly in recent years. It includes crop ecology, which is applied ecology in the ordinary sense, and ecological crop geography, which considers the effects of both physiological and economic factors on production and distribution of crop plants.[216] With this addition of "social" factors to the physical and physiological ones, the already complex environment becomes still more so, and the crop ecologist must integrate his observations and conclusions with additional fields. This phase of ecology is, as a whole, beyond our consideration here, but it is appropriate to emphasize that ecological principles are becoming a part of our way of thinking. They should undoubtedly be given even greater attention in these days of a planned economy, which affects us all.

Crop Ecology. The cultivated plant is as subject to ecological law as a native one; and, consequently, there is as much ecology to be studied in a corn, tobacco, or cotton field as there is in a forest. To be sure, largely by trial and error, the farmer has learned to grow crops so that they give a reasonable return for his labor. On the whole, however, this has been done at the expense of the soil as a natural resource. The natural fertility of most of our soils is largely depleted, erosion has ruined thousands of acres and reduced the productivity of many more, and water tables have been lowered to such an extent that crops in areas with rainfall sufficient for hardwood forest are suffering during dry spells as much as they would in grassland climate. Thanks to increased knowledge of fertilizers, the development of productive hybrid strains of various crop plants, and modern mechanized methods, our yields have steadily increased; but this cannot proceed indefinitely, especially since much of the increase in yield results in further depletion of the soil.

To counteract the downward trend of productivity, soil conservation and erosion control are receiving greater attention. Increased knowledge of crop ecology is imperative so that the highest yielding species will be grown under the proper conditions of cultivation and on the right sites. If possible, yields must be maintained at high levels at the same time that soils are improved rather than being depleted. The ecology of weeds, pests, and diseases must be studied so that the depredations of these products of cultivation may be held in check effectively. These things are not being neglected by agronomists and horticulturists, but there are special contributions that can be made if the investigator has the ecological point of view.

Land Use. At one time it was customary to clear all workable land for agriculture, permit plowland to revert to pasture only when it became unprofitable, and permit pasture in turn to revert to forest only under the same conditions. This procedure is not as common as it once was; perhaps it should be reversed completely. Perhaps the soundest ecological apportionment of the landscape would be represented by a minimum of carefully selected, skillfully operated

Fig. 182. *Once-fertile farm land that has been unnecessarily destroyed by surface erosion and gullying because of lack of concern (note that straight-row cultivation still prevails) and lack of understanding. Perhaps this area should have been put into forest long since. If it had been, it would still be valuable.*—U. S. Soil Conservation Service.

plowland with a maximum of natural vegetation. Where this natural vegetation consists of grassland, regulated pasture is an aspect of its normal development; where it consists of forest, it should be scrupulously protected against grazing, and whatever pastures are required should be handled with the same measure of skill that has been suggested for the plowland.[356]

Maintaining stands of natural vegetation provides areas for ecological comparison and diagnosis, insures that soil is being rebuilt and retained, provides organic matter, insures a regulation of moisture conditions that man cannot duplicate, and provides food and shelter for wildlife, which may be significant in reducing crop pests.

The planning of such land use should be, in so far as possible, based upon ecological principles as related to soil, topography, exposure, and drainage in terms of the climate and cultivated crops it will support. Special land-use problems arise on hilly land, which need not necessarily be unproductive. Ecological studies of hill-culture[263] are showing how some such lands may be used to grow orchards, vineyards, pasture, and other crops without depletion or erosion of the

Fig. 183. *A half-acre farm pond in West Virginia of the type being widely installed for food production and recreation.*—U. S. Soil Conservation Service.

soil.* Under proper conditions artificial fishponds can be a profitable investment. Successful operation of such ponds requires adequate knowledge of their ecology, including sizes and depths for different climates, drainage, amount of available water and necessary aeration, rate of silting under different conditions, fish food relations, kinds and amounts of fertilizer necessary, kinds of fish, and rate of stocking. Marshes might be retained and improved for muskrat production; but, again, the practical problems, largely ecological, have not been sufficiently explored. Stream margins create other land-use problems.

* Much of the following discussion of applied ecology in agriculture is adapted from an unpublished report by the Committee on Applied Ecology of the Ecological Society of America, 1944.

Fig. 184. *This eroded stream bank in Wisconsin was graded, laid with willow poles, and planted with a few willow sprouts. Only two seasons were required to produce the growth shown in the second picture where undercutting is effectively stopped and shelter is provided for wildlife.*—U. S. Soil Conservation Service.

Fig. 185. *Waste field-margins such as the fourteen-foot strip (1) abandoned because of root competition and erosion can be made useful.* (2) Lespedeza bicolor (*tall*) *and* L. sericea *planted in strips are holding the margin stable and producing food and cover for small game.*—U. S. Soil Conservation Service.

Usually they are grazed, with resulting accelerated run-off and bank cutting. The species that might appear under protection should be known, as well as the most desirable species for checking erosion. In many sections, planted hedges and field border plantings are being recommended on the unproductive margins of fields to reduce erosion and provide cover for wildlife. The ecology of the planted species must be known as well as its effects on the crop beside it. Also the ecology of the insects, birds, and mammals of these margins must be known. Are they desirable, beneficial, or are they harmful to desirable species?

Land Management. The operations by which land is prepared for crops, their planting, harvest, and use are known as *land management*.

Fig. 186. *A simple illustration of improper management. The amount of runoff on this slope means leaching and erosion. Certainly the rows should not have been put in up and down the hill, and perhaps, without terraces, clean cultivation should be ruled out on this field.* —U. S. Soil Conservation Service.

For greatest efficiency, good land management must parallel good land use. These are arts but, today, arts requiring all the help of science possible.[359]

A farm planted year after year to wheat or cotton does not, even with fertilizer, conform to the balances that occur in nature. Well-managed fields may seem to approach a condition of balance as a result of rests with rotation pasture, the use of legumes, and the addition of fertilizer. Yields may be high and sustained, soil may not erode, and all appear to be at its best. In terms of natural vegetation, however, our modern methods of land management may be questioned. Cultivation produces conditions similar to those in early stages of succession, conditions that in nature would be temporary and soon change in the direction of climax. We must have crops, but, if climax vegetation utilizes natural conditions most effectively—and that seems reasonable—are our methods of cultivation the best we can use for obtaining our crops? Is our method of deep plowing, with destruction of soil structure, best under all conditions? Should all

Fig. 187 (1). *An Indiana field after fall plowing showing severe erosion. Picture taken when it was decided to retire field to permanent pasture with contour furrows.*

crops be cultivated clean and all organic matter be turned into the soil? Might not mixed crops producing a complete cover as in nature be more desirable? Perhaps we have gone too far in producing unnatural conditions. The artificial environment of cultivation results in soil erosion, a modified soil flora and fauna, and changes in water relations. Also we have more diseases of crop plants and more insect pests than ever before.

These are ecological problems. Intelligent land use minimizes some of them. Practices like contour plowing, terracing, and strip cropping are moves in the direction of reducing them. With increased knowledge of the ecology of crop plants, especially in terms of natural vegetation, some of our methods of use and management may require revision.

Pasture Problems. If the same attention to management were given to pastures as to plowed land it would be a reversal of the usual point

Fig. 187 (2). *The next year, after gully-control work, this excellent planted pasture had taken over, the soil was stabilized and the field saved for long-continued usefulness.*—Both photos by U. S. Soil Conservation Service.

Fig. 188. *Deciduous forest farm wood lots, pastured (above) and not pastured (below), which illustrate the effects of browsing and trampling on reproduction and general forest condition.*—U. S. Forest Service.

of view, since pastures are so often on the poorest land and are given little or no attention. With the steady expansion of dairying, especially into sections of the country where adequate pastures do not produce themselves, the need for study of pasture ecology increases. The necessity for seeding is now widely accepted. Many species have been tested for palatability, yield, food value, and soil-building properties but there is still much to be learned about the management of planted pastures. Regional pasture ecology has not progressed as far as range ecology. There is much yet to be studied, tested, and put into practice. The implementation of such a program moves slowly in sections where pastures are not recognized as a crop to be managed like any other.

An illustration of the misconceptions regarding pasture is the common practice of including the farm wood lot in the pastured area, although it provides little more than browsing, which supplements feed during off seasons. To the ecologist, it is obvious that this is at the expense of seedlings and ground cover and that it will result in stand deterioration.[245] Silviculturists have shown that properly managed wood lots can yield as great a return as any average farm acreage, but the wood lot pasture persists. A study of maple groves in Ohio[358] showed that in three years the elimination of grazing resulted in an increased yield of maple syrup, worth more than twice what the rental for pasture would have been. At the same time, the condition of the stand was noticeably improved. As more such information is accumulated[116] it is to be hoped that its application will follow.

Regional pasture studies of an ecological nature must be continued so that both species and their culture can be recommended with confidence for climate, soils, and land management policies as they occur.

Insect Problems. The relationship between land-management practices and insect populations is inadequately known.[176] Whether insect pests will increase or decrease with strip-cropping or particular crop rotations cannot be said with certainty. Probably more complex are the relationships of insect populations to the birds and mammals that will appear in response to such conservation practices as cover crops, hedges, and field border plantings. Whenever the acreage of a cultivated species is increased extensively in an area, or a new species is introduced for special purposes such as erosion control, insects may appear with it or abruptly increase in numbers to pest proportions. Such relationships and innumerable others need more study. The possibilities for applied insect ecology in agriculture and forestry are almost unlimited.

Rodent Problems. Especially for range lands, ecological knowledge of rodents is still inadequate. In spite of this, rodent control has been attempted in these areas for years. More should be known of the effects upon rodent populations of kinds and degree of grazing as well as what effects the various rodent-control measures have on the condition of the range. With the latter, it should be possible to say what percentage of a rodent population can be destroyed by a control measure, how long before the surviving population will return to normal, and to what extent species move in from untreated areas. Complicating the above problems is the usually cyclical fluctuation of most rodent populations and the obvious desirability for adjusting control to these natural fluctuations. Other suggested ecological problems are the relationship of rodents to reseeding, succession, and climax in range land, and their numbers and effects upon orchards when managed with cover crops.

Weeds. The occurrence of weeds as a result of land use and their control by cultural practices have received far less attention than control by direct, physical means. Yet cultural control or control as a result of good land management is likely to be the most permanent and least costly. Certainly the weed problem has not been reduced by centuries of cultivation, mowing, and burning. Even modern "hormone" sprays, although they have come into wide use, are no panacea.[336] If progress is to be made, the autecology of the principal weed species must be studied in detail. If, then, the effects of various types of land use and management upon the occurrence of specific weed species is learned, there is a reasonable possibility that ecological controls could be used that would reduce the weed problem, under certain situations at least.

A recent development in applied ecology related to weeds, in the broadest sense, is the selective application of selective herbicides in the control or management of vegetation under power lines and on other rights-of-way.[141] When the ecology of the community is known, selective elimination of undesired species for a few years can result in an herb or shrub dominance that may be so stable that eliminating invaders and holding succession requires only a little further periodic effort.[310] Considering the effort required to keep modern right-of-way acreages clear, this method holds much promise. Note that it is not a blanket-spraying technique but an ecological approach to conversion of vegetation from one growth form to another.

Conservation

The problems of conservation are extremely diverse, including as they do such things as soil and soil water, wildlife of all kinds, and aesthetic considerations. All that we have discussed of applied ecology could be classified under the general heading of conservation. The field is so broad as to require specialists of all kinds in its management; but this, of all fields, requires training to see each problem in the light of others. Nowhere can the ecological point of view be more effectively applied.[397]

To illustrate the limited effectiveness of specialization without ecological appreciation, witness such operations as have been known to take place almost simultaneously on public lands: a road crew cutting a grade in a clay bank so as permanently to roil a trout stream that another crew is improving with dams and shelters; a silvicultural crew felling wolf trees and border shrubbery necessary for game food; a roadside cleanup crew burning all fallen oak fuel available for fireplaces that are being built by a recreation crew; a planting crew setting out pines in the only open fields available to deer and partridge; and a fire-line crew cutting and burning all hollow snags on a wildlife refuge.[234] Such conflicting activities have not been uncommon in the name of conservation. Some government agencies have spent millions for flooding marshes and improving them for wildlife, while other agencies were attempting to drain marshes of questionable agricultural value. Great dams have been built for reclamation purposes, but the watersheds above them have been ignored.[360] With continued lumbering and grazing, the reservoirs are silting in so rapidly that the usefulness of the dams promises to be short-lived. To assure integration of such activities may not require a "declaration of interdependence,"[397] but certainly the recognition of the interdependence of biological phenomena is necessary. This end will certainly be served if those responsible are ecologically trained or have an ecological point of view regardless of their special interests.

Fig. 189. *The deposition of silt and sand behind a dam in this fashion defeats its purpose of water storage and reduces the efficiency as a source of hydroelectric power.*—U. S. Soil Conservation Service.

Soil Conservation. The recognition of soil conservation as a national problem is of recent origin. The Soil Conservation Service was made a permanent bureau of the U. S. Department of Agriculture in 1935, although it originated as the Soil Erosion Service in the Department of the Interior in 1933. Since then great progress has been made in educating the public to the need for a continuous program of conservation, and soil conservation as a science has developed rapidly. The scope of the field and the problems involved have been admirably summarized in various publications.[23, 207]

Early publicity by soil conservationists was essentially a plea to save our irreplaceable land, a great deal of which was already permanently lost and much of which is in the process of being ruined. More recently, the emphasis has been upon rebuilding lands that have deteriorated. The modern philosophy considers soils, like forests, to be natural resources that are renewable and, therefore, subject to management that will give a sustained yield over an indefinite period of time.[265] Such a program is, of course, as justifiable as the original, which aimed primarily at erosion control. It indicates that the conservation program has been successful and is maturing.

Soil conservation is, therefore, more than erosion control. It also involves the retention of water, especially on slopes, and its utilization to best advantage. At the same time, it aims to maintain or increase soil fertility and productivity. Thus soil conservation is merely the practice of agriculture in the best possible way, and we have already suggested how the ecological approach to such problems is most likely to be successful.

Not all the various measures successfully introduced for erosion control and soil building are applicable everywhere but must be adjusted in terms of soil types and climate. However, certain generalizations can be made that have wide application and whose special use or desirability often must be determined by a knowledge of local ecology. Vegetative cover is the most effective means of checking erosion. This raises questions as to what cover is desirable or possible under different conditions, where it should be permanent, and when it should be of native vegetation. These problems are related to strip-cropping, gully control, cover crops, and decisions to cultivate hilly land, put it into pasture, or plant it to forest. It is now assumed that the control of erosion will pay dividends only when proper crop rotations and fertilizing practices are followed. The interrelationships must be known for every crop and region.

Much advance has been made in cultural practice. Contour plowing, in which cultivation follows lines of equal elevation, is becoming steadily more common. In many areas, strip-cropping is an additional control, in which clean-cultivated crops are planted between strips of cover crops, such as legumes, which retard runoff and hold soil. A further necessity on contoured slopes may be terraces, which are ridges so placed that they catch and hold water in a channel behind themselves and thus check runoff and cause water to soak in. In special instances, deep furrows are maintained (listing) in which water and snow are held, and crops are planted in the bottom of these troughs. Basin listing is done on some soils with special machinery that shapes these troughs with cross dams at regular intervals further to reduce runoff. It has been shown that wind erosion can be reduced by "stubble mulching," in which subsurface tillage keeps old organic debris on the surface. Windbreaks of various kinds are known to be effective also.

All these are examples of modern practices that are proving effective under special conditions. They are not by any means new, since they have been reported in various forms far back in history. It is their application in the light of modern knowledge that marks advance. The more complete the knowledge of all factors involved—crop, soil, climate—the greater the success of their application in the future. The research programs continue, and the investigations in progress are invariably ecological in nature. Studies include effects of cropping systems, crop rotations, handling of crop residues, and management in terms of runoff, yield, and soil properties.[263]

Some special problems of soil conservation still requiring a great

Fig. 190. *Aerial view showing strip-cropping of terraces that follow contours. Erosion is checked, much water is retained, and what runs off is directed to a sodded runaway channel. Such elaborate operations may require co-operation of several landowners. In this instance, two farms are involved.*—U. S. Soil Conservation Service.

deal of study are related to drainage of water-logged land and swamps, irrigation of lands with insufficient water, clearing of toxic salts from irrigated land and other lands not previously cultivated.

Water Supply. The conditions necessitating soil erosion control and the prevention of runoff of surface water are commonly reflected in the general water supply. In many agricultural areas with adequate rainfall, there are water problems that did not exist at the time of settlement. Where once streams and springs were abundant and flowed continuously, now they are intermittent, and summer water supplies are often low. In Ohio, the water table, as evident in well depths, is from 15 to 50 feet lower than originally. Floods appear to be more frequent and are certainly more destructive than before. On the credit side, there are now no malaria problems related to undrained swamps or typhoid epidemics resulting from contaminated

Fig. 191. *The type of dam and spillway being installed primarily for water conservation. When full, this reservoir extends fifteen miles upstream over an area of 10,000 acres. The flow from the dam can be controlled, thereby providing constant flow during dry periods and reducing danger of flooding with high water.*—U. S. Soil Conservation Service.

city water supplies.[356] The adverse conditions result partially from the removal of natural vegetation for agriculture. As much water falls today as before, but more of it runs off rapidly. Thus summer drought and spring floods are partially explainable.

There are other contributing factors. Roads, so important to the farmer for transportation, likewise serve to drain off water from his fields. This has been especially bad in the mid-western states where all roads were originally laid out in an east-west, north-south grid

Fig. 192. *A power-dam lake at the edge of a town in Minnesota as it appeared in 1926 when it was extensively used for fishing and recreation. By 1936 excessive silting had left only a small channel. Watersheds above the dam were improperly handled; timber was removed, slopes were cultivated, and few precautions were taken to prevent erosion.*—U. S. Soil Conservation Service.

pattern of blocks, which disregarded topography and provided a powerful system of artificial drainage. Also great drainage projects were instituted in the earlier days of agriculture, and these, too, served to speed the removal of water.

The trend in concern over surface water proceeded from drainage projects to those dealing with flood control. Such concern is still with us, and necessarily so, because of the destructiveness of floods to both property and land; but, a new trend is now apparent in the attempts to conserve, retain, and store water so that it may be available when needed, so that water tables may be held at higher levels, and so that flood waters may be controlled. To this end dams and reservoirs are being constructed and watersheds are being protected.[88]

A recent factor in the lowering of water tables is the great increase of use of water in industry and the rapid increase of air-conditioning. Much of the water used for the latter is wasted because it is not used for any other purpose. The lowering of the water table by using water for this purpose has caused much concern in large cities and has resulted in legislation aimed at controlling the use of this natural resource. Most large users drill their own wells, but this practice is being limited. In some cities, it is required that the water must be forced back into the earth at the levels from which it is drawn. Others have found ways to re-use water. Consumption in homes has also increased dramatically as may be judged by the fact that there are almost three times as many bathrooms today as in 1930. Many areas where the population has risen rapidly are hard put to meet water needs and must find new supplies or restrict their expansion.

Our water supply is a natural resource just as are the others we have discussed. When its availability is reduced, it affects agriculture, industry, fish and game, recreation, and perhaps home use. The trend is already in the direction of its conservation. Probably it will go further. Ecological problems of many kinds will arise in connection with control of water in streams and reservoirs, and the effects upon water table levels. Water is involved in every phase of applied ecology.

Another facet of the problem of water supply is its pollution by industrial waste and sewage. Here again, there are innumerable problems of an ecological nature. Their solution often requires the co-operation of engineers, chemists, bacteriologists, and limnologists. As always, when such specialists are drawn together, their success is greatest when they see their own fields in relation to the whole. This is the ecological approach.

Wildlife. Like soil, water, and forests, our wildlife constitutes a renewable natural resource, which, consequently, can be maintained, or even restored, while it is used, if the use is a wise one. All of these renewable resources are so intimately related that a program for the conservation of one must necessarily consider the others as well. This ecological point of view is fully appreciated by leaders in wildlife management. It is also realized that when man becomes the dominant organism the management of soil, water, forests, and grassland is inevitable—and wildlife, too, if it is to be preserved.

If wildlife management is to be successful, man must know the ecology of the species involved, whether they are fish, birds, or game animals. Life cycles must be known, as must breeding habits, food habits, and food chains, migration routes, preferred habitats, diseases, predators, population trends, and the carrying capacities of given habitats. Such complete information is not yet available. "In its present state, wildlife management is an effort to apply to urgent problems the ecological and biological data that are now available, always with the consciousness that existing tools, methods, and processes may have to be discarded as new and better information becomes available." [152] Ecological knowledge is still woefully incomplete for most of our wildlife, although information accumulates steadily. As it accumulates, programs of management increase in effectiveness.

The range of ecological problems related to wildlife management is tremendous. The complexity of management can perhaps be suggested by indicating some of the kinds of things that must be taken into consideration. It would seem that if food and cover are provided for an organism its needs should be satisfied. But, for many species, the feeding habits are inadequately known. Cover can be provided for some species but, under present conditions, frequently only in localized areas. If that is true, it is not uncommon for food problems to become complicated during the winter months when the species tends to become concentrated on these restricted areas. A population that is reasonable in summer may become excessive in winter and result in death by starvation for many individuals. Encouraging the increase of one species may be detrimental to another one; consequently, individual species must be studied in relation to others. In this connection, predation must be considered from an ecological standpoint.

Species whose numbers have declined to extremely low levels may be propagated under controlled conditions and then released, but the cost is often excessive. Others may be taken from areas of overpopulation and transported elsewhere to start a new population. Such activities have sometimes been successful but in other instances have failed because of factors that were not known or understood. The ecology of the species and of the region must be known. If it is known, there is a reasonable possibility that the species can be encouraged to increase naturally at much less expense and trouble. The problems related to overpopulations of protected species are no less complicated, the ideal being a condition in which natural propagation produces a constant population supportable by the environment and perhaps an excess sufficient to permit a reasonable take by the sportsman.

When it is realized that such problems and many more are in the process of solution for big game, birds of all kinds, fur animals, fish, and other wildlife, it should be apparent that there is much basic ecological work to be done that has possibilities of application. The mistakes that have been made in wildlife management have undoubtedly resulted more often from inadequate ecological information rather than from lack of appreciation of how such knowledge could be applied if it were available. Wildlife management is applied ecology, and it will progress as basic ecological knowledge becomes available and is integrated by wildlife ecologists.

Game refuges provide a safeguard against lack of knowledge and provide the opportunity for acquiring needed information. Particularly, they insure that scarce or disappearing species do not become extinct as some have in the past, for here they are protected and given every encouragement to increase. Usually such refuges do not result in the restoration of a vanishing population. They do, however, insure a continuous breeding stock from which restoration may be made, and they give excellent opportunity for the study of the species involved under relatively undisturbed conditions or under available conditions.[153] A few such refuges are still in near primitive condition and thus can provide much of the biological knowledge of habitat, vegetation, and wildlife that must be learned to manage other refuges and ultimately the general program of wildlife conservation. Other refuges provide the testing grounds for management procedures as knowledge accumulates.

Landscaping

The planning and planting of vegetation for home beautification or in public parks or gardens involves aesthetic considerations but like-

Fig. 193. *An unsightly, eroding road cut in Illinois and its stable appearance three years after planting with trees that blend with topography and native vegetation.*—U. S. Soil Conservation Service.

wise should be backed by an appreciation of the ecology of the species involved. If plantings are not made in terms of the requirements of the species used, they cannot be successful. Soil texture and structure must be considered as they affect water relations. Slope and exposure modify drainage and temperature just as they do in natural environments. Tolerance of shade, light, or extremes of temperature cannot be ignored when planning artificial combinations of species. Some species must be planted in moist places, some require full sunlight, some need to be partially shaded. Competition and all the other factors affecting natural communities operate among planted species as well. The same factors that limit the ranges of natural communities operate to limit the usable materials of landscape design for different sections of the country. Landscaping is, therefore, most successful when based upon ecological principles.

Fig. 194. *On such road-building projects erosion control must be given serious and prompt attention. These great fills have been stabilized by mechanical means and have likewise been planted. If aesthetic considerations have entered into the stabilization program, they are not yet apparent.*—U. S. Forest Service.

Natural landscaping is a recent development resulting from man's modern engineering activities, which drastically change topography, drainage, and vegetation when he constructs modern highways, dams, and airports. Great exposures of subsoil in cuts and fills require cover and replanting not only for aesthetic reasons but also to check erosion and slumping. It is to be expected that engineers should give first consideration to the efficiency of installation and use of a project under construction; but, when this has been the only concern, after effects on drainage and erosion have frequently created serious problems.

Not only has natural beauty been destroyed unnecessarily at times, but extensive expanses of bare soil, in fills and cuts, have been left for nature to recover and stabilize. The re-establishment of natural vegetation is often impossible before erosion and slumping cause disruption of drainage, road blocks, and similar difficulties. Consequently, stabilization must be provided for through artificial means and by seed-

Fig. 195. *The old and the modern manner of handling a road cut. Note the gradual back slope, seeded surface, and shallow, sodded runoff channel, all designed to check erosion.*—U. S. Soil Conservation Service.

ing and planting. The problem is intensified by the infertility of the subsoil, upon which few things will grow. Although the first concern should be stabilization, there should be consideration of succession and the possibility of harmonizing the developing vegetation with that of the surrounding terrain.

In addition to large cuts and fills along mountain highways, there are problems of maintaining road shoulders, ditches, and spillways. Certainly not all is known about the best species for such purposes under all conditions. Also the natural beauty destroyed by a new right-of-way need not be permanently lost. With a minimum of management it would seem that native species could be encouraged to provide cover and beauty, especially along the new express highways, which are increasing in number. It does not seem impossible that ecological knowledge applied in advance could prevent some erosion and drainage problems and save some of the destruction of natural vegetation. Certainly roadside ecology is worth considering both practically and aesthetically.

Plant Indicators

Elsewhere we have emphasized that plant communities give a better indication of the nature of environment than we can obtain by measurements of individual factors. The character and makeup of vegetation is an expression of the integrated effects of all factors operating in a habitat. When the relationships involved are well known, the vegetation becomes an indicator that can be interpreted or, in some instances, read like an instrument.

The practical use of plants as indicators is nothing new, for Pliny[206] wrote of selecting soil for wheatland by the natural vegetation it supported. More recently, in the settling of North America the pioneers used the principle widely in selecting their lands for agricultural purposes. With increasing knowledge, their selections became more effective as is indicated today by lands that have been abandoned and that have remained so. In any agricultural region, an experienced farmer knows the characteristics of soils and habitats supporting local peculiarities of vegetation, or often only a single indicator species.

Such practices and beliefs are usually the result of trial and error experiences, as well they must be, until the responses of a crop plant are tested under the conditions indicated by native vegetation. The knowledge has often been acquired after costly experience. If the requirements of an introduced plant are known and the characteristics of the habitats of native species are studied, the guessing may be reduced. Selection of native species as indicators of local conditions and fitting the ecological requirements of appropriate cultivated plants to these conditions involves ecological methods and thinking. Actually this is not easily accomplished because of our still limited knowledge of the ecology of both native and cultivated plants. It suggests the possibilities of the indicator method, however, in an applied field.[205]

The scope of possible uses of indicators[344] involves much of the entire field of ecology,[79] which necessarily limits the discussion here. Many of the possibilities of their application we have considered in other connections. Consequently, only certain practical aspects, in which they have been successfully applied or might be further expanded will be discussed.

It may sometimes be difficult to recognize or select indicator species. Those with restricted distributions and those tolerating only narrow ranges of habitat conditions should be most useful. Such plants should show responses to minor habitat differences. Thus it follows that similar local conditions in different climatic areas would probably support different indicators. Also the same species might not always be indicative of the same things throughout its range. Differences in geological or cultural history might make it necessary to interpret the significance of an indicator since it need not always be the same. It is rather generally agreed that a group of species or a whole community is more reliable as an indicator than a single species and that dominants, especially of the climax,[79] or at least characteristic species[49] are more useful indicators than lesser species. Above all, application of the method cannot be successful without judgment, good sense, and interpretation in terms of each situation.

Agricultural Indicators. That crop centers and types of agriculture are correlated with climate and climax vegetation[417] is obvious. The agricultural areas of North America follow a pattern very similar to that of a map of natural vegetation.[365] The northeastern conifer region suggests general agriculture at the lower altitudes and latitudes where the land is level and soil is deep. In the transition from boreal to deciduous forest, white pine-red pine-jack pine forests are on sandy soils, which are, in general, undesirable for agriculture, while the northern hardwoods-hemlock forest indicates the best soils for cultivation. The range of the deciduous forest formation marks the best agricultural region of the east with the greatest diversity of crops. Away from the southern Appalachian and Ohio Valley center, as the associations become less complex and oak and hickory become relatively more important, so also does agriculture become more specialized.

On the prairie, both tall and mixed grasses indicate fertile and productive land for cereals, hay, and fodder. Likewise, the natural grass cover provides valuable grazing facilities. The short grass area indicates productive soil whose cultivated crops are limited by moisture. The most favorable sections can be dry-farmed, but otherwise irrigation is necessary for cultivation. As a result, the land is most widely used for grazing.

Vegetation indicating general land use has been given more attention in the western United States than elsewhere.[367] Subalpine vegetation indicates a growing season too short for cultivated crops, steep slopes, and poor agricultural soil. The montane zone also has a short season with cool weather but permits some cultivation if the land is not too rough. Piñon-juniper in the woodland zone indicates productive soil if irrigation is possible, but chaparral indicates inferior agricultural land under almost any circumstances.

Plant indicators of land use in the arid regions of the West are rather well known because of several intensive studies in different areas. Irrigation is necessary everywhere except on the best soils in the sagebrush areas of the northern portion of the Great Basin. Elsewhere, in addition to the need for irrigation, native species indicate other necessities or precautions.[364] Table 12, although specifically applicable only to the Sonoran Desert region of Arizona and southeastern California, illustrates the principles involved.

These generalizations indicate how natural vegetation may be useful in determining regional land use. It is the details of local conditions as indicated by native species that need more study. If the equivalent cultivated and native species were known for different soils, sites, and exposures, it would be possible to state with confidence which fields should be cultivated and which should be put to pasture or wood lot, as well as which crops should be grown in a particular field. The more complete such knowledge is, the more effectively land can be used, and the more certainly land values can be fixed for sale and taxation.

Land evaluation on an ecological basis has been made use of at various times, and a simple illustration will serve to indicate the pos-

Table 12. *Potentialities of Lands for Crop Production as Indicated by the Principal Plant Communities of the Southwestern Desert (after Sampson*[344]*).*

Vegetation	Predominant species	Probable success under irrigation
Creosote bush.......	*Larrea divaricata*........	Successful where native cover is luxuriant; of doubtful success on lands of rock outcrop or with rock layers or hardpan
Desert sage........	*Atriplex polycarpa*......	Successful where native cover is luxuriant; of low value on hardpan soil
Mesquite and chamiso..........	*Prosopis glandulosa* *Atriplex canescens*.......	Partly successful; special crops on level tracts
Chamiso...........	*Atriplex canescens*.......	Successful
Mesquite thicket.....	*Prosopis glandulosa*.....	Successful when salts leached out
Seep weed.........	*Dondia intermedia*.......	Not successful; much abandoned farm land on this cover. Successful when salts leached out
Saltbush and arrowweed.......	*Atriplex lentiformis* *Pluchea sericea*.........	Successful when drained
Pickleweed.........	*Allenrolfea occidentalis*...	Successful when drained and leached
Saltgrass...........	*Distichlis stricta*.........	Successful when drained and leached
Yucca-cactus........	*Yucca mohavensis* *Ferocactus acanthodes* *Opuntia bigelovii*.......	Partly successful; land usually too steep or soil too rocky
Giant cactus-paloverde........	*Carnegiea gigantea* *Cercidium torreyanum*....	Successful when drained

sibilities. Not long ago the construction of dams for water control in the upper Mississippi River necessitated legal action to fix the value of much lowland that would be flooded when the project was completed. One of the basic questions involved the establishment of criteria for determining which acreages were cultivatable and which were not. It was possible to show by means of the natural vegetation, regardless of whether the land had or had not been cultivated, which areas were only rarely flooded and, therefore, desirable agriculturally, which flooded frequently, and which were always too wet for cultivation. Once this was worked out it could be applied generally throughout the area. The information was used effectively for establishing equitable land values in several court proceedings.

Fig. 196. *These productive fields and orchards in Hurricane Valley, Utah, irrigated from the big ditch at left, are bordered on all sides by sagebrush desert. Knowledge of natural vegetation and soil gained from such projects makes possible confident statements of probable success or failure when others are to be established.*—U. S. Forest Service.

Fig. 197. *Death of shrubs and a browse line in a pasture as indicators of too heavy grazing by cattle.*—U. S. Forest Service.

Range and Pasture Indicators. The use of plants as indicators is basic to range management.[391] A knowledge of the important indicator plants and the application of their meaning to handling of grazing land has become fundamental to successful management. Plant indicators are used to judge the condition of the range and particularly to

Fig. 198. *Winter range* (Atriplex nuttallii) *in Colorado, so badly overgrazed that there is practically no vegetation left and gullying is serious on all the slopes. Such depletion is obvious to anyone, but recognition of the onset of these conditions should be possible for those who know the indicators.*—U. S. Forest Service.

recognize signs of deterioration or improvement under certain usages. They are used to determine the kind, degree, and time of grazing, and for determining the grazing capacity of a range. When the plants present are considered in conjunction with soil conditions and the climax, the previous use of the range can be interpreted and its potential usefulness under proper management can be predicted.

Misuse of range lands is obvious in late stages, but it is difficult to recognize when it first begins and should be corrected. Among the indicators that must be watched for are thinning of cover and a lowered vitality of the principal species, replacement of good forage plants by inferior ones, close grazing of species that ordinarily would not be preferred, and, with this, accelerated erosion.[392] It is also highly desirable that the slow successional changes in species composition resulting from grazing under a certain system be recognized. Usually if these are in the direction of climax, they are advantageous. If they show an increase of forbs or of unpalatable species, management practices must be corrected before the trend becomes serious.

In each grazing region, the significant indicators must be known and interpreted. Often selected species can be used and checked upon to simplify evaluations. Likewise, restricted areas, selected on the basis of experience, may be used for observation as representative of the general conditions on a range as a whole.

Range management is obviously applied ecology in which indicators play an important part. The more completely the ecology of the species and communities is known under grazing conditions, the more readily their responses can be interpreted and the more effective management practices can be.

Forest Site Indicators. In forestry, as in agriculture, the indicator significance of one group of plants must be interpreted and applied to an entirely different group of plants. Since forest indicators are commonly herbs or shrubs, there is often some difficulty in translating their meaning to apply to trees. In the broadest sense, forest indicators are site indicators, but rarely do they suggest more than a portion of the several factors that contribute to site. Physical or chemical characteristics of soil, moisture relationships, aeration, or erosion may be indicated by some species. With these and others the probable development of a particular stand can be interpreted. Still others may indicate the past history of vegetation on the site or the probable successional trend to be anticipated in the future.

It is fundamental to indicator interpretation that the successional trends of a region be thoroughly understood for every type of habitat. Only when an indicator is considered in relation to the stage of succession concerned can its meaning be at all clear.

The use of subordinate or dependent species as indicators of site quality has been attempted under various conditions since such a system for classifying forest types was set up in Finland.[70] This system assumes that, since communities of similar structure occupy similar sites, it is possible to judge a site and the nature of the dominants from the ground cover alone. Thus recognition of the herbs, mosses, and lichens on the forest floor with an estimate of their relative proportions might suffice for evaluation of the stand and the quality of the site on which it grows.

Perhaps the most comprehensive attempt to apply the method in North America was made in the Adirondack Mountain area.[184] Elsewhere smaller areas with fewer communities have been studied. Although special phases of the method have proved useful in certain situations, as a whole it has found limited application. Although herbs undoubtedly affect the dominants by modifying soil structure and water relations, and likewise through competition with seedlings of dominant species, there are arguments against the validity of information based on herbs alone, particularly since they derive water and nutrients from different soil horizons than do the dominant trees. It is, therefore, suggested that all the lesser woody vegetation should also be included. There is evidence that the same herbaceous species predominate on more that one soil type, and, therefore, their significance is questioned. Often the indicator types are of limited extent, and several may be present within a single stand. Interpretation, then, becomes difficult in terms of management. Undoubtedly, the foresters' not uncommon lack of familiarity with lesser vegetation and frequent inclination to ignore it entirely have been factors in limiting the testing and application of the method in American forests.

Because the subordinate vegetation changes after lumbering or fire and because height of trees, the commonest criterion of site, cannot then be known, it is desirable that some relatively simple means of evaluation of site be available that can be applied at any time. Of the various approaches to this problem through physical measurement of the soil, one is outstanding for its simplicity.[87] After extensive testing of numerous variables it was found that the site index of southern pines can be accurately determined if only the depth of the A horizon and the soil type are known. Using the xylene equivalent (determined like moisture equivalent) of the B horizon (determined in advance for the soil type) and the depth of the A horizon, a positive statement of site quality can be made whether the land is in forest, cultivated, or abandoned and regardless of slope or exposure. This would seem to be the most promising approach to recognition of site quality. Once these two factors are known for the soils of an area, they can be recorded like a soils map, which then becomes a map of site index to be interpreted for management purposes.

Innumerable indicators, other than site indicators, are used in forestry. Relicts are particularly useful, and successional indicators are applied regularly. Special instances have been suggested elsewhere. It is appropriate to emphasize that indicator applications are invariably successful when the ecology of the region and the species is known.

Human Ecology

No textbook of ecology would be complete without emphasizing that the principles presented are as applicable to man as to any other organism and that men have relationships to environment, to each other, and to other organisms that will be understood best if considered on an ecological basis. This section on applied ecology has given some idea of how man is using his accumulated knowledge of the operation of natural phenomena. In some instances he has used it well. Yet, at the very time that he has become the dominant organism in a complex system, it may be questioned if this knowledge is really being used to his best advantage. In spite of his dominance, resulting from increased numbers, knowledge, and technocracy, he is not freeing himself from dependence upon the system over which he is gaining control. There is, then, a real need for the greatest possible understanding of the system as a whole, and this includes man's place in it, both as to his potentials and limitations.

Human ecology has attracted much interest in recent years as the public in general has become aware and appreciative of the concepts and values of ecology. There is a growing realization that man is subject to ecological laws. The very fact that the term ecology now frequently appears without definition in popular magazines and newspapers is evidence of progress in this direction.

Scholars and investigators in fields outside ecology have adopted the term and some have effectively applied ecological methodology and thinking to their areas of interest. As might perhaps be expected, anthropologists have used the ecological approach to great advantage and this may have stimulated attempts to apply it elsewhere. Although the social ecology of animals has been given much attention[6] there have been only a few advocates of ecological methods in the analysis of man's social behavior.[2] A few sociologists have thought in terms of human ecology for some years, but recently there has been an upsurge of interest that gives promise of a wider introduction of ecological philosophy in an area where it may be used to advantage. The recent appearance of textbooks with "human ecology" in the title[180, 313, 439] is of real interest, but their content suggests that there is not a close agreement as to what constitutes the subject. Sociology is so broad a field and ecology is so inclusive that disparity in approaches to human ecology may be inevitable.

It may even be that sociology, with anthropology and archaeology to supplement it, is inadequate for interpreting human ecology. Certain phases of man's social behavior must certainly fall in the province of the psychologist. Furthermore, if the functioning of human communities is to be understood, economics may become a consideration in human ecology. Perhaps the major objective of human ecology should be the synthesizing of ideas from several disciplines[361] so that an advantage will accrue to all the contributing specialties as a result of transfer of ideas and better understanding of accomplishments and their meaning.

Some ideas of human ecology as expressed by a sociologist[271] interested in communities seem particularly pertinent here. The scope of human ecology is so great that it must have a synoptic view of plant, animal, and human communities; since all are interrelated and governed by the same principles involved in competition, symbiosis, succession, balance, and optimal population. Approached in this fashion, the laws, processes, and structure of human population are seen to be subservient to the more comprehensive laws of ecology, since the latter are the determiners of regional economic and social types. When the arrangement and spatial adaptations of populations are considered, such ecological processes as aggregation, mobility, specialization, distance, and succession are excellent bases of evaluation. They permit the establishment of ecological indices for the measurement of types and trends of social mobility, distance, dominance, and change.

Finally, let us return to phases of this discussion that have been touched upon earlier. It is hoped that, in this last chapter, enough practical aspects of ecology have been suggested to show its wide applicability. Furthermore, the aim has been to show that its application is necessary if man is to continue to enjoy the full benefits of his environment upon which he is dependent, in which he is a factor, and over which he is a dominant. We have suggested that people with a wide variety of interests have concerned themselves with the general subject of human ecology. Among plant ecologists, Dr. Paul B. Sears is outstanding for his efforts in behalf of applied ecology and, particularly, human ecology. As a conclusion to this section it is, therefore, entirely proper that we quote one of his chapter headings from "Life and Environment,"[353] which reads, "The social function of ecology is to provide a scientific basis whereby man may shape the environment and his relations to it, as he expresses himself in and through his culture patterns."

General References

C. C. Adams. General Ecology and Human Ecology.

H. H. Bennett. *Soil Conservation.*

F. E. Clements. *Plant Indicators: The Relation of Plant Communities to Processes and Practice.*

E. A. Colman. *Vegetation and Watershed Management.*

I. N. Gabrielson. *Wildlife Conservation.*

E. H. Graham. *Natural Principles of Land Use.*

C. E. Kellogg. *The Soils That Support Us.*

K. H. W. Klages. *Ecological Crop Geography.*

P. B. Sears. *Life and Environment.*

H. L. Shantz. *Natural Vegetation as an Indicator of the Capabilities of Land for Crop Production in the Great Plains Area.*

L. A. Stoddart and A. D. Smith. *Range Management.*

J. W. Toumey and C. F. Korstian. *Foundations of Silviculture upon an Ecological Basis.*

References Cited

1. Aamodt, O. S. War among plants. *Turf Culture,* **2**: 240-244, 1942.

2. Adams, C. C. General ecology and human ecology. *Ecology,* **16**: 316-335, 1935.

3. Aikman, J. M. Native vegetation of the shelterbelt region. In *Possibilities of Shelterbelt Planting in the Plains Region* (pp. 155-174). Washington, D. C., Govt. Printing Office 1935.

4. ——, and Smelser, A. W. The structure and environment of forest communities in central Iowa. *Ecology,* **19**: 141-150, 1938.

5. Allard, H. A. Length of day in relation to the natural and artificial distribution of plants. *Ecology,* **13**: 221-234, 1932.

6. Allee, W. C. *Animal Aggregations. A Study in General Sociology. Chicago*: Univ. of Chicago Press, 1931. 431 pp.

7. Anderson, D. B. Relative humidity or vapor pressure deficit. *Ecology,* **17**: 277-282, 1936.

8. Anderson, L. E. The distribution of *Tortula pagorum* in North America. *Bryol.,* **46**: 47-66, 1943.

9. Anderson, P. J., and Rankin, W. H. Endothia canker of chestnut. *Cornell Univ. Agr. Exp. Stat. Bull.* 347: 530-618, 1914.

10. Anderson, R. M. Effect of the introduction of exotic animal forms. *Proc. 5th Pacific Sci. Congr.,* Vol. 1: 769-778, 1933.

11. Archibald, E. E. A. The specific character of plant communities. *Jour. Ecol.,* **37**: 260-288, 1949.

12. Ashbel, D. Frequency and distribution of dew in Palestine. *Geogr. Rev.,* **39**: 291-297, 1949.

13. Ashby, E. Statistical ecology. II. A reassessment. *Bot. Rev.,* **14**: 222-234, 1948.

14. Bailey, L. H., Rothacker, J. S., and Cummings, W. H. A critical study of the cobalt chloride method of measuring transpiration. *Pl. Physiol.,* **27**: 563-574, 1952.

15. Ball, John. Climatological diagrams. *Cairo Sci. Jour.,* **4**: no. 50 n.v., 1910.

16. Bauer, H. L. Moisture relations in the chaparral of the Santa Monica mountains, California. *Ecol. Monog.,* **6**: 409-454, 1936.

17. ——. The statistical analysis of chaparral and other plant communities by means of transect samples. *Ecology,* **24**: 45-60, 1943.

18. Baver, L. D. *Soil Physics.* New York: John Wiley & Sons, Inc., 1940. 370 pp.

19. Beard, J. S. Climax vegetation in

tropical America. *Ecology,* **25**: 127-158, 1944.

20. ——. The classification of tropical American vegetation-types. *Ecology,* **36**: 89-100, 1955.

21. BEAVEN, G. F., and OOSTING, H. J. Pocomoke Swamp: A study of a cypress swamp on the eastern shore of Maryland. *Bull. Torr. Bot. Cl.,* **66**: 367-389, 1939.

22. BEDFORD, THE DUKE OF, and PICKERING, S. U. Effect of one crop upon another. *Jour. Agric. Sci.,* **6**: 136-151, 1914.

23. BENNETT, H. H. *Soil Conservation.* New York: McGraw-Hill Book Co., 1939. 993 pp.

24. BERGER-LANDEFELDT, U. Über den Wasserverbrauch von Pflanzenverbänden. *Planta,* **37**: 6-11, 1949.

25. BERNARD, M. Precipitation. In *Physics of the Earth IX: Hydrology,* pp. 32-55. New York: McGraw-Hill Book Co., 1942.

26. BILLINGS, W. D. The structure and development of old field shortleaf pine stands and certain associated physical properties of the soil. *Ecol. Monog.,* **8**: 437-499, 1938.

27. ——. The plant associations of the Carson Desert Region, Western Nevada. *Butler Univ. Bot. Stud.,* **7**: 89-123, 1945.

28. ——. Vegetation and plant growth as affected by chemically altered rocks in the western Great Basin. *Ecology,* **31**: 62-74, 1950.

29. ——. The shadscale vegetation zone of Nevada and eastern California in relation to climate and soils. *Am. Midl. Nat.,* **42**: 87-109, 1949.

30. ——, and DREW, W. B. Bark factors affecting the distribution of corticolous bryophitic communities. *Am. Midl. Nat.,* **20**: 302-330, 1938.

31. BJORKMAN, E. The ecological significance of the ectotrophic mycorrhizal association in forest trees. *Svensk Bot. Tidskr.,* **43**: 223-262, 1949.

32. BLUMENSTOCK, D. I., and THORNTHWAITE, C. W. Climate and the world pattern. In *Climate and Man,* pp. 98-127. (*See* No. 410.)

33. BÖCHER, T. W. Phytogeographical studies of the Greenland flora. *Meddel. om Grønland* **104** (3): 1-56. 1933.

34. BOND, G., and SCOTT, G. D. An examination of some symbiotic systems for fixation of nitrogen. *Ann. Bot.,* **19**: 67-77, 1955.

35. BOOTH, W. E. Tripod method of making chart quadrats. *Ecology,* **24**: 262, 1943.

36. BORCHERT, J. R. The climate of the central North American grassland. *Ann. Assoc. Amer. Geogr.,* **40**: 1-39, 1950.

37. BORMAN, F. H. The statistical efficiency of sample plot size and shape in forest ecology. *Ecology,* **34**: 474-487, 1953.

38. BOURDEAU, P. F. A test of random versus systematic sampling. *Ecology,* **34**: 499-512, 1953.

39. BOUYOUCOS, G. J. Directions for making mechanical analyses of soils by the hydrometer method. *Soil Sci.,* **42**: 225-230, 1936.

40. ——. and MICK, A. H. An electrical resistance method for the continuous measurements of soil moisture under field conditions. *Mich. Agr. Exp. Stat. Tech. Bull.* 172, 1940. 38 pp.

41. BOYCE, S. G. The salt spray community. *Ecol. Monog.,* **24**: 29-67, 1954.

42. Boyko, H. On forest types of the semi-arid areas at lower altitudes *Palestine Jour. Bot.*, R. *Ser.*, **5**: 1-21, 1945.

43. Boynton, D., and Reuther, W. A way of sampling soil gases in dense subsoils and some of its advantages and limitations. *Proc. Soil Sci. Soc. Amer.* **3**: 37-42, 1938.

44. Braun, E. Lucy. Physiographic ecology of the Cincinnati region. *Ohio State Univ. Bull.* 20: no. **34**: 116-211, 1916.

45. ——. The undifferentiated deciduous forest climax and the association segregate. *Ecology*, **16**: 514-519, 1935.

46. ——. The differentiation of the deciduous forest of the eastern United States. *Ohio Jour. Sci.*, **41**: 235-241, 1941.

47. ——. *Deciduous Forests of Eastern North America.* Philadelphia: The Blakiston Co., 1950. 596 pp.

48. ——. The phytogeography of unglaciated eastern United States and its interpretation. *Bot. Rev.*, **21**: 297-375, 1955.

49. Braun-Blanquet, J. *Plant Sociology: the Study of Plant Communities.* (Trans., rev., and ed. by G. D. Fuller and H. S. Conard.) New York: McGraw-Hill Book Co., 1932. 439 pp.

50. Briggs, L. J., and Shantz, H. L. The wilting coefficient for different plants and its indirect determination. *U. S. Dept. Agr., Bureau of Plant Industry Bull.* 230, 1912.

51. Bromley, S. W. The original forest types of southern New England. *Ecol. Monog.*, **5**: 61-89, 1935.

52. Brown, Dorothy. *Methods of Surveying and Measuring Vegetation.* Bucks, England: Commonwealth Agricultural Bureau. 1954. 223 pp.

53. Bruner, W. E. The vegetation of Oklahoma. *Ecol. Monog.*, **1**: 99-188, 1931.

54. Buell, M. F., and Cain, R. L. The successional role of southern white cedar, *Chamaecyparis thyoides*, in south-eastern North Carolina. *Ecology*, **24**: 85-93, 1943.

55. ——, and Cantlon, J. E. A study of two communities of the New Jersey pine barrens and a comparison of methods. *Ecology*, **31**: 567-586, 1950.

56. ——, and Gordon, W. E. Hardwood-conifer forest contact zone in Itasca Park, Minn. *Am. Midl Nat.*, **34**: 433-439, 1945.

57. Burkholder, P. The role of light in the life of plants. *Bot. Rev.*, **2**: 1-52, 97-172, 1936.

58. Byram, G. M., and Jemison, G. M. Solar radiation and forest fuel moisture. *Jour. Agr. Res.*, **67**: 149-176, 1943.

59. Cain, S. A. Concerning certain phytosociological concepts. *Ecol. Monog.*, **2**: 475-505, 1932.

60. ——. Studies on virgin hardwood forest: II. A comparison of quadrat sizes in a quantitative phytosociological study of Nash's Woods, Posey County, Indiana. *Am. Midl. Nat.*, **15**: 529-566, 1934.

61. ——. Studies of virgin hardwood forest: III. Warren's Woods, a beech-maple climax forest in Berrien County, Mich. *Ecology*, **16**: 500-513, 1935.

62. ——. The composition and structure of an oak woods, Cold Spring Harbor, Long Island, with special attention to sampling methods. *Am. Midl. Nat.*, **17**: 725-740, 1936.

63. ——. The species-area curve. *Am. Midl. Nat.*, **19**: 573-581, 1938.

64. ——. The climax and its complexities. *Am. Midl. Nat.,* **21**: 146-181, 1939.

65. ——. Pollen analysis as a paleoecological research method. *Bot. Rev.,* **5**: 627-654, 1939.

66. ——. Sample-plot technique applied to alpine vegetation in Wyoming. *Am. Jour. Bot.,* **30**: 240-247, 1943.

67. ——. *Foundations of Plant Geography.* New York: Harper & Brothers, 1944. 556 pp.

68. ——. A biological spectrum of the flora of the Great Smoky Mountains National Park. *Butler Univ. Bot. Studies,* **7**: 1-14, 1945.

69. ——. Life-forms and phytoclimate. *Bot. Rev.,* **16**: 1-32, 1950.

70. Cajander, A. K. Theory of forest types. *Acta Forestalia Fennica,* **29**: 1-108, 1926.

71. Carpenter, J. R. The grassland biome. *Ecol. Monog.,* **10**: 617-684, 1940.

72. Chandler, R. F., Jr. Cation exchange properties of certain forest soils in the Adirondack section. *Jour. Agr. Res.,* **59**: 491-505, 1939.

73. Chaney, R. W. Tertiary forests and continental history. *Bull. Geol. Soc. Amer.,* **51**: 469-488, 1940.

74. ——. Tertiary centers and migration routes. *In* Origin and development of natural floristic areas with special reference to North America (a symposium). *Ecol. Monog.,* **17**: 139-148, 1947.

75. Chapman, H. H. Is the longleaf type a climax? *Ecology,* **13**: 328-334, 1932.

76. Church, J. E. Snow and snow surveying. In *Physics of the Earth IX: Hydrology,* pp. 83-148. New York: McGraw-Hill Book Co., 1942.

77. Clapham, A. R. The form of the observational unit in quantitative ecology. *Jour. Ecol.,* **20**: 192-197, 1932.

77a. Clausen, J. *Stages in the Evolution of Plant Species.* Ithaca: Cornell University Press, 1951. 206 pp.

78. Clements, F. E. *Plant Succession: An Analysis of the Development of Vegetation.* Carnegie Inst. Wash. Publ. 242, 1916. 512 pp.

79. ——. *Plant Indicators: The Relation of Plant Communities to Processes and Practice.* Carnegie Inst. Wash. Pub. 290, 1920. 388 pp.

80. ——. The relict method in dynamic ecology. *Jour. Ecol.,* **22**: 39-68, 1934.

81. ——. Experimental ecology in the public service. *Ecology,* **16**: 342-363, 1935.

82. ——. Nature and structure of the climax. *Jour. Ecol.,* **24**: 252-284, 1936.

83. ——, and Shelford, V. E. *Bio-ecology.* New York: John Wiley & Sons, Inc., 1939. 425 pp.

84. Clinton, G. P., and McCormick, F. A. Dutch elm disease, *Graphium ulnii. Conn. Agr. Exp. Stat. Bull.* 389: 301-752, 1936.

85. Coile, T. S. Soil samplers. *Soil Sci.,* **42**: 139-142, 1936.

86. ——. Some physical properties of the B horizons of Piedmont soils. *Soil Sci.,* **54**: 101-103, 1942.

87. ——. Relation of soil characteristics to site index of loblolly and shortleaf pine in the lower Piedmont region of North Carolina. *Duke Univ. School of Forestry Bull.* 13, 1948. 78 pp.

88. Colman, E. A. *Vegetation and Watershed Management.* New York: Ronald Press, 1953. 412 pp.

89. ——, and HENDRIX, T. M. The fiberglas soil-moisture instrument. *Soil Sci.*, **67**: 425-438, 1949.

90. CONARD, H. S. The plant associations of Central Long Island. *Am. Midl. Nat.*, **16**: 433-516, 1935.

91. ——. Plant associations on land. *Am. Midl. Nat.*, **21**: 1-27, 1939.

92. ——. *The Background of Plant Ecology*. Ames: Iowa State Press, 1951. 238 pp. (a transl. of *The Plant Life of the Danube Basin*, by Anton Kerner, 1863).

93. COCK, D. B, and ROBESON, S. B. Varying hare and forest succession. *Ecology*, **26**: 406-410, 1945.

94. COOPER, A. W. Sugar pine and western yellow pine in California. *U. S. Dept. Agr., Forest Service Bull.* 690, 1906.

95. COOPER, W. S. The climax forest of Isle Royale, Lake Superior, and its development. *Bot. Gaz.*, **55**: 1-44, 115-140, 189-235, 1913.

96. ——. Redwoods, rainfall and fog. *Plant World*, **20**: 179-189, 1917.

97. ——. *The Broad-Sclerophyll Vegetation of California*. Carnegie Inst. Wash. Publ. 319, 1922. 124 pp.

98. ——. The fundamentals of vegetational change. *Ecology*, **7**: 391-413, 1926.

99. ——. Seventeen years of successional change upon Isle Royale, Lake Superior. *Ecology*, **9**: 1-5, 1928.

100. ——. A third expedition to Glacier Bay, Alaska. *Ecology*, **12**: 61-96, 1931.

101. ——. The problem of Glacier Bay, Alaska; a study of glacier variations. *Geogr. Rev.*, **27**: 37-62, 1937.

102. ——, and FOOT, HELEN. Reconstruction of a late Pleistocene biotic community in Minneapolis, Minnesota. *Ecology*, **13**: 63-73, 1932.

103. COTTAM, G., and CURTIS, J. T. A method for making rapid surveys of woodlands by means of pairs of randomly selected trees. *Ecology*, **30**: 101-104, 1949.

104. ——. Correction for various exclusion angles in the random pairs method. *Ecology*, **36**: 767, 1955.

105. ——, and HALE, B. W. Some sampling characteristics of a population of randomly dispersed individuals. *Ecology*, **34**: 741-757, 1953.

106. COULTER, J. M., BARNES, C. R., and COWLES, H. C. A *Textbook of Botany. Vol. III. Ecology* (revised by Fuller, G. D.), 1-499, New York: American Book Co., 1931.

107. COWLES, H. C. The ecological relations of the vegetation on the sand dunes of Lake Michigan. *Bot Gaz.*, **27**: 95-117, 167-202, 281-308, 361-391, 1899.

108. ——. The physiographic ecology of Chicago and vicinity. *Bot. Gaz.*, **31**: 73-108, 145-181, 1901.

109. COX, H. J. Thermal belts and fruit growing in North Carolina. *Mo. Weath. Rev.* Suppl. 19, 1923.

110. CROCKER, R. L. Soil genesis and the pedogenic factors. *Quart. Rev. Biol.*, **27**: 139-168, 1952.

111. CULBERSON, W. L. The corticolous communities of lichens and bryophytes in the upland forests of northern Wisconsin. *Ecol. Monog.*, **25**: 215-231, 1955.

112. CURTIS, J. T. A prairie continuum in Wisconsin. *Ecology*, **36**: 558-566, 1955.

113. ——, and MCINTOSH, R. P. The interrelations of certain analytic

and synthetic characters. *Ecology,* **31**: 434-455, 1950.

114. ——. An upland forest continuum in the prairie-forest border region of Wisconsin. *Ecology,* **32**: 476-496, 1951.

115. DACHNOWSKI, A. P. Peat deposits and their evidence of climatic change. *Bot. Gaz.,* **72**: 57-89, 1921.

116. DAMBACH, C. A. A ten-year ecological study of adjoining grazed and ungrazed woodlands in northeastern Ohio. *Ecol. Monog.,* **14**: 255-270, 1944.

117. DANSEREAU, P. L'erablière Laurentienne. II. Les successions et leurs indicateurs. *Canad. Jour. Res., C,* **24**: 235-291, 1946.

118. ——, and SEGADAS-VIANNA, F. Ecological study of the peat bogs of eastern North America. I. Structure and evolution of vegetation. *Canad. Jour. Bot.,* **30**: 490-520, 1952.

119. DAUBENMIRE, R. F. Exclosure technique in ecology. *Ecology,* **21**: 514-515, 1940.

120. ——. Vegetational zonation in the Rocky Mountains. *Bot. Rev.,* **9**: 325-393, 1943.

121. ——. Temperature gradients near the soil surface with reference to techniques of measurement in forest ecology. *Jour. Forest.,* **41**: 601-603, 1943.

122. ——. The life-zone problem in the northern intermountain region. *Northwest Sci.,* **20**: 28-38, 1946.

123. ——. *Plants and Environment.* New York: John Wiley and Sons, 1947. 424 pp.

124. ——. Forest vegetation of northern Idaho and adjacent Washington, and its bearing on concepts of vegetation classification. *Ecol. Monog.,* **22**: 301-330, 1952.

125. ——. Alpine timberlines in the Americas and their interpretation. *Butler Univ. Bot. Stud.,* **11**: 119-136, 1954.

126. DAVID, F. N., and MOORE, P. G. Notes on contagious distributions in plant populations. *Ann. Bot.,* **18**: 47-54, 1954.

127. DAVIS, J. H. The natural features of southern Florida, especially the vegetation, and the everglades. *Fla. Dept. of Conserv. Geol. Bull.* **25**. 311 pp. 1943.

128. DAVIS, R. O. E., and BENNETT, H. H. Grouping of soils on the basis of mechanical analysis. *U. S. Dept. Agr. Circ.* 419, 1927. 14 pp.

129. DECANDOLLE, A. L. *Géographie Botanique Raisonée.* Paris: 1855. 1365 pp.

130. DEEVEY, E. S. Pollen analysis and history. *Am. Scientist.* **32**: 39-53, 1944.

131. DILLER, O. D. The relation of temperature and precipitation to the growth of beech in northern Indiana. *Ecology,* **16**: 72-81, 1935.

132. DIMBLEBY, G. W. A simple method for the comparative estimate of soil water. *Plant and Soil,* **5**: 143-154, 1954.

133. DOMIN, K. Is the evolution of the earth's vegetation tending toward a small number of climatic formations? *Acta Bot. Bohemica,* **2**: 54-60, 1923.

134. DOUGLASS, A. E. *Climatic Cycles and Tree Growth. A study of the annual rings of trees in relation to climate and solar activity. Carnegie Inst. Wash. Publ.* 289: 1-127, 1919.

135. ——. Vol. II., *ibid.,* 1-166, 1928.

136. ——. Vol. III. *Climatic Cycles and Tree Growth; A study of cycles,* 1-171, 1936.

137. Drude, O. *Handbuch der Pflanzengeographie*. Stuttgart: J. Engelhorn, 1890. 582 pp.

138. Dyksterhuis, E. J. The vegetation of the western Cross Timbers. *Ecol. Monog.*, **18**: 325-376, 1948.

139. Eggler, W. A. The maple-basswood forest type in Washburn County, Wisconsin, *Ecology*, **19**: 243-263, 1938.

140. Egler, F. E. Arid southeast Oahu vegetation, Hawaii. *Ecol. Monog.*, **17**: 383-435, 1947.

141. ——. Our disregarded rights-of-way—ten million unused wildlife acres. *Trans. 18th No. Amer. Wildlife Conf., 1953:* 148-157, 1953.

142. Ellison, L. A comparison of methods of quadratting shortgrass vegetation. *Jour. Agr. Res.*, **64**: 595-614, 1942.

143. Erdtman, G. *An Introduction to Pollen Analysis*. Waltham, Mass.: Chronica Botanica Co., 1943. 239 pp.

144. Fenton, E. W. The influence of rabbits on the vegetation of certain hill-grazing districts of Scotland. *Jour. Ecol.*, **28**: 438-449, 1940.

145. Fenton, G. R. The soil fauna: with special reference to the ecosystem of forest soil. *Jour. Ecol.*, **16**: 76-93, 1947.

146. Ferguson, T. P., and Bond, G. Observations on the formation and function of the root nodules of *Alnus glutinosa* (L.) Gaertn. *Ann. Bot.*, **17**: 175-188, 1953.

147. Flint, R. F., and Deevey, E. S. Radiocarbon dating of late-Pleistocene events. *Amer. Jour. Sci.*, **249**: 257-300, 1951.

148. Flowers, S. Vegetation of the Great Salt Lake Region. *Bot. Gaz.*, **95**: 353-418, 1934.

149. Freeland, R. O. Apparent photosynthesis in some conifers during winter. *Plant Physiol.*, **19**: 179-185, 1944.

150. Fuller, H. J. Carbon dioxide concentration of the atmosphere above Illinois forest and grassland. *Am. Midl. Nat.*, **39**: 247-249, 1948.

151. Funke, G. The influence of *Artemisia absinthium* on neighboring plants. An essay of experimental sociology. *Blumea*, **5**: 281-293, 1943.

152. Gabrielson, I. N. *Wildlife Conservation*. New York: The Macmillan Company, 1941, 249 pp.

153. ——. *Wildlife Refuges*. New York: The Macmillan Company, 1943. 257 pp.

154. Garner, W. W. Photoperiodism. In Duggar, *Biological Effects of Radiation*. Vol. II, 677-713, New York: McGraw-Hill Book Co., 1936.

155. Garner, W. W. Recent work on photoperiodism. *Bot. Rev.*, **3**: 259-275, 1937.

156. ——, and Allard, H. A. Effect of the relative length of day and night and other factors of the environment on growth and reproduction in plants. *Jour. Agr. Res.*, **18**: 553-606, 1920.

157. Garren, K. H. Effects of fire on vegetation of the southeastern United States. *Bot. Rev.*, **9**: 617-654, 1943.

158. Geiger, R. *The Climate Near the Ground*. (transl. M. N. Stewart). Cambridge: Harvard Univ. Press, 1950. 482 pp.

159. Gleason, H. A. The structure and development of the plant association. *Bull. Torrey Bot. Cl.*, **43**: 463-481, 1917.

160. ——.The individualistic concept of the plant association. *Bull.*

Torrey Bot. Cl., **53**: 7-26, 1926.; *Amer. Midl. Nat.*, **21**: 92-110, 1939.

161. Glinka, K. D. *The Great Soil Groups of the World and Their Development* (Engl. transl. by C. F. Marbut). Ann Arbor, Mich.: Edwards Brothers, 1927. 150 pp.

162. Glock, W. S. *Principles and Methods of Tree-Ring Analysis.* Carnegie Inst. Wash. Publ. 486: 1-100, 1937.

163. Good, R. *The Geography of the Flowering Plants.* London: Longmans, Green and Co. 1947.

164. Goodall, D. W. Quantitative aspects of plant distribution. *Biol. Rev. Cambridge Philos. Soc.*, **27**: 194-245, 1952.

165. ——. Objective methods for the classification of vegetation. II. Fidelity and indicator value. *Australian Jour. Bot.*, **1**: 434-456, 1953.

166. Gordon, W. E. Nomograms for conversion of psychrometric data. *Ecology*, **21**: 505-508, 1940.

167. Graham, E. H. *Natural Principles of Land Use.* New York: Oxford University Press, 1944. 274 pp.

168. Griggs, R. F. The edge of the forest in Alaska and the reasons for its position. *Ecology*, **15**: 80-96, 1934.

169. ——. The timberlines of northern America and their interpretation. *Ecology*, **27**: 275-289, 1946.

170. Grisebach, A. H. R. *Die Vegetation der Erde nach ihrer klimatischen Anordnung.* Leipzig: W. Engelmann, 1872. 2 vol., 603 and 635 pp.

171. Grosenbaugh, L. R. Plotless timber estimates—new, fast, easy. *Jour. Forestry*, **50**: 32-37, 1952.

172. Haeckel, E. Ueber Entwicklungsgang und Aufgabe der Zoologie. *Jenaischer Zeitschr. für Naturwiss.* **5**: 353-370, 1869.

173. Haldane, J. S., and Graham, J. I. *Methods of Air Analysis.* London: Charles Griffin, 1935. 177 pp.

174. Hall, T. F., and Penfound, W. T. Cypress-gum communities in the Blue Girth Swamp near Selma, Alabama. *Ecology*, **24**: 208-217, 1943.

175. Hansen, H. P. Postglacial forest succession, climate and chronology in the Pacific northwest. *Trans. Am. Phil. Soc.* **37**: 1-130, 1947.

176. Hanson, H. C. Ecology in agriculture. *Ecology*, **20**: 111-117, 1939.

177. ——. Fire in land use and management. *Am. Midl. Nat.*, **21**: 415-434, 1939.

178. Harshberger, J. W. *Phytogeographic Survey of North America.* New York: G. E. Stechert & Company, Inc., 1911.

179. Hatch, A. B. The physical basis of mycotrophy in *Pinus*. *Black Rock Forest Bull.*, **6**: 1937. 168 pp.

180. Hawley, A. H. *Human Ecology. A theory of community structure.* New York: Ronald Press, 1950. 456 pp.

181. Hawley, Florence. *Tree-Ring Analysis and Dating in the Mississippi Drainage.* Chicago: University of Chicago Press, 1941. 110 pp.

182. Hayes, G. L., and Kittredge, J. Comparative rain measurements and raingage performances on a steep slope adjacent to a pine stand. *Trans. Amer. Geophys. Union*, **30**: 295-301, 1949.

183. Heiberg, S. O., and Chandler, R. F. A revised nomenclature of

forest humus layers for the northeastern United States. *Soil Sci.*, **52**: 87-99, 1941.

184. HEIMBURGER, C. *Forest Type Studies in the Adirondack Region.* Cornell Univ. Agr. Exp. Sta. Mem. 165:1-122, 1934.

185. HENDERSON, L. J. *The Fitness of the Environment.* New York: The Macmillan Company, 1913. 317 pp.

186. HENDRICKS, B. A. Effect of forest litter on soil temperature. *Chronica Botanica,* **6**: 440-441, 1941.

187. HENDRIX, T. M. Calibration of fiberglas soil moisture units. *Soils Sci.,* **71**: 419-428, 1951.

188. HOFMAN, J. V. The establishment of a Douglas fir forest. *Ecology,* **1**: 49-53, 1920.

189. HOPKINS, B. A. A new method for determining the type of distribution of plant individuals. *Ann. Bot.,* **18**: 213-228, 1954.

190. HUFFAKER, C. B. Vegetational correlations with vapor pressure deficit and relative humidity. *Am. Midl. Nat.,* **28**: 486-500, 1942.

191. HUMBOLDT, A. VON. *Ideen zu einer Geographie der Pflanzen nebst einem Naturgemälde der Tropenländer.* Tübingen: 1807. 182 pp.

192. HUMM, H. J. Bacterial leaf nodules. *Jour. N. Y. Botanical Garden,* **45**: 193-199, 1944.

193. HUMPHREYS, W. J. *Fogs and Clouds.* Baltimore: Williams and Wilkins Co., 1926.

194. ——. *Ways of the Weather.* Lancaster, Pa.: Jaques Cattell Press, 1942. 400 pp.

195. ILVESSALO, Y. Vegetationsstatistische Untersuchungen über die Waldtypen. *Acta Forest. Fenn.,* **20**: 1-70, 1922.

196. JACOBS, M. R. The effect of wind sway on the form and development of *Pinus radiata. Austr. Jour. Bot.,* **2**: 35-51, 1954.

197. JACCARD, P. The distribution of the flora in the alpine zone. *New Phytol.,* **11**: 37-50, 1912.

198. JENNY, H. *Factors of Soil Formation.* New York: McGraw-Hill Book Co., 1941. 281 pp.

199. ——, and COWAN, E. W. The utilization of adsorbed ions by plants. *Science,* **77**: 394-396, 1933.

199a. JOLLY, G. M. Theory of sampling. In *Methods of Surveying and Measuring Vegetation,* by Dorothy Brown, pp. 8-18. 1954.

199b. JONES, G. N. *A Botanical Survey of the Olympic Peninsula, Washington.* U. of Wash. Publ. in Biol. **5**: 1-286, 1936.

200. JUST, T. Geology and plant distribution. *Ecol. Monog.,* **17**: 127-137, 1947.

201. KEARNEY, T. H. Report on a botanical survey of the Dismal Swamp region. *U. S. Nat. Herb., Contr.* **5**: 321-585, 1901.

202. ——, BRIGGS, L. J., SHANTZ, H. L., MCLANE, J. W. and PIEMEISEL, R. L. Indicator significance of vegetation in Tooele Valley, Utah. *Jour. Agr. Res.,* **1**: 365-417, 1914.

203. KEEVER, CATHARINE. Causes of succession on old fields of the Piedmont, North Carolina. *Ecol. Monog.,* **20**: 229-250, 1950.

204. ——. Present composition of some stands of the former oak-chestnut forest in the southern Blue Ridge Mountains. *Ecology,* **34**: 44-54, 1953.

205. KELLEY, A. P. Plant indicators of soil types. *Soil Sci.,* **13**: 411-423. 1922.

206. KELLOGG, C. E. Development and

Significance of the Great Soil Groups of the United States. *U. S. Dept. Agr. Misc. Pub.* 229, 1936.

207. ——. *The Soils That Support Us.* New York: The Macmillan Company, 1941. 370 pp.

208. KENOYER, L. A. A study of Raunkiaer's law of frequency. *Ecology,* **8**: 341-349, 1927.

209. ——. Ecological notes on Kalamazoo County, Michigan based on the original land survey. *Paps. Mich. Scad. Sci., Arts and Letters,* **11**: 211-217, 1930.

210. ——. Forest distribution in southwestern Michigan as interpreted from the original land survey (1826-32). *Paps. Mich Acad. Sci., Arts and Letters,* **19**: 107-111, 1934.

211. KIMBALL, H. H. Intensity of solar radiation at the surface of the earth and its variations with latitude, altitude, season and time of the day. *Mo. Weath. Rev.,* **63**: 1-4, 1935.

212. KINCER, J. B. Climate and weather data for the United States. In *Climate and Man.* 685-699. (*See* No. 410.)

213. KITTREDGE, J. Forests and water aspects which have received little attention. *Jour. For.,* **34**: 417-419, 1936.

214. ——. *Forest Influences.* New York: McGraw-Hill Book Co., 1948. 394 pp.

215. ——. Influences of forests on snow in the ponderosa-sugar pine-fir zone of the central Sierra Nevada. *Hilgardia,* **22**: 1-96, 1953.

216. KLAGES, K. H. W. *Ecological Crop Geography.* New York: The Macmillan Company, 1942. 615 pp.

217. KLYVER, F. D. Major plant communities in a transect of the Sierra Nevada mountains of California. *Ecology,* **12**: 1-17, 1931.

218. KORSTIAN, C. F., and BRUSH, W. D. Southern white cedar. *U. S. Dept. Agr. Tech. Bull,* 251, 1931.

219. ——, and COILE, T. S. Plant competition in forest stands. *Duke Univ. School of Forestry Bull.* 3, 1938. 125 pp.

220. KRAMER, P. J. Photoperiodic stimulation of growth by artificial light as a cause of winter killing. *Plant Physiol.,* **12**: 881-883, 1936.

221. ——. Species differences with respect to water absorption at low temperatures. *Am. Jour. Bot.,* **29**: 828-832, 1942.

222. ——. Soil moisture in relation to plant growth. *Bot. Rev.,* **10**: 525-559, 1944.

223. ——. *Plant and Soil-Water Relationships.* New York: McGraw-Hill Book Co., 1949. 347 pp.

224. ——. Water relation to plant growth. *In* The Physical and economic foundation of natural resources. I. Photosynthesis-Basic features of the process. 28-32, 1952. *U. S. Congr. Interior and Insular Affairs Comm.*

225. ——, RILEY, W. S., and BANNISTER, T. T. Gas exchange of cypress knees. *Ecology,* **33**: 117-121, 1952.

226. KULCZYŃSKI, S. Zespoły roślin w Pieninach.—Die Pflanzenassoziationen der Pieninen. *Polon. Acad. des Sci. et Lettres, Cl. des Sci. Math. et Nat. Bull. Internatl.* sér. B (Suppl. II) 1927: 57-203, 1937. (n.v.)

227. KURZ, H. and DEMAREE, D. Cypress buttresses and knees in relation to water and air. *Ecology,* **15**: 36-41, 1934.

228. LARSON, L. T., and WOODBURY, T. D. Sugar pine. *U. S. Dept. Agr. Bull.* 426, 1916. 40 pp.

229. LAWRENCE, D. B. Some features of the vegetation of the Columbia River Gorge with special reference to asymmetry in trees. *Ecol. Monog.*, **9**: 217-257, 1939.

230. ——. Glacier fluctuation for six centuries in southeastern Alaska and its relation to solar activity. *Geogr. Rev.*, **40**: 191-223, 1950.

231. ——. Estimating dates of recent glacier advances and recession rates by studying tree growth layers. *Trans. Amer. Geophys. Union,* **31**: 243-248, 1950.

232. LEE, C. H. Total evaporation for Sierra Nevada watersheds by the method of precipitation and runoff differences. *Trans. Amer. Geophys. Union,* Pt. I, pp. 50-66, 1941.

233. LEIGHLY, JOHN. Climatology since the year 1800. *Trans. Amer. Geophys. Union,* **30**: 658-672, 1949.

234. LEOPOLD, A. Conservation economics. *Jour. Forest.* **32**: 537-544, 1934.

235. LEWIS, F. J. The vegetation of Alberta. II. The swamp moor and bog forest. *Jour. Ecol.,* **16**: 18-70, 1928.

236. LIBBY, W. F. *Radiocarbon Dating.* Chicago Univ. Press. 2nd ed. 1955. 167 pp.

237. LINDSEY, A. A. Testing the line-strip method against full tallies in diverse forest types. *Ecology,* **36**: 485-495, 1955.

238. LITTLE, S., and MOORE, E. B. The ecological role of prescribed burns in the pine-oak forests of southern New Jersey. *Ecology,* **30**: 223-233, 1949.

239. LIVINGSTON, B. E. A single index to represent both moisture and temperature conditions as related to plant growth. *Physiol. Research* no. 9: 421-440, 1916.

240. ——. Atmometers of porous porcelain and paper, their use in physiological ecology. *Ecology,* **16**: 438-472, 1935.

241. ——, and KOKETSU, R. The water-supplying power of the soil as related to the wilting of plants. *Soil Sci.,* **9**: 469-485, 1920.

242. ——, and SHREVE, F. *The Distribution of Vegetation in the United States, as Related to Climatic Conditions.* Carnegie Inst. Wash. Publ. 284, 1921. 590 pp.

243. LULL, H. W., and ELLISON, L Precipitation in relation to altitude in central Utah. *Ecology,* **31**: 479-484, 1950.

244. LUNDEGARDH, H. *Environment and Plant Development.* (transl. by E. Ashby) London: Edward Arnold and Co., 1931. 330 pp.

245. LUTZ, H. J. Effect of cattle grazing on vegetation of a virgin forest in northwestern Pennsylvania. *Jour. Agr. Res.,* **41**: 561-570, 1930.

246. ——. The vegetation of Heart's Content, a virgin forest in northwestern Pennsylvania. *Ecology,* **11**: 1-29, 1930.

247. ——. Origin of white pine in virgin forest stands of northwestern Pennsylvania. *Ecology,* **16**: 252-256, 1935.

248. ——. Determinations of certain physical properties of forest soils: I. Methods utilizing samples collected in metal cylinders. *Soil Sci.,* **57**: 475-487, 1944.

249. ——. Determination of certain physical properties of forest soils: II. Methods utilizing loose samples collected from pits. *Soil Sci.,* **58**: 325-333, 1944.

250. ——, and CHANDLER, R. F. *Forest Soils.* New York: John Wiley and Sons. 514 pp.

251. LYON, C. Tree ring width as an index of physiological dryness in New England. *Ecology,* **17**: 457-478, 1936.

252. MACKINNEY, A. L. Effects of forest litter on soil temperature and soil freezing in autumn and winter. *Ecology,* **10**: 312-322, 1929.

253. MCCUBBIN, W. A. Preventing plant disease introduction. *Bot. Rev.,* **12**: 101-139, 1946.

254. MCDERMOTT, R. E. Effects of saturated soil on seedling growth of some bottomland species. *Ecology,* **35**: 36-41, 1954.

255. ——. Seedling tolerance as a factor in bottomland timber succession, *Mo. Agr. Expt. Stat. Res. Bull.,* **557**: 1-11, 1954.

256. MCDOUGALL, W. B. *Plant Ecology.* Philadelphia: Lea & Febiger 1949, 4th ed. 234 pp.

257. ——, and JACOBS, M. C. Tree mycorhizas from the central Rocky Mountain region. *Am. Jour. Bot.,* **14**: 258-266, 1927.

258. MARBUT, C. F. A scheme for soil classification. *First Internatl. Congr. Soil Sci.* (1927) *Proc. and Paps.* 4, 1-31. 1928.

259. ——. Soils of the United States. In *Atlas of American Agriculture. Pt. III.* 98 pp. Washington, D. C.: U. S. Dept. Agr. Bur. Chem. and Soils, 1935.

260. MASON, H. L. The principles of geographic distribution as applied to floral analysis. *Madroño,* **8**: 209-226, 1936.

261. MATZKE, E. B. Effect of street lights in delaying leaf-fall in certain trees. *Am. Jour. Bot.,* **23**: 446-452, 1936.

262. MEINZER, O. E. (ed.) *Hydrology.* New York: McGraw-Hill Book Co., 1942. 712 pp.

263. MENDELL, F. H., and AIKMAN, J. M. Soil and water conservation. In *Present Status and Outlook of Conservation in Iowa.* Rep. of the Sonservation Comm. Ia. Acad. Sci. **51**: 87-96, 1944.

264. MERRIAM, C. H. *The Geographic Distribution of Animals and Plants in North America.* (U. S. Dept. Agr. Yearbook.) Washington, D. C.: Govt. Printing Office, 1894, 203-214.

265. MICKEY, K. B. *Man and the Soil.* Chicago: International Harvester Co., 1945. 110 pp.

266. MIDDLETON, W. E. K., and SPILHAUS, A. F. *Meteorological Instruments.* Toronto: Univ. of Toronto Press, 1953. 3rd ed., rev. 286 pp.

267. MOORE, P. G. A test for non-randomness in plant populations. *Ann. Bot.,* **17**: 57-62, 1953.

268. MOSS, E. H. The vegetation of Alberta. IV. The poplar association and related vegetation of central Alberta. *Jour. Ecol.,* **20**: 380-415, 1932.

269. ——. Forest communities in northwestern Alberta. *Canad. Jour. Bot.,* **31**: 212-252, 1953.

270. MUENSCHER, W. C. *Weeds.* New York: The MacMillan Co., 1955. 2nd ed. 560 pp.

271. MUKERJEE, R. *Man and His Habitation. A Study in Social Ecology.* New York: Longmans, Green & Co., 1940, 313 pp.

272. MULLER, C. H. The association of desert annuals with shrubs. *Am. Jour. Bot.,* **40**: 53-59, 1953.

273. NELSON, T. C. Chestnut replacement in the southern highlands. *Ecology,* **36**: 352-353, 1955.

274. NICHOLS, G. E. The vegetation of northern Cape Breton Island.

Nova Scotia. *Trans. Conn. Acad. Arts and Sci.*, **22**: 249-467, 1918.

275. ——. A working basis for the ecological classification of plant communities. *Ecology*, **4**: 11-23, 154-180, 1923.

276. ——. The hemlock-white pine-northern hardwood region of eastern North America. *Ecology*, **16**: 403-422, 1935.

277. OLIVER, W. R. B. New Zealand epiphytes, *Jour. Ecol.*, **18**: 1-51, 1930.

278. OLMSTED, C. E. Experiments on photoperiodism, dormancy, and leaf age and abscission in sugar maple. *Bot. Gaz.*, **112**: 365-393, 1951.

279. OLMSTED, L. B., ALEXANDER, L. T., and MIDDLETON, H. E. A pipette method of mechanical analysis of soils based on improved dispersion procedure. *U. S. Dept. Agr. Tech. Bull.* 170, 1930. 22 pp.

280. OOSTING, H. J. An ecological analysis of the plant communities of Piedmont, North Carolina. *Am. Midl. Nat.*, **28**: 1-126, 1942.

281. ——. The comparative effect of surface and crown fire on the composition of a loblolly pine community. *Ecology*, **25**: 61-69, 1944.

282. ——. Botanical notes on the flora of East Greenland. In *The Coast of Northeast Greenland, The Louise A. Boyd Expeditions of* 1937 and 1938, pp. 225-269. Am. Geogr. Soc. Spec. Publ. 30, 1948.

283. ——. Ecological processes and vegetation of the maritime strand in the southeastern United States. *Bot. Rev.*, **20**: 226-262, 1954.

284. ——, and ANDERSON, L. E. Plant succession on granite rock in eastern North Carolina. *Bot. Gaz.*, **100**: 750-768, 1939.

285. ——, and BILLINGS, W. D. Edapho-vegetational relations in Ravenel's Woods. *Am. Midl. Nat.*, **22**: 333-350, 1939.

286. ——, and BILLINGS, W. D. Factors effecting vegetational zonation on coastal dunes. *Ecology*, **23**: 131-142, 1942.

287. ——, and BILLINGS, W. D. The red fir forest of the Sierra Nevada: Abietum magnificae. *Ecol. Monog.*, **13**: 261-274, 1943.

289. ——, and BILLINGS, W. D. A comparison of virgin spruce-fir forest in the northern and southern Appalachian system. *Ecology*, **32**: 84-103, 1951.

290. ——, and KRAMER, P. J. Water and light in relation to pine reproduction. *Ecology*, **27**: 47-53, 1946.

291. ——, and REED, J. F. Ecological composition of pulpwood forests in northwestern Maine. *Am. Midl. Nat.*, **31**: 182-210, 1944.

292. ——, and REED, J. F. Virgin spruce-fir forest in the Medicine Bow Mountains, Wyoming. *Ecol. Monog.*, **22**: 69-91, 1952.

293. PALPANT, E. H., and LULL, H. W. Comparison of four types of electrical resistance instruments for measuring soil moisture. *Forest Serv. U.S.D.A. Occas. Pap.* **128**, 1953.

294. PARISH, S. B. Vegetation of the Mohave and Colorado deserts of southern California. *Ecology*, **11**: 481-499, 1930.

295. PARKER, K. W. Application of ecology in the determination of range condition and trend. *Jour. Range Mgt.*, **7**: 14-23, 1954.

296. PAULEY, S. S. and PERRY, T. O. Ecotypic variation of the photoperiodic response in Populus. *Jour. Arnold Arboretum*, **35**: 167-188, 1954.

297. PEARSE, K., PECHANEC, J. F., and PICKFORD, G. D. An improved pantograph for mapping vegetation. *Ecology,* **16**: 529-530, 1935.

298. PECHANEC, J. F., and STEWART, G. Sagebrush-grass range sampling studies: size and structure of sampling unit. *Jour. Amer. Soc. Agron.,* **32**: 669-682, 1940.

299. PECK, M. E. A preliminary sketch of the plant regions of Oregon. I. Western Oregon. *Am. Jour. Bot.,* **12**: 69-91, 1925.

300. PENFOUND, W. T. A study of phytosociological relationships by means of aggregations of colored cards. *Ecology,* **26**: 38-57, 1945.

301. ——. Southern swamps and marshes. *Bot. Rev.,* **18**: 413-446, 1952.

302. ——, and HOWARD, J. R. A phytosociological study of an evergreen oak forest in the vicinity of New Orleans, Louisiana. *Am. Midl. Nat.,* **23**: 165-174, 1940.

303. ——, and MACKANESS, F. P. A note concerning the relation between drainage pattern, bark conditions and the distribution of corticolous bryophytes. *Bryol.,* **43**: 168-170, 1940.

304. PENMAN, H. L. The dependence of transpiration on weather and soil conditions. *Jour. Soil Sci.,* **1**: 74-89, 1949.

305. PHILLIPS, J. Succession, development, the climax, and the complex organism: An analysis of concepts. *Jour. Ecol.,* **22**: 554-571; **23**: 210-246, 488-508, 1931.

306. PIEMEISEL, R. L. Causes affecting change and rate of change in a vegetation of annuals in Idaho. *Ecology,* **32**: 53-72, 1951.

307. PLATT, R. B. An ecological study of the mid-Appalachian shale barrens and of the plants endemic to them. *Ecol. Monog.,* **21**: 269-300, 1951.

308. ——, and WOLF, J. N. General uses and methods of thermistors in temperature investigations, with special reference to a technique for high sensitivity contact temperature measurement. *Plant Physiol.,* **25**: 507-512, 1950.

309. POLUNIN, N. Botany of the Canadian eastern Arctic. III. Vegetation and ecology. *National Museum of Canada Bull.,* **104**. Ottawa: King's Printer and Controller of Stationery. 1948. 304 pp.

310. POUND, C. E., and EGLER, F. E. Brush control in southeastern New York: fifteen years of stable tree-less communities. *Ecology,* **34**: 63-73, 1953.

311. POTZGER, J. E., and KELLER, C. O. The beech line in northwestern Indiana. *Butler Univ. Bot. Studies,* **10**: 108-113, 1952.

312. PRESTON, F. W. The commonness, and rarity, of species. *Ecology,* **29**: 254-283, 1948.

313. QUINN, J. A. Human Ecology. New York: Prentice-Hall. 1950. 561 pp.

314. Radiocarbon dates. *Sci.,* **114**: 291-296, 1951; **116**: 409-414, 1952; **119**: 135-140, 1954; **122**: 954-960, 1955.

315. RAUNKIAER, C. *The Life Forms of Plants and Statistical Plant Geography; Being the Collected Papers of C. Raunkiaer.* Oxford: Clarendon Press, 1934. 632 pp.

316. RAUP, H. M. Recent changes of climate and vegetation in southern New England and adjacent New York. *Jour. Arnold Arboretum,* **18**: 79-117, 1937.

317. ——. Botanical problems in boreal America. *Bot. Rev.,* **7**: 147-248, 1941.

318. Rayner, M. C. The significance of the mycorhizal habit in plant crops. *Proc. Soc. Appl. Bact.*, 1949, **45**: 1949.

319. Reed, John Frederick. *Root and Shoot Growth of Shortleaf and Loblolly Pines in Relation to Certain Environmental Conditions.* Duke Univ. School of Forestry Bull. 4, 1939, 52 pp.

320. Reed, J. F., and Cummings, R. W. Soil reaction-glass electrode and colorimetric methods for determining *p*H values of soils. *Soil Sci.*, **59**: 97-104, 1945.

321. Rice, E. L., and Kelting, R. W. The species-area curve. *Ecology*, **36**: 7-11, 1955.

322. ——, and Penfound, W. T. An evaluation of the variable-radius and paired-tree methods in the blackjack-post oak forest. *Ecology*, **36**: 315-320, 1955.

333. Richards, L. A. A pressure-membrane extraction apparatus for soil solution. *Soil Sci.*, **51**: 377-386, 1941.

334. ——. Soil moisture tensiometer materials and construction. *Soil Sci.*, **53**: 241-248, 1942.

335. Richards, P. W. *The Tropical Rain Forest.* Cambridge Univ. Press, 1952. 450 pp.

336. Robbins, W. W., Crafts, A. S., and Raynor, R. N. *Weed Control: a textbook and manual.* New York: McGraw-Hill Book Co., 1952. 503 pp.

337. Robinson, P. The distribution of plant populations. *Ann. Bot.*, **18**: 35-46, 1954.

338. Rogers, H. T., Pearson, R. W., and Pierre, W. H. The source and phosphotase activity of exoenzyme systems of corn and tomato roots. *Soil Sci.*, **54**: 353-366, 1942.

339. Rubel, E. Ecology, plant geography and geo-botany; their history and aim. *Bot. Gaz.*, 84: 428-439, 1927.

340. ——. Plant communities of the world. In *Essays in Geobotany*, pp. 263-290. Berkeley, Calif.: Univ. of Calif. Press, 1936.

341. Russell, E. J. (revised by E. W. Russell) *Soil Conditions and Plant Growth.* London: Longmans, Green and Co., 1950, 635 pp.

342. ——, and Appleyard, A. The atmosphere of the soil; its composition and causes of variation. *Jour. Agr. Sci.*, **7**: 1-48, 1915.

343. Rycroft, H. B. Random sampling of rainfall. *Jour. So. Afr. Forest Assoc.*, **18**: 71-81, 1949.

344. Sampson, A. W. Plant indicators —concept and status. *Bot. Rev.*, **5**: 155-206, 1939.

345. Schimper, A. F. W. *Plant Geography upon a Physiological Basis.* (Transl. by W. R. Fisher.) Oxford: Clarendon Press, 1903, 839 pp.

346. Schouw, J. F. *Grundzüge einer allgemeinen Pflanzengeographie.* Berlin, 1823. 524 pp.

347. Schreiner, O., and Reed, H. S. The production of deleterious excretions by roots. *Bull. Torr. Bot. Cl.* **34**: 279-301, 1907.

348. Schumacher, F. X., and Chapman, R. A. *Sampling Methods in Forestry and Range Management.* Duke Univ. School of Forestry Bull. 7, rev. ed. 1948. 221 pp.

349. Scofield, C. S. The measurement of soil water. *Jour. Agr. Res.*, **71**: 375-402, 1945.

350. Scorer, R. S. Smog. *Sci. Progr.*, **42**: 396-405, 1954.

351. Sears, P. B. The natural vegetation of Ohio. *Ohio Jour. Sci.*, **25**: 139-149; **26**: 128-146, 139-231, 1925-26.

352. ——. Climatic interpretation of post-glacial pollen deposits in North America. *Bull. Amer. Meteorol. Soc.*, **19**: 177-185, 1938.

353. ——. *Life and Environment.* New York: Teachers College, Columbia University, 1939. 175 pp.

354. ——. Postglacial vegetation in the Erie-Ohio area. *Ohio Jour. Sci.*, **41**: 225-234, 1941.

355. ——. Xerothermic theory. *Bot. Rev.*, **8**: 708-736, 1942.

356. ——. The ecological basis of land use and management. *Proc. 8th Am. Sci Congr.*, **5**: 223-233, 1942.

357. ——. History of conservation in Ohio. In *The History of the State of Ohio.* VI: *Ohio in the Twentieth Century*—pp. 219-240, Columbus, Ohio. Ohio State Archaeological Society, 1942.

358. ——. Grazing versus maple syrup. *Science,* **98**: 83-84, 1943.

359. ——. Man and nature in the modern world. In *Education for Use of Regional Resources* (Rept. of Gatlinburg Conference II, sponsored by Committee on Southern Regional Studies and Education of the American Council in Education), 1944. Chp. 3: 25-44.

360. ——. Importance of ecology in the training of engineers. *Science.* **106**: 1-3, 1947.

361. ——. Human Ecology: a problem in synthesis. *Sci.*, **120**: 959-963, 1954.

362. Shanks, R. E. Forest composition and species association in the beech-maple forest region of western Ohio. *Ecology,* **34**: 455-466, 1953.

363. ——. Plotless sampling trials in Appalachian forest types. *Ecology,* **35:** 237-244, 1954.

364. Shantz, H. L. Natural vegetation as an indicator of the capabilities of land for crop production in the Great Plains area. *U. S. Dept. Agr. Bur. Pl. Ind. Bull.* 201, 1-100, 1911.

365. ——. Plants as soil indicators. In *Soils and Men,* pp. 835-860. (*See* No. 409.)

366. ——. *Fire as a Tool in the Management of the Brush Ranges of California.* Calif. Divis. Forestry, Spec. Publ., 1947, 156 pp.

367. ——, and Zon, R. The physical basis of agriculture: Natural vegetation. In *Atlas of American Agriculture.* (Pt. I, Sect. E. 29 pp.) Washington, D. C.: U. S. Dept. Agr., 1924.

368. Shelford, V. E. (editor). *Naturalist's Guide to the Americas.* Baltimore: Williams & Wilkins Company, 1926. 761 pp.

369. Sherman, L. K., and Musgrave, G. W. Infiltration. In *Hydrology,* O. E. Meinzer, ed.) pp. 244-258. New York: McGraw-Hill Book Co., 1942.

370. Shields, Lora M. Nitrogen sources of seed plants and environmental influences affecting the nitrogen cycle. *Bot. Rev.*, **19**: 321-376, 1953.

371. Shirley, H. L. Light as an ecological factor and its measurement. *Bot. Rev.*, **1**: 355-381, 1935.

372. ——. Reproduction of upland conifers in the Lake States as affected by root competition and light. *Am. Midl. Nat.*, **33**: 537-612, 1945.

373. ——. Light as an ecological factor and its measurement, II. *Bot. Rev.*, **11**: 497-532, 1945.

374. Shreve, F. A map of the vegetation of the United States. *Geog. Rev.*, 3: 119-125, 1917.

375. ——. The plant life of the Sonoran Desert. *Sci. Mo.*, **42**: 195-213, 1936.

376. ——. The desert vegetation of North America. *Bot. Rev.*, **8**: 195-246, 1942.

377. ——. and WIGGINS, I. L. Vegetation and flora of the Sonoran Desert. Vol. I. (F. Shreve) Vegetation of the Sonoran Desert. *Carnegie Inst. Wash. Publ.*, **591**, 1951. 192 pp.

378. SIGAFOOS, R. S. Frost action as a primary physical factor in tundra plant communities. *Ecology*, **33**: 480-487, 1952.

379. SINCLAIR, J. G. Temperatures of the soil and air in a desert. *Mo. Weath. Rev.*, **50**: 142-144, 1922.

380. SMALL, J. *pH and Plants.* New York: D. Van Nostrand Company, Inc., 1946. 216 pp.

381. SMILEY, F. J. *A Report upon the Boreal Flora of the Sierra Nevada of California.* Univ. of Calif. Publ. in Botany 9, 1921. 423 pp.

382. SMITH, A. Seasonal subsoil temperature variations. *Jour. Agr. Res.*, **44**: 421-428, 1932.

383. SMITH, A. D. A discussion of the application of a climatological diagram, the hythergraph, to the distribution of natural vegetation types. *Ecology*, **21**: 184-191, 1940.

384. SOMMERS, G. F., and HAMNER, K. C. Phototube-type integrating light recorders: a summary of performance over a five-year period. *Plant Physiol.*, **26**: 318-330. 1951.

385. SPURR, S. H. A new definition of silviculture. *Jour. Forest.*, **43**: 44, 1945.

386. ——. *Aerial Photographs in Forestry.* New York: Ronald Press, 1948. 340 pp.

387.——, and CLINE, A. C. Ecological forestry in central New England. *Jour. Forest.*, **40**: 418-420, 1942.

388. STAKMAN, E. C., and CHRISTENSEN, C. M. Aerobiology in relation to plant disease. *Bot. Rev.*, **12**: 205-253, 1946.

389. STERN, W. L., and BUELL, M. F. Life-form spectra of New Jersey pine barrens forest and Minnesota jack pine forest. *Bull. Torrey Bot. Cl.*, **78**: 61-65, 1951.

390. STEWART, G. and HUTCHINGS, S. S. The point-observation-plot (square-foot density) method of vegetation survey. *Jour. Amer. Soc. Agron.*, **28**: 714-722, 1936.

391. STODDART, L. A., and SMITH, A. D. *Range Management.* New York: McGraw-Hill Book Co., 2nd ed., 1955. 433 pp.

392. TALBOT, M. W. Indicators of southwestern range conditions. *U. S. Dept. Agr. Farmers' Bull.* 1782, 1937. 35 pp.

393. TANSLEY, A. G. The classification of vegetation and the concept of development. *Jour. Ecol.*, **8**: 118-149, 1920.

394. ——. The use and abuse of vegetational concepts and terms. *Ecology*, **16**, 284-307 1935.

395. ——. *The British Islands and Their Vegetation.* Cambridge University Press, 1939. 930 pp.

396. ——. The early history of modern plant ecology in Britain. *Jour. Ecol.*, **35**: 130-137, 1947.

397. TAYLOR, W. P. What is ecology and what good is it? *Ecology*, **17**: 333-346, 1936.

398. THORNTHWAITE, C. W. The climates of North America. *Geog. Rev.*, **21**: 633-654, 1931.

399. ——. Atmospheric moisture in relation to ecological problems. *Ecology*, **21**: 17-28, 1940.

400. ——. An approach toward a rational classication of climate. *Geogr. Rev.*, **38**: 55-94, 1948.

401. Tippett, L. H. C. *Random Sampling Numbers. Tracts for Computers XV.* Cambridge University Press, 1927.

402. Tisdale, E. W., and Zappetini, G. Halogeton studies on Idaho ranges. *Jour. Range Mgt.*, **6**: 225-236, 1953.

403. Toumey, J. W., and Kienholz, R. *Trenched Plots under Forest Canopies.* Yale Univ. School of Forestry Bull. 30, 1931. 31 pp.

404. ——, and Korstian, C. F. *Foundations of Silviculture upon an Ecological Basis.* New York: John Wiley & Sons, Inc., 1947, 2nd ed. 468 pp.

405. Transeau, E. N. Forest centers of eastern North America. *Am. Nat.* **39**: 875-889, 1905.

406. ——. The prairie peninsula. *Ecology*, **16**: 423-437, 1935.

407. ——, Sampson, H. C., and Tiffany, L. H. *Texbook of Botany.* New York: Harper and Brothers, 1940. 812 pp.

408. Trewartha, G. T. *An Introduction to Weather and Climate.* New York: McGraw-Hill Book Co., 3rd ed, 1954. 402 pp.

409. U. S. Department of Agriculture. *Soils and Men.* (U. S. Dept. Agr. Yearbook). Washington, D. C.: Gov. Printing Office, 1938. 1232 pp.

410. U. S. Department of Agriculture. *Climate and Man.* (U. S. Dept. Agr. Yearbook). Washington, D. C.: Gov. Printing Office, 1941, 1248 pp.

411. U. S. Weather Bureau. *Cloud Forms According to the International System of Classication.* Washington, D. C.: Gov. Printing Office, 1928.

412. Veihmeyer, F. J. Evaporation from soils and transpiration. *Trans. Am. Geophysical Union* (19th Ann. Meeting), 612-619, 1938.

413. Vestal, A. G. Minimum areas for different vegetations. *Ill. Biol. Monog.*, **30**: 1-129, 1949.

414. Waksman, S. A. *Soil Microbiology.* New York: Wiley, 1952. 356 pp.

415. ——. *Humus: Origin, Chemical Composition, and Importance in Nature.* Baltimore: Williams & Wilkins Company, 1936. 494 pp.

416. Walker, R. B. The ecology of serpentine soils. II. Factors affecting plant growth on serpentine soils. *Ecology*, **35**: 259-266, 1954.

417. Waller, A. E. Crop centers of the United States. *Jour. Amer. Soc. Agron.*, **10**: 49-83, 1918.

418. Ward, H. B., and Powers, W. E. *Weather and Climate.* Evanston, Ill., 1942. 112 pp.

419. Warming, E. *Oecology of Plants.* (Transl. by P. Groom and I. B. Balfour.) Oxford: Clarendon Press, 1909. 422 pp.

420. Weaver, J. E. Replacement of true prairie by mixed prairie in eastern Nebraska and Kansas. *Ecology*, **24**: 421-434, 1943.

421. ——. *North American Prairie.* Lincoln, Nebr.: Johnsen Publishing Co., 1954. 348 pp.

422. ——, and Clements, F. E. *Plant Ecology.* New York: McGraw-Hill Book Co., 1938 (2nd ed.). 601 pp.

423. Wells, B. W. Plant communities of the coastal plain of North Carolina and their successional relations. *Ecology*, **9**: 230-242, 1928.

424. ——. A new forest climax: the salt spray climax of Smith Island, North Carolina. *Bull. Torr. Bot. Cl.*, **66**: 629-634, 1939.

425. ——, and SHUNK, I. V. The vegetation and habitat factors of the coarser sands of the North Carolina coastal plain. *Ecol. Monog.*, **1**: 465-521, 1931.

426. WELM, H. G. The influence of forest cover on snow-melt. *Trans. Amer. Geophys. Union*, **29**: 546-556, 1948.

427. WENT, F. W. The dependence of certain annual plants on shrubs in California deserts. *Bull. Torr. Bot. Cl.*, **69**: 100-114, 1942.

428. WHITTAKER, R. H. A criticizm of the plant association and climatic climax concepts. *Northwest Sci.*, **25**: 17-31, 1951.

429. ——. A consideration of climax theory: the climax as a population pattern. *Ecol. Monog.*, **23**: 41-78, 1953.

430. ——. Vegetation of the Great Smoky Mountains. *Ecol. Monog.*, **26**: 1-80, 1956.

431. WISLER, C. O., and BRATER, E. F. *Hydrology.* New York: John Wiley and Sons, 1949. 419 pp.

432. WODEHOUSE, R. P. *Pollen Grains, Their Structure, Identification and Significance in Science and Medicine.* New York: McGraw-Hill Book Co., 1935. 574 pp.

433. WOLFENBARGER, D. O. Dispersion of small organisms. Distance dispersion rates of bacteria, spores, seeds, pollen, and insects; incidence rates of diseases and injury. *Am. Midl. Nat.*, **35**: 1-152, 1946.

434. WOODBURY, A. M. Distribution of pigmy conifers in Utah and Northeastern Arizona. *Ecology*, **28**: 113-126, 1947.

435. WOODIN, H. E., and LINDSEY, A. A. Juniper-pinyon east of the continental divide, as analyzed by the line-strip method. *Ecology*, **35**: 473-489, 1954.

436. WORK, R. A. Stream-flow forecasting from snow surveys. *U. S. D. A. Soil Conserv. Serv. Circ.*, **914**: 1-16, 1953.

437. WRIGHT, A. H., and WRIGHT, A. A. The habitats and composition of the vegetation of Okefinokee Swamp, Georgia. *Ecol. Monog.*, **2**: 190-232, 1932.

438. ZEUNER, F. E. *Dating the Past, an Introduction to Geochronology.* London: Methuen and Co., 3rd ed, 1952. 495 pp.

439. ZIPF, G. K. *Human Behavior and the Principles of Least Effort. An Introduction to Human Ecology.* Cambridge, Mass.: Addison-Wesley Press, 1949. 573 pp.

Index

Page numbers in bold face type indicate illustrative material